Lecture Notes in Electrical Engineering

Volume 1080

The book series *Lecture Notes in Electrical Engineering* (LNEE) publishes the latest developments in Electrical Engineering—quickly, informally and in high quality. While original research reported in proceedings and monographs has traditionally formed the core of LNEE, we also encourage authors to submit books devoted to supporting student education and professional training in the various fields and applications areas of electrical engineering. The series cover classical and emerging topics concerning:

- Communication Engineering, Information Theory and Networks
- Electronics Engineering and Microelectronics
- Signal, Image and Speech Processing
- Wireless and Mobile Communication
- Circuits and Systems
- Energy Systems, Power Electronics and Electrical Machines
- Electro-optical Engineering
- Instrumentation Engineering
- Avionics Engineering
- Control Systems
- Internet-of-Things and Cybersecurity
- Biomedical Devices, MEMS and NEMS

For general information about this book series, comments or suggestions, please contact leontina.dicecco@springer.com.

To submit a proposal or request further information, please contact the Publishing Editor in your country:

China

Jasmine Dou, Editor (jasmine.dou@springer.com)

India, Japan, Rest of Asia

Swati Meherishi, Editorial Director (Swati.Meherishi@springer.com)

Southeast Asia, Australia, New Zealand

Ramesh Nath Premnath, Editor (ramesh.premnath@springernature.com)

USA, Canada

Michael Luby, Senior Editor (michael.luby@springer.com)

All other Countries

Leontina Di Cecco, Senior Editor (leontina.dicecco@springer.com)

**** This series is indexed by EI Compendex and Scopus databases. ****

Nitasha Hasteer · Seán McLoone · Manju Khari ·
Purushottam Sharma
Editors

Decision Intelligence Solutions

Proceedings of the International Conference on Information Technology, InCITe 2023, Volume 2

Springer

Editors
Nitasha Hasteer
Dy. Director (Academics) and Head - IT
Amity School of Engineering
and Technology
Noida, Uttar Pradesh, India

Seán McLoone
Centre for Intelligent Autonomous
Manufacturing Systems
Queen's University Belfast
Belfast, UK

Manju Khari
School of Computer and Systems Sciences
Jawaharlal Nehru University
Delhi, India

Purushottam Sharma
Department of Information Technology
Amity University Uttar Pradesh
Noida, Uttar Pradesh, India

ISSN 1876-1100 ISSN 1876-1119 (electronic)
Lecture Notes in Electrical Engineering
ISBN 978-981-99-5993-8 ISBN 978-981-99-5994-5 (eBook)
https://doi.org/10.1007/978-981-99-5994-5

This Springer imprint is published by the registered company Springer Nature Singapore Pte Ltd.
The registered company address is: 152 Beach Road, #21-01/04 Gateway East, Singapore 189721, Singapore

Paper in this product is recyclable.

Preface

Decision intelligence is a discipline that combines data science, machine learning, and behavioral science to help people and organizations make better decisions. It involves the use of advanced analytics and decision-making techniques to generate insights and recommendations that are grounded in data and aligned with human values and preferences. Decision intelligence applies a systematic approach to decision-making that considers multiple options, evaluates trade-offs, and incorporates feedback and learning to continuously improve decision-making over time. The ultimate goal of decision intelligence is to improve the quality and speed of decision-making while minimizing risks and uncertainties. It can be applied to a wide range of decision domains, including healthcare, education, governance and business.

This book is a compilation of research work carried out by esteemed experts in the area of decision intelligence. The objective of this compilation is to provide relevant and timely information to those who are striving to contribute to this domain by utilizing the latest technology. The content of this book is based on the research papers accepted during the 3rd International Conference on Information Technology (InCITe 2023) held at Amity University, Uttar Pradesh, Noida, India, on 2 and 3 March 2023. It is a conglomeration of research papers covering interdisciplinary research and in-depth applications of AI & machine learning for decision making, intelligent system design & modeling, intelligent networks and security, image processing and cognition systems, software engineering and quality management. The content would serve as a rich knowledge repository on information and communication technologies, neural networks, fuzzy systems, natural language processing, datamining and warehousing, big data analytics, cloud computing, security, social networks and intelligence, decision-making and modelling, information systems and IT architectures. We are hopeful that it will prove to be of high value to graduate students, researchers, scientists, practitioners and policymakers in the field of information technology.

We thank the management of Amity University, Uttar Pradesh, who believed in us and provided the opportunity for publishing the book. We are grateful to Science and Engineering Research Board, Department of Science and Technology (DST) and Defence Research and Development Organization (DRDO), Government of India, for their valuable support. We are thankful to the advisory committee of InCITe 2023 for the continuous support and guidance. Special thanks to all the national and international reviewers who helped us in selecting the best of the works as different chapters of the book. We wish to acknowledge and thank all the authors and co-authors of different chapters who cooperated with us at every stage of publication and helped us to sail through this mammoth task. We owe our sincere gratitude to all our family members and friends who helped us through this journey of publishing a book. Our appreciation goes to each and every individual who supported us in this endeavour. Last but not least, we are grateful to the editing team of Springer who provided all guidance and support to us in the compilation of the book and also shaped it into a marketable product.

Nitasha Hasteer
Manju Khari
Seán McLoone
Purushottam Sharma

Contents

About the Editors

Dr. Nitasha Hasteer is a Professor & Head of the Information Technology Department and Dy. Director (Academics) at Amity School of Engineering and Technology, Amity University, India. She has 22 years of teaching, research, and administrative experience in academics and industry. She holds a Ph.D. in Computer Science and Engineering, and her interest areas are machine learning, cloud computing, crowdsourced software development, software project management, and process modeling through multi-criteria decision-making techniques. She has published more than 60 papers in various journals and international conferences in these areas and has guided many postgraduate students in their dissertations. She has been on the editorial board of many international conferences and obtained funding from government agencies such as the Science and Engineering Research Board, Department of Science Technology (DST), Defence Research and Development Organization (DRDO), Indian National Science Academy (INSA), and Council of Scientific & Industrial Research (CSIR) for organizing conferences in the area of Information Technology. She is a member of the International Association of Computer Science & Information Technology (IACSIT), Singapore, the Institution of Engineering and Technology (IET), UK, and various other professional bodies.

Dr. Manju Khari is Professor at Jawaharlal Nehru University (JNU), New Delhi. Before joining JNU, she worked with Netaji Subhas University of Technology, East Campus, under Govt. Of NCT Delhi. Her Ph.D. is in Computer Science and Engineering from the National Institute of Technology Patna, and she received her master's degree in Information Security from Guru Gobind Singh Indraprastha University, Delhi, India. She is also the co-author of two books, published by NCERT of XI and XII, and co-editor of 10 edited books. She has delivered expert talks, has been a guest speaker at international conferences, and member of the reviewer/technical program committee at various international conferences. Besides this, she has been associated with many research activities such as being Associate Editor/Guest Editor of Springer, Wiley, and Elsevier books and reviewer for various international journals.

Dr. Seán McLoone is Professor of Applied Computational Intelligence and Director of the Centre for Intelligent Autonomous Manufacturing Systems (i-AMS) at Queen's University Belfast. Prof McLoone's research interests are in the general area of intelligent systems with a particular focus on data-based modeling and analysis of dynamical systems. This encompasses techniques ranging from classical system identification, fault diagnosis, and statistical process control to modern artificial intelligence and biologically inspired adaptive learning algorithms and optimization techniques. His research has a strong application focus, with many projects undertaken in collaboration with the industry in areas such as process monitoring. control, optimization, time series prediction, and in-line sensor characterization. At a professional level, Prof. McLoone is a Chartered Engineer, a Fellow of the Institute of Engineering Technology (IET), and a Senior Member of the Institute of Electrical and Electronic Engineers (IEEE). He is a member of the IFAC Technical Committees on Computational Intelligence in Control (TC3.2), and Modelling, Identification, and Signal Processing (TC1.1) and has previously served as a Member of the Accreditation Board of Engineers Ireland and as a non-executive Director on the Board of Directors of Irish Manufacturing Research Ltd. He currently serves as an Associate Editor for the 'Transactions of the Institute of Measurement and Control' and as a member of the Editorial Board of 'Engineering Applications of Artificial Intelligence'.

Dr. Purushottam Sharma is working as a Professor in the Department of Information Technology, Amity School of Engineering & Technology at Amity University Uttar Pradesh. He has more than 17 years of experience in research, academia, and industry. His research interest includes artificial intelligence, data analytics, temporal data mining, and high-performance networks. Dr. Sharma has published more than 70 research papers in SCI, ESCI, Scopus-indexed journals, and reputed international conferences. He has multiple technical patents in his name. He is a Cisco certified Instructor Trainer ITQ (Instructor Trainer Qualification) and has also obtained CCNA, CCNA(R&S) Global Certificate from Cisco System USA. He has delivered lectures on Networking related subjects in e-Learning mode to 29 African Countries using the Pan-African Network at Amity University Uttar Pradesh.

Dual Band Open Slot and Notch Loaded Bandwidth Enhanced Microstrip Patch Antenna for IoT/WiMAX/WLAN Applications

Ramesh Kumar Verma, Dheeraj Tripathi, Dukhishyam Sabat, Maninder Singh, and Akhilesh Kumar

Abstract The demand of multiband microstrip antennas is increasing due to the technological advancement in modern communication systems. This work presents a dual band microstrip patch antenna of enhanced bandwidth (BW). The intended antenna is designed by creating one open rectangular shape slot and four notches. The open slot is loaded at the top of radiation patch while four notches are loaded at each corner. The lower band of intended antenna is resonating from 2.50 to 2.75 GHz while upper band is resonating from 4.12 to 5.87 GHz. The lower band of antenna exhibits bandwidth of 250 MHz (9.52%) while upper band of antenna exhibits bandwidth of 1750 MHz (35.04%). The intended antenna resonates at 2.64 GHz in lower band with −18.9 dB return loss (RL) while at frequencies 4.43 GHz and 5.5 GHz in upper band with −14.82 dB and −31.19 dB return loss. The intended antenna shows peak gain of 4.07, 4.62 and 4.9 dB at resonating frequencies. The intended antenna can be convenient for IoT (in ISM), WiMAX and WLAN utilizations.

Keywords Dual band · Open slot · Notches · IoT · WiMAX · WLAN

R. K. Verma
Department of CSE, IMS Engineering College, Ghaziabad, UP, India

D. Tripathi
Department of ECE, Noida Institute of Engineering and Technology, Greater Noida, UP, India

D. Sabat · A. Kumar (✉)
Department of ECE, National Institute of Technology, Assam Silchar, India
e-mail: akhilesh31017@gmail.com

D. Sabat
e-mail: dukhishyam_rs@ece.nits.ac.in

M. Singh
Department of ECE, Motilal Nehru National Institute of Technology Allahabad, Prayagraj, UP, India

N. Hasteer et al. (eds.), *Decision Intelligence Solutions*, Lecture Notes in Electrical Engineering 1080,
https://doi.org/10.1007/978-981-99-5994-5_1

1 Introduction

In order to improve functionality of the system at a lower buildup cost and smaller size, antenna with greater performance are built for portable wireless systems. Multi-band antennas improve the system's functionality while simultaneously shrinking the size and cost of the device [1, 2]. Researchers have therefore been driven to develop unique multiband antennas as a result of the increasing expansion of mobile communication networks [3]. Mobile communication systems are in great demand and have seen rapid expansion in recent years [4]. IoT is a developing network of machines, gadgets, and things that can all connect to the Internet wirelessly and communicate with one another. There is a wide variety of wired and wireless connectivity methods available for IoT devices. IoT protocols typically operate at 2.4 GHz, 5 GHz, 4.33 GHz and 915 MHz in the ISM band [5].

For the designing of dual band antennas, two arc-shaped strips [6], fractal ring with BW of 700 and 680 MHz [7], split-ring resonator with BW of 120 and 3132 MHz [8], dual-square ring with BW of 100 and 200 MHz [9], rectangular rings-based with BW of 450 and 450 MHz [10], pentagonal shaped CDRA with BW of 210 and 1820 MHz [11], patch of asymmetrical I-shape with BW of 790 and 1670 MHz [12] and ring CDRA with BW of 780 and 2490 MHz [13] are used. A dual band CRLH unit cell loaded antenna is presented for WLAN/WiMAX resonates at 2.5 and 5.8 GHz with 22.8% and 10.8% bandwidth [14] while a ring antenna is presented using ECL (Electric inductive capacitive) resonator for 3.74 and 5.1 GHz frequency with BW of 500 and 860 MHz [15]. Another antenna for 5G application having ellipse shaped ring is designed with 100 and 500 MHz bandwidth [16] while elliptical-ring CPW fed antenna of compact size is proposed with bandwidth of 4.1% and 9.4% at 2.60 and 3.48 GHz for WiMAX [17]. However, triangular shape antenna is presented for RFID application resonating at 0.912 and 2.45 GHz [18]. Other dual band antenna using U-shape and circular ring slots is proposed for GSM and WiMAX having 14.1% and 6.4% bandwidth [19] while monopole fan shape patches using shorting pins dual band antenna is presented for ISM band showing 3.4% and 3.8% bandwidth [20]. Apart from these, L-shaped dual band antenna having 4.13% and 8.82% bandwidth is proposed for 2.45 and 5.125 GHz [21].

This study presents a dual band, improved bandwidth rectangular patch antenna. The lower band of intended antenna is resonating from 2.50 to 2.75 GHz while upper band is resonating from 4.12 to 5.87 GHz. The lower band of antenna exhibits bandwidth of 250 MHz (9.52%) while upper band of antenna exhibits bandwidth of 1750 MHz (35.04%). The intended antenna is designed by creating one open rectangular shape slot and four notches. The open slot is created at the top of radiation patch while four notches are loaded at each corner. The intended antenna can be useful for IoT (in ISM), WiMAX and WLAN utilizations [22]. The Mentor Graphics IE3D is used for designing of intended antenna [23].

2 Antenna Design

The following set of equations is used for the determination of length (L) and width (W) of patch for the given resonating frequency (f_r), dielectric constant (ε_r) and thickness (h) [24]:

$$W = \frac{c}{2f_r}\sqrt{\frac{2}{\varepsilon_r + 1}} \tag{1}$$

$$\varepsilon_{reff} = [\left(\frac{1}{2}\right) * (\varepsilon_r + 1) + \left(\frac{1}{2}\right) * (\varepsilon_r - 1)\left(1 + \frac{12\text{h}}{W}\right)^{-0.5}] \tag{2}$$

The extension of the length (ΔL),

$$\frac{\Delta L}{h} = (0.412)\frac{(\varepsilon_{reff} + 0.3)(\frac{W}{h} + 0.264)}{(\varepsilon_{reff} - 0.258)(\frac{W}{h} + 0.8)} \tag{3}$$

Actual patch length,

$$L = \frac{0.5c}{f_r\sqrt{\varepsilon_{reff}}} - 2\Delta L \tag{4}$$

The ground size is calculated as [25]:

$$L_g = L + 6\text{h} \tag{5}$$

$$W_g = W + 6\text{h} \tag{6}$$

3 Design Specifications and Proposed Geometry

The geometry of intended antenna has been represented in Fig. 1(a) while design procedure is exhibited in Fig. 1(b), 1(c) and 1(d). The geometrical parameters and specification of intended antenna are exhibited in Table 1. The presented antenna is designed using FR-4 substrate. For design frequency 2.45 GHz, the size of the patch are determined as Lp = 29 mm (length) and Wp = 37 mm (width). However, the ground size (length and width) of intended antenna are Lg = 39 mm and Wg = 47 mm respectively. The patch of proposed antenna has been loaded with one open rectangular slot of size 10 mm × 15 mm on the top of patch and four rectangular notches of size 5 mm × 10 mm each on four corners. The intended antenna is excited by 50Ω microstrip line feed via feed strip of size 4 mm × 5 mm at bottom of the patch [26].

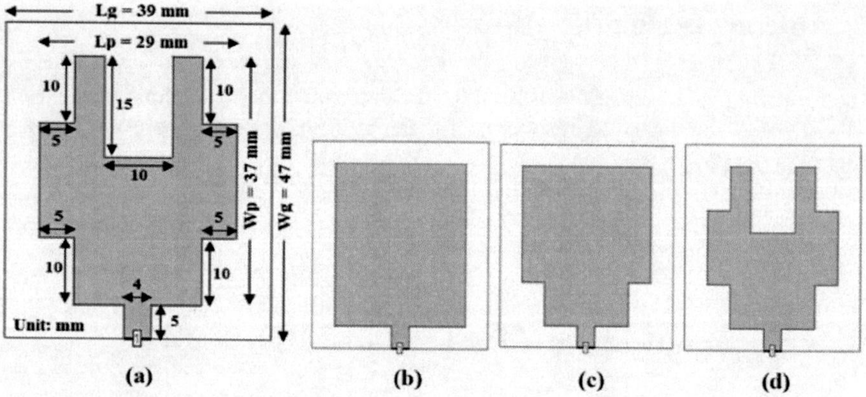

Fig. 1 Proposed geometry of antenna and design steps

Table 1 Specifications of antenna

Parameter	Values	Parameter	Values
Frequency	2.45 GHz	Corner notch length	10 mm
Dielectric constant	4.4	Corner notch width	5 mm
Thickness (h)	1.6 mm	Upper middle open slot length	15 mm
Wp (Patch width)	37 mm	Upper middle open slot width	10 mm
Lp (Patch length)	29 mm	Feed width	4 mm
Wg (Ground width)	47 mm	Feed length	5 mm
Lg (Ground length)	39 mm		

4 Procedure of Antenna Design

Initially, a conventional antenna (Antenna-1) has been designed with rectangular ground of size 39 × 47 mm^2 and a rectangular patch of size 29 × 37 mm^2 on a FR-4 substrate as exhibited in Fig. 1(b). This antenna resonates at 2.45 GHz between 2.37–2.56 GHz, 3.84 GHz between 3.6–4.04 GHz and 5.29 GHz between 4.97–5.54 GHz with impedance bandwidth of 190 MHz (7.71%), 440 MHz (11.52%) and 570 MHz (10.85%) respectively. The return losses of −12.20, −18.18 and −21.82 dB at 2.45, 3.84 and 5.29 GHz are observed. In next step, antenna embedded with notches (Antenna-2) has been designed with two rectangular notches of size 10 × 5mm^2 at both lower corners as exhibited in Fig. 1(c). This antenna is also resonating in triple band at 2.64 GHz between 2.50–2.75 GHz, 4.27 GHz between 4.0–4.56 GHz and 5.39 GHz between 4.94–5.67 GHz with return losses of –30.96, −16.01 and −16.29 dB at 2.64, 4.27 and 5.39 GHz. The BW of 250 MHz (9.52%), 560 MHz (13.08%) and 730 MHz (13.76%) is obtained. Finally, antenna (Antenna-3) is designed with loading two other rectangular notches of size 10 × 5mm^2 at both upper corners and one open rectangular slot of size 10 mm × 15 mm on the top of patch as exhibited in Fig. 1(d). Now, this antenna (proposed) resonates at dual band

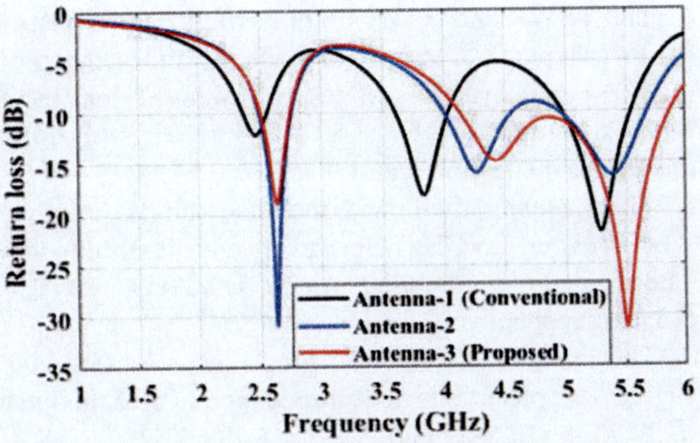

Fig. 2 RL plots of different antennas

with improved bandwidth between 2.50–2.75 GHz and 4.12–5.87 GHz resonating at 2.64, 4.43 and 5.5 GHz exhibiting bandwidth of 250 MHz (9.52%) and 1750 MHz (35.04%) respectively. At 2.64, 4.43, and 5.5 GHz, return losses of −18.9, −14.82 and −31.19 dB are noted. The RL plots of these antennas are exhibited in Fig. 2.

5 Results and Discussion

The presented dual band antenna is resonating with improved bandwidth in lower band from 2.50–2.75 GHz and upper band from 4.12–5.87 GHz. The BW of proposed antenna is achieved 250 MHz (9.52%) in lower band and 1750 MHz (35.04%) in upper band. The RL of −18.9 dB at 2.64 GHz in lower band, −14.82 dB at 4.43 GHz and −31.19 dB at 5.5 GHz in upper band is obtained. Return loss graph of presented dual band antenna is exhibited in Fig. 3. The lower band 2.50–2.75 GHz is applicable for WiMAX and upper band 4.12–5.87 GHz is applicable for ISM and WLAN.

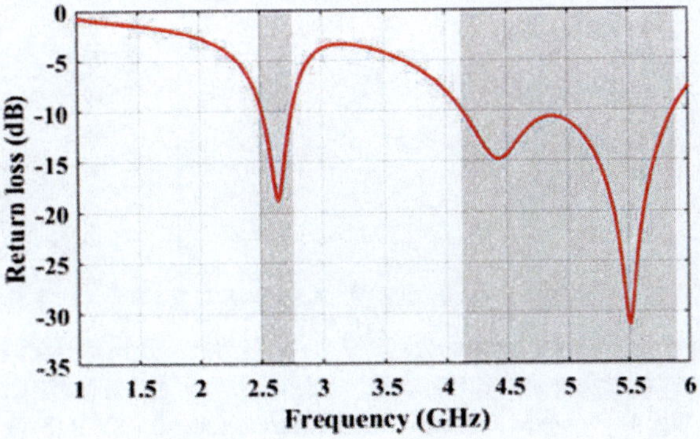

Fig. 3 RL plot of dual band antenna

The VSWR and peak gain plots of presented antenna is exhibited in Fig. 4(a) and 4(b) while peak directivity and antenna efficiency plots is exhibited in Fig. 4(c) and 4(d) respectively. The VSWR of proposed dual band antenna is 1.256, 1.444 and 1.057 at frequency 2.64, 4.43 and 5.5 GHz while simulated peak gain of presented antenna is 4.07, 4.62 and 4.90 dB at 2.64, 4.43 and 5.5 GHz respectively.

The presented dual band antenna exhibits directivity of 4.13, 4.56 and 4.90 dB at 2.64, 4.43 and 5.5 GHz respectively and maximum directivity of 5.08 dB at 5.75 GHz. The efficiency of intended antenna is 80.93%, 79.29% and 81.94% at 2.64, 4.43 and 5.5 GHz respectively.

The input impedance (Z) graph of presented dual band antenna is exhibited in Fig. 5. The observed input impedance of intended antenna is Z = 61.24–j5.81Ω, Z = 39.74 + j12.88Ω and Z = 47.4–j0.652Ω at 2.64, 4.43 and 5.5 GHz respectively. Figure 6 represents the 2D radiation pattern of intended dual band antenna for phi = 0° and phi = 90° at 2.64, 4.43 and 5.5 GHz. The comparison of intended antenna and published work is exhibited in Table 2.

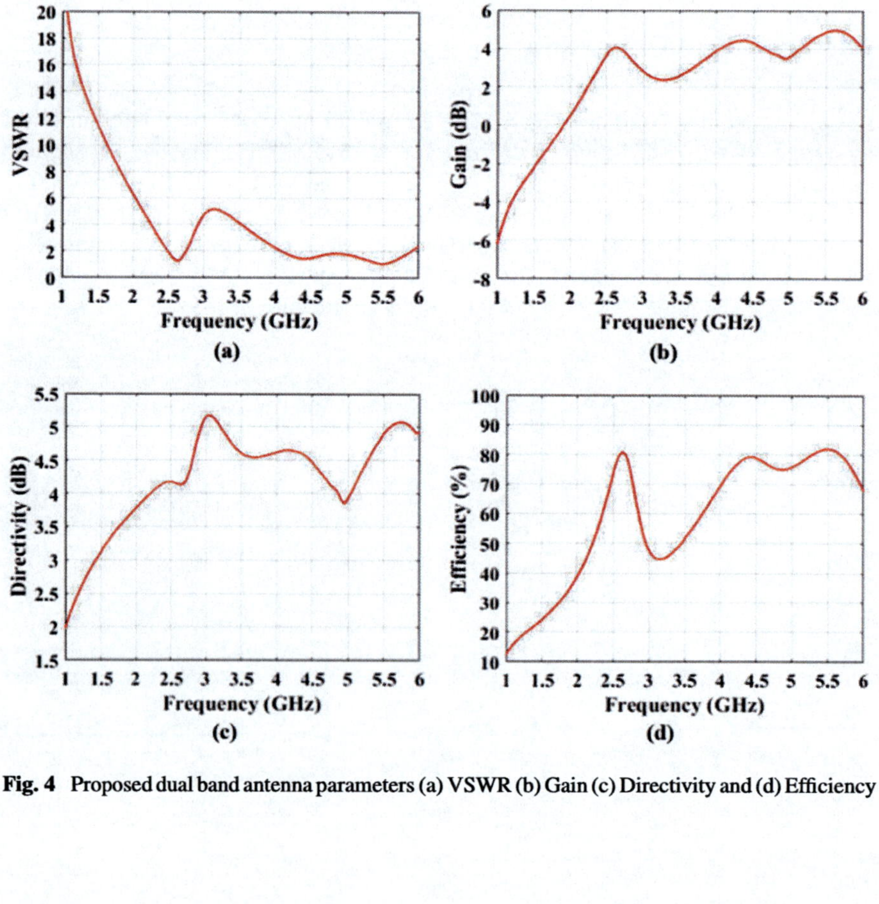

Fig. 4 Proposed dual band antenna parameters (a) VSWR (b) Gain (c) Directivity and (d) Efficiency

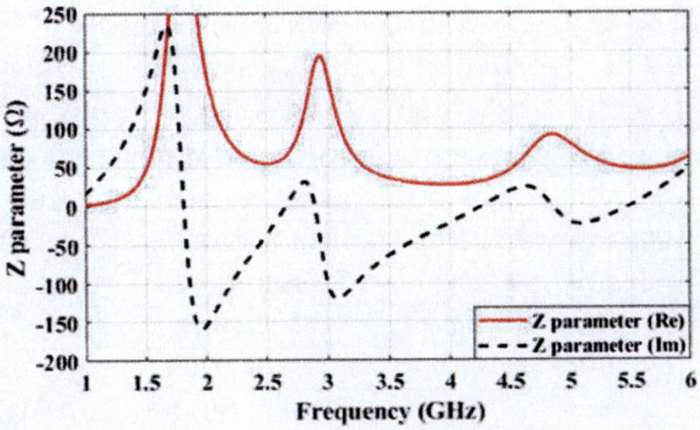

Fig. 5 Input impedance (Z) plot of proposed dual band antenna

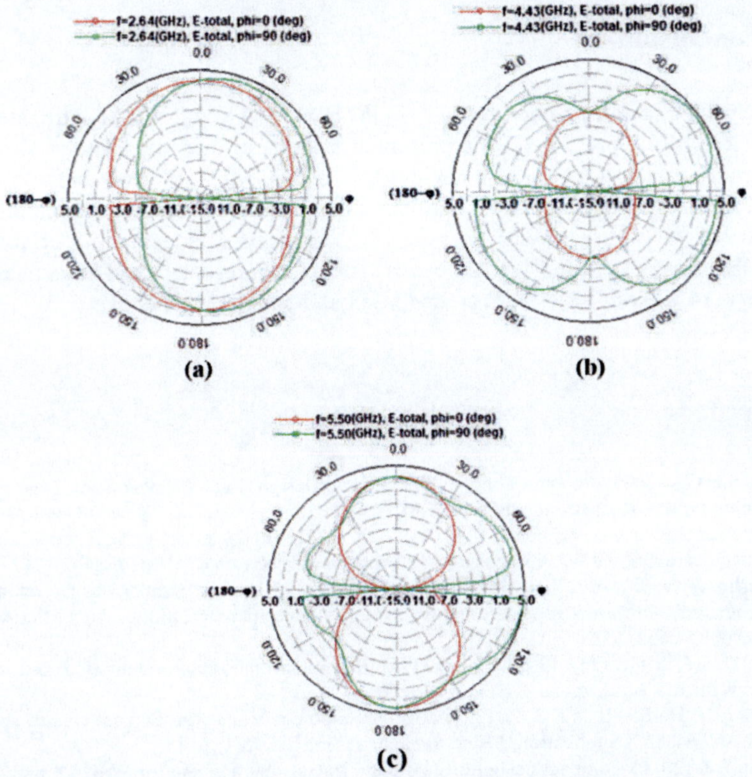

Fig. 6 2D radiation pattern at frequencies (a) 2.64 GHz (b) 4.43 GHz and (c) 5.5 GHz

Table 2 Performance comparison with published work

Ref	Size (mm^2)	Bandwidth	Gain (dB)	Efficiency
[14], 2018	44 × 70	22.8% and 10.8%	1.5, 4.4	93%
[15], 2018	30 × 30	500 and 860 MHz	1.23, 1.57	82%
[16], 2020	180 × 60	100 and 500 MHz	2.4, 6.1	Not reported
[17], 2019	30 × 33	100 and 330 MHz	2.39, 1.75	88%
[18], 2018	66 × 80	207 and 700 MHz	2.24, 5.32	Not reported
[19], 2019	48 × 48	14.1% and 6.4%	NR	82, 86%
[20], 2016	48 × 48	3.4% and 3.8%	−0.5, 2.45	35.8, 61.9%
Proposed	39 × 47	250 and 1750 MHz	4.07, 4.62, 4.9	80.93, 79.29, 81.94%

6 Conclusion

A patch antenna of dual band for IoT (in ISM), WiMAX and WLAN utilization has been successfully presented. The presented antenna of dual band shows simulated peak gain of 4.07, 4.62 and 4.9 dB and BW of 250 MHz (9.52%) and 1750 MHz (35.04%) resonating at 2.64, 4.43 and 5.5 GHz. The lower band is resonating from 2.50 to 2.75 GHz covers WiMAX band while upper band is resonating from 4.12 to 5.87 GHz covers ISM and WLAN bands. The RL of presented dual band antenna is −18.9, −14.82 and −31.19 dB at 2.64, 4.43 and 5.5 GHz respectively.

References

1. Balanis CA (2005) Antenna Theory, Analysis and Design. John Wiley & Sons, New York
2. Kumar G, Ray K.P.: Broadband Microstrip Antenna. Artech House, Norwood, MA (2003)
3. Verulkar S, Khade A, Trimukhe MA, Gupta RK (2022) Dual band split ring monopole antenna structures for 5G and WLAN applications. Progress In Electromagnetics Research C 122:17–30
4. Mishra B, Verma RK, Yashwanth N, Singh R (2022) A review on microstrip patch antenna parameters of different geometry and bandwidth enhancement techniques. Int J Microw Wirel Technol 14:652–673
5. Kulkarni P, Srinivasan R (2021) Compact polarization diversity patch antenna in L and WiMAX bands for IoT applications. AEU-Int J Electron C 136:1–8
6. Yoon JH, Ha SJ, Rhee YC (2015) A novel monopole antenna with two arc-shaped strips for WLAN/WiMAX application. J Electromagnetic Eng Sci 15(1):6–13
7. Naji DK (2016) Compact design of dual-band fractal ring antenna for WiMAX and WLAN applications. Int J Electromag Appl 6(2): 42–50
8. Christydass SPJ, Gunavathi N (2021) Dual-band complementary split-ring resonator engraved rectangular monopole for GSM and WLAN/WiMAX/5G sub-6 GHz band. Progress In Electromagnetics Res. C 113:251–263
9. Swain BR, Sharma AK (2019) An investigation of dual-band dual-square ring (DSR) based microstrip antenna for WiFi/WLAN and 5G-NR wireless applications. Progress In Electromagnetics Res M 86:17–26
10. Singla G, Khanna R, Parkash D (2019) CPW fed rectangular rings-based patch antenna with DGS for WLAN/NII applications. Int J Microw Wirel Technol 11:523–531

11. Rai C, Singh A, Singh S, Singh AK, Verma RK (2022) Dual-band and dual polarized inverted pentagonal shaped hybrid cylindrical dielectric resonator antenna for wireless applications. Wireless Pers Commun 124:2121–2139
12. Verma RK, Srivastava DK, Tripathi RP, Rajpoot V (2022) Wide dual band asymmetrical I-shape rectangular microstrip patch antenna for PCS/UMTS/WiMAX/IMT applications. Wireless Pers Commun 122:1577–1598
13. Rai C, Singh S, Singh AK, Verma RK (2022) Design and analysis of dual-band circularly polarized hybrid ring cylindrical dielectric resonator antenna for wireless applications in C and X-band. Wireless Pers Commun 126:1383–1401
14. Li H, Zheng Q, Ding J, Guo G (2018) Dual-band planar antenna loaded with CRLH unit cell for WLAN/iMAX application. IET Microwaves Antennas Propag 12(1):132–136
15. Daniel RS, Pandeeswari R, Raghavan S (2018) Dual-band monopole antenna loaded with ELC metamaterial resonator for WiMAX and WLAN applications. Appl Phys A Mater Sci Process 124:1–7
16. Arya AK, Kim SJ, Kim S (2020) A dual-band antenna for LTE-R and 5G lower frequency. Progress In Electromagnetics Res. Letters 88:113–119
17. Dattatreya, G., Naik, K.K.: A low volume flexible CPW-fed elliptical-ring with split-triangular patch dual-band antenna. Int J RF Microw Comput Aided Eng 29:1–9 (2019)
18. Dehmas M, Azrar A, Mouhouche F, Djafri K, Challal M (2018) Compact dual band slotted triangular monopole antenna for RFID applications. Microw Opt Technol Lett 60:432–436
19. Gangwar SP, Gangwar K, Kumar A (2019) Dual band modified circular ring shaped slot antenna for GSM and WiMAX applications. Microw Opt Technol Lett 61(12):2752–2759
20. Tak J, Kang DG, Choi J (2016) A compact dual-band monopolar patch antenna using TM01 and TM41 modes. Microw Opt Technol Lett 58(7):1699–1703
21. Mishra, A., Ansari, J.A., Kamakshi, K., Singh, A., Aneesh, M., Vishvakarma, B.R.: Compact dual band rectangular microstrip patch antenna for 2.4/5.12 GHz wireless applications. Wireless Networks **21**(2), 347–355 (2015)
22. Tripathi D, Srivastava DK, Verma RK (2021) Bandwidth enhancement of slotted rectangular wideband microstrip antenna for the application of WLAN/WiMAX. Wireless Pers Commun 119:1193–1207
23. Zeland Software, Inc., IE3D Simulation Software Version 9.0 and Mentor Graphics IE3D Simulation Software Version 15.30
24. Verma RK, Srivastava DK, Mishra B (2022) Circuit theory model-based analysis of triple-band stub and notches loaded epoxy substrate patch antenna for wireless applications. Int J Commun Syst 35:1–17
25. Verma RK (2022) Bandwidth enhancement of an inverted F-shape notch loaded rectangular microstrip patch antenna for wireless applications in L and S-band. Wireless Pers Commun 125:861–877
26. Yadav A, Singh P, Verma RK, Singh VK (2023) Design and comparative analysis of circuit theory model based slot loaded printed rectangular monopole antenna for UWB applications with notch band. Int J Commun Syst 36(3):1–15

Carbon Nanotubes as Interconnects: A Short Review on Modelling and Optimization

Gaurav Mitra, Vangmayee Sharda, and Ruchi Sharma

Abstract Carbon nanotube (CNT) interconnects have a substantial influence on the performance and power loss of integrated circuits (ICs). When copper's conductivity declines considerably owing to side effects in future generations of technology, nanotubes, which are rolled-up sheets of carbon one atom thick, hold great potential for fixing some of the most pressing interconnect issues. Carbon nanotubes possess a number of remarkable properties, including very high mechanical strength, stability, and broad electron mean-free paths. This literature review examines the physical circuit design and modelling of carbon nanotubes. At a tolerable working temperature (100 °C), single-wall (SWCNT) and multi-wall carbon nanotube (MWCNT) interconnects are compared to Cu interconnects for possible future improvement. These models accurately represent the size, temperature dependence, and numerous electron photonic scattering mechanisms of quantum conductance. Utilizing a hybrid structure comprised of Cu, SWCNTs, and MWCNTs yields the better network gain. In addition to the delay they provide to important channels, the power they squander, the distortion and vibration they cause each other, and their vulnerability to electron transfer, interconnects are regarded as one of the most significant challenges facing Giga scale integration.

Keywords Carbon Nanotubes · SWCNT · MWCNT · Deep Submicron (DSM) Technology Node

G. Mitra (✉) · R. Sharma
ECED, Bharati Vidyapeeth College of Engineering, New Delhi, Delhi, India
e-mail: gaurav.mitra@bharatividyapeeth.edu

R. Sharma
e-mail: ruchi.sharma@bharatividyapeeth.edu

V. Sharda
Amity University, Uttar Pradesh, Noida, India
e-mail: vsharma3@amity.edu

N. Hasteer et al. (eds.), *Decision Intelligence Solutions*, Lecture Notes in Electrical Engineering 1080,
https://doi.org/10.1007/978-981-99-5994-5_2

1 Introduction

Carbon interconnects considered for more than 65% of the capacity of high-performance circuits at 130-nm technology applications. As a result, their dynamic power dissipation is more than that of circuits [2]. As a result, Giga-scale systems are severely constrained since almost all modern processors have power constraints [3]. Interconnects have a significant and growing role in the delays of essential lines. Numerous variables make connection issues worse when integrated circuits (IC) go towards the nanoscale zone. Copper's resistivity grows with its cross-sectional area. Hence larger copper objects are more challenging to pass electricity through than smaller ones (35 nm at ambient temperature). The scattering of electrons at the wire surfaces increases as the diameter of the wire decreases. When electrons scatter off the surfaces of wires, they can either retain all of their longitudinal momenta or lose some of it depending on the quality of the surface [4].

Additionally, there is a link between the cross-sectional area and the average dimensions since electrical cables are polycrystalline. The narrower wire width or thickness is equal to the typical grain diameter. The grain barriers that electrons must traverse grows as wire cross sections become smaller. As a result, the MFP becomes smaller overall in scaled copper cables. The substance used as a barrier to keep copper is responsible for the third contributor to the increase in resistivity. Copper wires lose some effective resistance due to the poor conductivity of the barrier materials unless the barrier size and cross-sectional wire area are reduced proportionally [5]. Creating durable size of barriers below a few nm ranges will be difficult. Extremely large electron MFPs have been seen on quantum-scaled wires, even though scaling down standard wires' cross-sections increases resistance and adds extra scattering. The constrained scattering phase space explains this. Electrons can only travel in one dimension in a quantum wire. Therefore, only backward electron scattering is possible, which needs a sharp, significant shift in momentum. Because of this, luminescence wires do not encounter the same number of electron scatterings as conventional wires, where electrons may undergo several scatterings at modest angles and backscattering.

CNTs, compared to the copper MFP of 35 nm, provide a good illustration of a quantum wire for which micrometre-scale electron MFPs have been seen [6]. Depending on its chirality, a single-wall nanotube (SWNT) is a graphite tube that ranges in size from 0.5 to a few nanometers and can either be metallic or semiconducting. The diameters of MWNTs, on the other hand, range from a few to more than 100 nm. The remarkable features of CNTs originate from their one-dimensional nanotube structure, strong sp2 bonding, and the odd band structure of graphene. CNTs have robust mechanical bonding [1] due to the importance of the sp2 connection in graphene over the sp3 coupling in a diamond.

2 Literature Reviews on Carbon Interconnect

CNTs may conduct very high current densities due to their mechanical bonding and the minimal interaction among electrons and compounds. Interconnect applications in gig-scale systems are up-and-coming because of the enormous MFP, and electro-migration resistance was seen in CNTs. Therefore, extensive research is being conducted to create CNT interconnect technology. Increasing self-assembled CNTs using chemical vapour deposition (CVD) on plates with pre-patterned catalysts has advanced significantly during the last ten years since no post-growth processing is required. The blend of top-down and bottom-up methods promises CNT integration on a broad scale. By using the gas flow or by adding an electric field during CVD, the directed development of sparse SWCNTs has been observed. A significant hurdle still exists in the large-scale and high-yield development of dense arrays of CNTs. CNT chirality has not been controlled, and only one-third of SWCNTs are conductive [7].

Since high-quality CNTs are frequently generated at temperatures above 600 °C, this in situ growing approach poses considerable challenges with CMOS compatibility. Recent research on creating nanotubes between 420°C and 480°C has been encouraging. Another appealing method is post-growth assembly because it isolates metal and semiconducting nanotubes. For example, SWCNTs floated in a solution on a multi-electrode array are sorted by electrocardiographs, providing an AC signal electric field. When an AC electrical influence is generated, metallic SWCNTs, which are more polarised, are preferentially deposited upon the substrate between the electrodes. The metallic SWCNTs that were deposited are likewise straight. Other methods of separation include filtration and ultraconservatives. Connections with a low resistance to single SWCNTs or vertically bundled MWCNTs have been described in the literature [8]. Connections with a low resistance to single SWCNTs or vertically bundled MWCNTs have been described in the literature [8].

Constructing trustworthy, low-resistance connections to horizontal CNT bundles is a challenge that must be surmounted. The performance of CNT interconnects must be accurately predicted before they can be used in IC. However, these substantial challenges must be overcome. First, it is possible to quantify the maximum performance improvement that CNT interconnects can provide, which would aid in determining their practicality. Such modelling would also make it possible to decide which CNT interconnect applications are most promising, generating significant guidance for developing CNT technology appropriate for diverse interconnect uses [9]. This literature review will continue in the following format. In the following paragraphs, we will discuss the circuit modelling for carbon nanotubes. Afterwards, we will discuss optimizing the connections between carbon nanotubes. After that, we'll discuss the algorithm and methods behind carbon interconnects and draw some conclusions.

Circuit Modelling for a Carbon Nanotube

Figure 1 depicts an equivalent-circuit model of a CNT. An analogous transmission line serves as the primary component of this approach. CNT is described in the solid frame, while a similar transmission line is shown in the dashed frame.

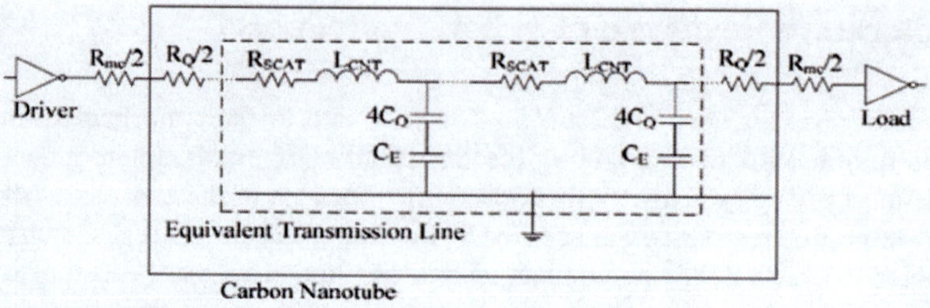

Fig. 1 An equivalent transmission model of SWCNT [23]

Number of Conduction Channels

Each conductor in a graphite tube might be usable for conduction [10], depending on how full or empty the tube is. The contribution of each conduction band at low voltage may be shown to be proportional to the probability that its lowest state is occupied, and this can be done with the use of the Landauer formula.

Resistance Components

Within the ballistic system, if there is no dispersion at the contact or over the length of a graphite shell, each conduction circuit yields one quantum conductance. The interaction resistance, which may reach hundreds of k if the cables are of low quality, is caused by scattering at the junctions. Numerous studies have recorded interaction resistances on the order of k or even hundreds, signaling there should be no fundamental restriction on reducing contact resistance to a point where it may be replaced by quantum resistance. If the electron's round-trip distance is greater than or equal to the mean free path (MFP) associated with the scattering mechanism, the electron will be dispersed by the defect or other phonon scattering mechanism. Because interconnect applications demand exceedingly long nanotubes, these scatterings are especially crucial [11].

Kinetic and Magnetic Inductances

A traditional description of an inductor is the resistances to changes in current that result from Faraday's law ($1/2\ LI^2$). This energy is stored as a magnetic field. Mass, which opposes changes in velocity and generates kinetic energy ($1/2\ mv^2$), is the mechanical analogue of inductance. A non-zero mass carrier moving with kinetic energy is what we mean when we talk about electric current [12]. Electron kinetic energy and magnetic field potential energy are added together to get the overall energy of an electric current. The latter is often disregarded in standard cables since it is orders of magnitude less than the former. In a quantum wire, the electrons' kinetic energy may be higher than the energy of the magnetic field [13]. The decreasing carrier density with increasing cross-sectional dimensions is proof of this phenomenon in classical mechanics. To retain the same current, the rate of electron flow must increase. The kinetic energy develops exponentially due to the quadratic connection between velocity and kinetic energy. In contrast, the magnetic inductance of a wire is mostly determined by its closeness to a return channel and only to a little extent by its cross-sectional area. In order to calculate electron velocity, the word "kinetic inductance" must be specified.

Quantum and Electrostatic Capacitances

The Pauli exclusion dictates that the number of electrons in states above the Fermi level must be increased for a quantum wire to undergo electric fluctuations. A quantum capacitance may also be established in parallel with the electrostatic capacitance. 300 ml is the electrostatic capacitance per conduction channel. Nevertheless, capacitance is often significantly more remarkable. Therefore, quantum capacitance can be ignored in most real applications [14].

Carbon Nanotube Interconnects Optimization

Depending on their scale, energy and signals are distributed across the IC via interconnects, categorized as local, semi-global, or global. These connectors are governed by standards that are inconsistent with one another. This section will discuss the three hierarchical levels of interconnect and why comparing their performance is essential.

Local Interconnects

When we look at the model of global interconnects, the biggest obstacle to Giga scale integration is because their length needs to scale with technology. The fact is that in contrast to global interconnects, local interconnects' sizes decrease as technology progresses. They are seen as less complicated and neglected. Local interconnects. How- ever, become more significant for several reasons as the industrial sector approaches the nanoscale age. The consequences of size, nanoscale lithography differences, polishing, and electro-migration will be detrimental to local interconnects. Although short in isolation, local interconnects aggregate together to form a much more extended network, slow significant traffic and waste energy. For example, more than fifty per cent of the electricity is located in a local connection. Most cutting-edge circuit, design, and technology solutions can only considerably mitigate the global interconnection problem. Network-on-chip, wave pipe lining, three-dimensional integration, and, most crucially, numerous core designs are a few examples [15].

One of the major challenges for local interconnects is the conflict between the need for signal and power interconnects. Since the capacitance of a connection of minimal size needs to be higher than a conventional logic gate, which is only 10–20 gate pitches wide [16], capacitors, not resistance, are the fundamental issue for local signal interconnects. Designers favor narrow angles for these signal interconnects to reduce latency, power consumption, and the dynamic change in delay caused by different switching patterns. Transport substantial currents while facilitating electricity interconnections. Designers would prefer thicker interconnects if doing so reduced reactance and power density within the IC and increased its resistance to distortion and electromagnetic interference in the power supply. Competition between signal and power interconnect requirements is one of the key issues for local interconnects.

Signal routing and power hookups are performed on the first few metal layers. It must have a consistent thickness. This causes costly power and signal connection in terms of cable space, delay, and cross-talk. The problem is exacerbated by the thickness variance generated by chemical mechanical polishing (CMP). Even greater nominal thicknesses are necessary to accommodate for interpretation and provide a high yield [17]. This article demonstrates how CNTs may be used as signal and power interconnects to solve some of the most pressing issues with local interconnects by physically separating the optimization of the two types of interconnects.

Signal Local Interconnects
Given that CNT technology makes it possible to tailor the thickness of a connection to its particular purpose and needs, it has the potential to transform the interconnect industry. Chemistry. The diameter of CNTs, for instance, is not a function of metal removal time or chemical mechanical polishing (CMP), but rather of the catalyst material used in their creation. According to the International Technology Roadmap for Semiconductors (ITRS), the vertical capacitance of signal interconnects may be decreased by a factor of 4. CAD tools can only be utilised to programme into thin SWCNT interconnects short interconnects with modest drivers and loads [18]. Similar approaches were used in prior polysilicon connection systems; thus, this technology seems familiar. Moreover, if the density and number of CNT layers are tailored to the dimensions, driver, and load sizes, thin SWCNT may increase performance. For this reason, CNT connection development would need to be closely managed [19].

Power Local Interconnects
Nanotubes with a diameter of just one gate pitch are needed at the initial stage of the connection level to supply energy to all logic gates. The minimum feature size is 16 times the smallest feasible gate pitch. As a result of quantum and connecting resistances that, for such short lengths, should not grow with the length, CNTs may experience considerable losses. With the advancement of technology, both copper and SWNT package resistivities have risen. Gate pitch decreases as scaling decreases transistor size.

As a result, as quantum resistance becomes more prevalent, the resistance of one-gate-pitch-long SWCNT packages will grow as technology advances. Copper wires' higher resistance results from their smaller cross-sectional areas, which enhances electron scattering at wire surfaces and grain boundaries [20]. Even if faultless connections can be achieved to all metallic nanotubes, copper-like resistance requires more than one metallic nanotube per 4 nm^2 of cross-sectional size. In order to reach this density, very tightly packed SWCNTs with over 80% metallic tubes would be necessary (considering a diameter of 0.5 nm and an inner tube spacing of 0.25 nm).

Carbon Interconnects Algorithms and Techniques
Global Interconnects

The use of CNTs as power and signal interconnect at the international level is covered in this subsection. Worldwide interconnects for signals. SWNT bundles have a higher conductivity than copper wires, which reduces the delay. Global interconnects, however, typically operate in the weak resistance-inductance-capacitance

(RLC) zone because of their huge cross-sectional dimensions [21]. The RLC area, where delay progressively reduces with decreasing resistance, may be reached by a significant drop in their resistance. Because mutual inductance has a wide-ranging impact and numerous lines might collectively create sound on an output side if they all switch simultaneously, crosstalk noise can also get quite big.

Power Global Interconnects
The two most important international power distribution network benchmarks are simultaneous switching noise (SSN) and DC voltage drop, which translates to IR drop. Metallic nanotubes may number in the thousands inside densely packed SWCNT bundles. Kinetic inductance for such bundles is low in compared to magnetic inductance because magnetic capacitance is proportional to the distance to the return channel, but kinetic inductance reduces linearly with the number of metallic nanotubes. CNT bundles have magnetic inductances that are equivalent to copper interconnects. Further, the inductance of the link is often substantially higher than that of on-chip interconnects. Therefore, there is a minor difference in SSN between copper and CNT [22].

3 Conclusions

CNTs are compared to copper wires, a conventional electrical conductor, for use in the transfer of power and signals at an on-chip temperature. CNTs with a thickness of one gate pitch are required for power distribution on the metallic's ground level. Because quantum resistance does not rely on length, CNT bundles have a very high resistivity for their compact sizes. At the first connection level, power and ground must be delivered through copper wires. A novel routing possibility at the second level is the use of long nanotubes for power distribution. After the ITRS, 50 nm-diameter MWCNTs used for power distribution may provide a sheet resistance 2.3 times less than copper. Even if flawless connections can be produced, the thickness of metallic SWCNTs must be more than 1 per 3 nm^2 to distribute power globally. Due to the fact that power grids need very short nanotubes, the contact resistance of power interconnects is particularly sensitive. SWCNT bundles can only surpass copper wires worldwide in terms of resistance and delay if they are tightly packed and include more than 40% metallic nanotubes. As only 40 % of SWCNTs with random chirality are deemed metallic, the necessity for a large concentration of metallic nanotubes would entail the management of chirality. Due to their length, global signal interconnections may survive contact resistance. Large-diameter MWCNTs are advantageous for global movement interconnects due to their ability to drastically decrease resistance. The capacity to build constant low-resistance connections is required for the integration of such interconnect networks. These methods must be compatible with the copper connection technology.

References

1. Naeemi A, Meindl JD: Review of Materials Research, pp 255–275 (2009)
2. Magen N, Kolodny A, Weiser U, Shamir N: (2004)
3. Sakurai T: Perspectives on power-aware electronics. In: Proceedings IEEE International Solid State Circuits Conference, pp 26–29. IEEE (2003)
4. Nagaraj, N.S., Bonifield, T., Singh, A., Bittlestone, C., Narasimha, U.: BEOL variability and impact on RC extraction. Proc. Des. Autom. Conf., 42nd, pp 758–59 (2005)
5. Steinhogl W, Schindler G, Steinlesberger G, Traving M, Engelhardt M (2005) Comprehensive study of the resistivity of copper wires with lateral dimensions of 100 nm and smaller. J Appl Phys 97:23706–23713
6. Rossnagel SM, Kuan TS (2004) Alteration of Cu conductivity in the size effect regime. J Vac Sci Technol B 22:240–287
7. Mann D, Javey A, Kong J, Dai WQ (2003) H: Ballistic transport in metallic nanotubes with reliable Pd ohmic contacts. Nano Lett 3:1541–1585
8. Mceuen PL, Fuhrer MS, Hongkun P (2002) Single-walled carbon nanotube electronics. Nanotechnol. IEEE Trans 1:78–85
9. Graham AP, Duesberg GS, Hoenlein W, Kreupl F, Liebau M (2005) How do carbon nanotubes fit into the semiconductor roadmap? Appl Phys A 80:1141–1151
10. Naeemi A, Meindl JD (2007) Design and performance modeling for single-walled carbon nanotubes as local, semiglobal, and global interconnects in gigascale integrated systems. Electron. Devices IEEE Trans 54:26–37
11. Nihei M, Kondo D, Kawabata A, Sato S, Shioya H: Low-resistance multi-walled carbon nanotube vias with parallel channel conduction of inner shells. In: Proceedings IEEE International Interconnect Tech. Conference, pp 234–270 (2005)
12. Vajtai R, Bingqing W, Joon Y, Anyuan J, Biswas, CKS: Building and testing organized architectures of carbon nanotubes. Nanotechnol. IEEE Trans **2**, 355–61 (2003)
13. Nieuwoudt A, Massoud Y (2006) Understanding the impact of inductance in carbon nanotube bundles for VLSI interconnect using scalable modeling techniques. Nanotechnol. IEEE Trans 5:758–765
14. Cao A, Baskaran R, Frederick MJ, Turner K, Ajayan PM, Ramanath G (2003) Direction-selective and length-tunable in-plane growth of carbon nanotubes. Adv Mater 15:1105–1114
15. Ural A, Li Y, Dai H (2002) Electric-field-aligned growth of single-walled carbon nanotubes on surfaces. Appl Phys Lett 81:3464–3466
16. Huang L, Cui X, White B, O'brien SP: Long and oriented single-walled carbon nanotubes grown by ethanol chemical vapor deposition. J. Phys. Chem. B **108**, 16451–56 (2004)
17. Hersam MC (2008) Progress towards monodisperse single-walled carbon nanotubes. Nat Nano 3:387–394
18. Naeemi A, Meindl JD (2008) Performance modeling for single- and multi-wall carbon nanotubes as signal and power interconnects in gigascale systems. Electron. Devices IEEE Trans 55:2574–2582
19. Datta S (2005) Quantum Transport: Atom to Transistor. Cambridge Univ. Press, Cambridge, UK/New York
20. Dresselhaus MS, Dresselhaus G, Avouris P: Carbon Nanotubes: Synthesis, Structure, Properties, and Applications. Springer (2001)
21. Naeemi A, Meindl JD (2005) Impact of electron-phonon scattering on the performance of carbon nanotube interconnects for GSI. Electron. Device Lett. IEEE 26:476–478
22. Javey A, Guo J, Paulsson M, Wang Q, Mann D (2004) High-field quasi ballistic transport in short carbon nanotubes. Phys Rev Lett 92:106804–106804
23. Li , Yin W-Y, Mao J-F: Modeling of carbon nanotube interconnects and comparative analysis with cu interconnects. In: Proceedings of Asia-Pacific Microwave Conference (2006).

AutoML Based IoT Application for Heart Attack Risk Prediction

N. Aishwarya, D. Yathishan, R. Alageswaran, and D. Manivannan

Abstract Machine Learning is used worldwide, for many applications, healthcare is also one such application. Machine learning can be crucial in determining whether or not there will be locomotors abnormalities, heart ailments, and other conditions. If foreseen far in advance, such information can offer crucial intuitions to professionals, who can then modify their detection and approach as per patient. The real-time data from IoT based healthcare systems is obtained, which is further processed to predict the potential risk of cardiovascular diseases. Developing a machine learning algorithm to predict potential heart diseases in people is focussed in this paper. In this paper, a comparative analysis of various classifiers, including decision trees, logistic regression, SVM, and random forests is conducted. Also, an ensemble classifier that performs hybrid classification by using both strong and weak classifiers is proposed because it can have a large number of training and validation samples. Current classifier and proposed new classifiers, such as Ada-boost, Grid Search CV is analysed. The accuracy outcome can be improved by using Auto ML. Auto ML provides its suggestions on which algorithm is to be used to get the best accuracy outcome. From the results obtained, it is evident that Auto ML can reduce the time taken to perform data analytics and it act s as a more accurate method. Along with Real-time IoT data observation and auto ML, the paper provides a complete aspect to a healthcare device.

Keywords Auto ML · Grid Search CV · Cardiovascular diseases · Eval ML

N. Aishwarya · D. Yathishan · R. Alageswaran (✉) · D. Manivannan
Sastra Deemed University, Thanjavur, India

N. Aishwarya
e-mail: aishwarya.narasimhan95@gmail.com

D. Manivannan
e-mail: dmv@cse.sastra.edu

N. Hasteer et al. (eds.), *Decision Intelligence Solutions*, Lecture Notes in Electrical Engineering 1080,
https://doi.org/10.1007/978-981-99-5994-5_3

1 Introduction

Globally, 12 million people die from heart disease each year, as per the World Health Organization. The leading causes of illness and fatality among large group of people is cardiovascular diseases. Cardiovascular illness is expected to be one of the most pressing issues in the data analysis field. The worldwide burden of cardiovascular disease has risen dramatically in recent years. Numerous research has been conducted in an attempt to identify the most important risk factors for heart disease and accurately assess the total risk. Heart disease is often known as the "silent killer" since it kills people without any visible symptoms.

Making decisions on lifestyle changes for high-risk patients depends critically on the early diagnosis of heart disease, which also lowers complications. A significant amount of data generated by IoT in the health care sector can be used to make predictions and judgments with the help of machine learning. The fundamental goal of this research is to establish a model for predicting the development of cardiovascular disease.

The project's goal is to help patients with heart disease determine if their discomfort is caused by the heart or not between appointments. As a result of the obtained data, doctors can remotely monitor and diagnose patients' suffering if it is not essential. This allows patients to avoid stress, dread, and unnecessary travel to the hospital.

Furthermore, the research aims to establish the best categorization system for diagnosing heart illness in a patient. The government or any of the healthcare related organisations can use this process of data collection, data analysis and predictive analysis to foresee the risk of a person having cardiovascular disease in the near future.

The prospect of offering enhanced services for managing human health and related issues were examined by Zhao, Wang, and Nakahira in 2011 [1]. They as well as provided a study direction for medical technology on IoT. Various health monitoring related sensing equipment and protocols were scrutinized, and few problems that are to be resolved have been brought forward. One of the causes of fatalities from heart disease is that the hazards are either not recognised or are recognised until much later. However, this issue can be solved and risk can be predicted early on using machine learning techniques. Support Vector Machines (SVM), Neural Networks, Decision Trees, Regression, and Naive Bayes classifiers are some of the methods utilised for these prediction issues [2]. Neural networks, decision trees, Naive Bayes, and associative classification are effective at envisioning cardiac disease, according to analytical studies on data mining approaches [3]. Compared to standard classifiers, associative classification produces excellent accuracy and strong pliability, even when processing unstructured data [4]. Decision tree classifiers are easy to use and accurate, according to a comparison of classification techniques [3]. The best algorithm was discovered to be Naive Bayes, which was then supervened by neural networks and decision trees. Data extraction approaches like ANN, time series, clustering, and classification methods were used to predict heart attract based on 15 parameters [5]. In order to acquire better accuracy, an additional effort of feature choosing has been proposed [6].Data from the dataset was harvested using the K-means clustering algorithm, and clusters were extracted using the Maximum peak Frequent Item set Algorithm (MAFIA) to make predictions of heart disease based on the weights given to various parameters. Frequent patterns with numbers over a threshold point were shown to be reliable at identifying myocardial lesion [7]. An additional feature selection step has been proposed to increase accuracy [6]. When taking into account in particular continuous time series measurements, they enforced certain common data extraction techniques, such as anomaly detection, prediction, and decision making. Finally, they identified a number of significant obstacles for data extraction techniques in health monitoring systems based on their analysis of the literature. Alemdar, Ersoy, et al. (2010) looked into several issues with wireless sensor networks in the healthcare industry [8]. It has been established that SVM-based classifiers yield exceptionally accurate results when distinguishing heartbeats. Through Particle Swarm Optimization (PSO), the features have been improved. Using Particle Swarm Optimisation, the classifier's performance was strengthened [10]. In cutting-edge research, a variety of techniques were employed to predict heart disease risks with good accuracy; yet, some algorithms focussing on classification identifies heart disease threat with suboptimal correctness. The major cutting-edge research that achieves high accuracy uses a mixed approach that includes categorization algorithms. In this research, we are engrossed on achieving an IoT based solution along with the implementation of Machine Learning classifier algorithm like Ada Boost, and Grid Search CV. Further, we apply Auto ML to identify the suggestions and run the best algorithm provided by the Artificial Intelligence to compare with the human selected Algorithms.

2 Proposed System Architecture

The real-time data from the IoT sensors are transmitted to a database in the cloud server through a microcontroller and a gateway. Max 30,100 is used as an integrated pulse oximeter and heart rate monitoring sensor.The cloud server runs both Machine Learning and Auto ML and sends the best and accurate output to the display unit by retracing the data transfer path. The data of the patient regarding Blood pressure, oxygen levels and heart rate along with their personal details like age, weight etc. is obtained. As the data transfer occurs over the lightweight protocol, there is no overhead in transmitting the message or data through the network. The gateway acts as an intermediate between the cloud server and the microcontroller in transferring the data. Modem acts as a gateway. An OLED is used as display as it is compatible with the microcontroller we use. The data is also sent as a notification to the hospitals for further reference to the doctors. Figure 1 indicates the Architecture of proposed system.

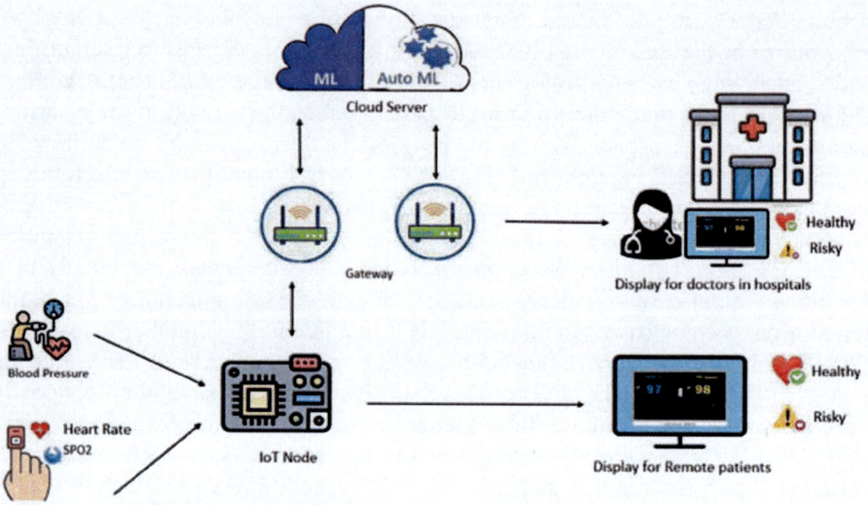

Fig. 1 Architecture of Proposed System

3 Methodology Auto ML

Automated machine learning is considered a convenient and comprehensive approach to solve and troubleshoot problems related to machine learning algorithms and models. Automatic machine learning provides end-to-end automation of ML algorithms and models. The Flow diagram for a traditional ML model is shown in Fig. 2. It is designed to perform automated data analysis to obtain precise and accurate results. An automated machine learning algorithm frees data scientists because it cleans and aggregates data and trains models automatically. With its technical attribute of automated functionality, AutoML automatically gathers data, extracts meaningful information, and detects partial data throughout the process as shown in Fig. 3. In addition, it optimizes the learning and functionality of the appropriate algorithm, automatically stores data, and detects leaks and misconfigurations. It ensures the accuracy and precision of the result, thereby eliminating the risk of bias. Additionally, since data scientists don't have to clean or collect data, organizations can use their skills to solve more severe and urgent problems.

Eval ML
EvalML is a Python-based open-source AutoML module that automates a substantial part of the machine learning process and allows us to assess which machine learning pipeline works best for a particular batch of data. It creates and optimises machine learning pipelines based on predefined goal functions. It can automatically do feature selection, model construction, hyper-parameter tweaking, cross-validation, and so on. It provides a comprehensive set of tools for understanding models. It is paired

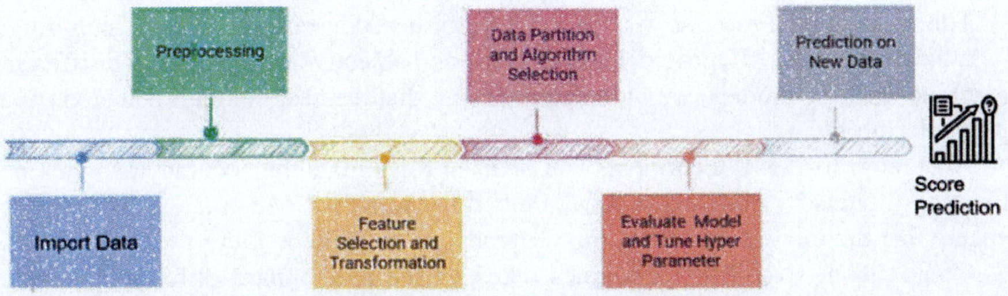

Fig. 2 Flow Diagram of Traditional ML

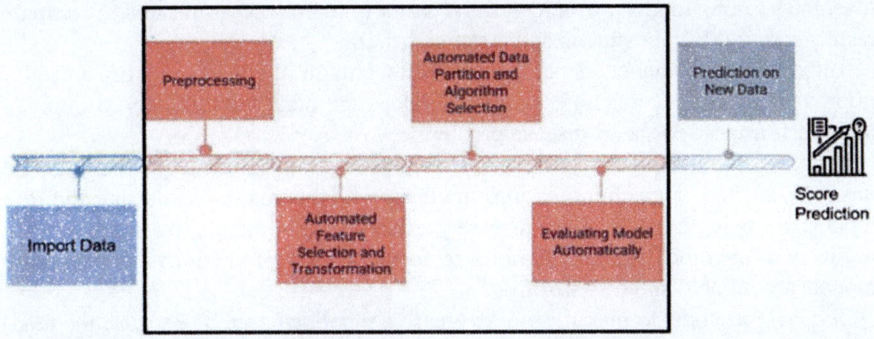

Fig. 3 Flow Diagram of AUTO ML

with Feature tools, an automated feature engineering framework, and a composed automated prediction engineering framework.

Eval ML plays an important role in heart disease prediction by allowing machine learning models to be evaluated and compared in a systematic and objective way. The performance of machine learning models can be measured using a variety of metrics, such as accuracy, precision, recall, and F1 score, and Eval ML techniques can be used to determine which models perform best for a given problem.

In the context of heart disease prediction, Eval ML techniques can be used to compare different machine learning models and determine which one provides the most accurate predictions. For example, different models may be trained on different sets of features or different algorithms may be used, and Eval ML techniques can be used to determine which approach is most effective for heart disease prediction.

In addition to comparing models, Eval ML techniques can also be used to validate the performance of a machine learning model by comparing its predictions to actual outcomes. This allows us to determine the accuracy of the model and to identify any sources of error or bias that may be present.

Eval ML allows machine learning models to be evaluated, compared, and validated in a systematic and objective way, which helps to ensure that the models being used are accurate and reliable.

There are several reasons to use Eval ML for heart disease prediction, including:

Objectivity: Eval ML provides a systematic and objective way to compare different machine learning models, which helps to ensure that the best model is selected for a particular problem.

Accuracy: Eval ML techniques can be used to validate the accuracy of machine learning models by comparing their predictions to actual outcomes. This helps to ensure that the models are providing accurate and reliable predictions.

Comparison: Eval ML techniques can be used to compare different machine learning models, algorithms, and feature sets to determine which approach is most effective for heart disease prediction.

Reproducibility: Eval ML techniques provide a standardized way of evaluating machine learning models, which makes it easier to reproduce results and to compare results across different studies and research groups.

Improved performance: By using Eval ML techniques, you can identify areas for improvement in your machine learning models and make adjustments to improve their performance for heart disease prediction.

Eval ML provides a comprehensive and objective way to evaluate and compare the performance of machine learning models, which helps to ensure that the best models are being used for heart disease prediction. Additionally, by validating the accuracy of the models, Eval ML helps to ensure that the predictions made by these models are reliable and trustworthy.

EvalML attempts to maximise or decrease a pipelined search. Because the feedback from the pipelines leads to model optimization. Regression, binary, and multi-class classification are just a few of the many supervised learning tasks that EvalML can handle. These are some of the data tests performed by EvalML.

- By feeding the model data during training, detecting target leaks.
- Identifies invalid datatypes
- Class imbalance
- Redundant features including constant columns and null columns.
- Identifies columns which are not useful for modelling.

4 Results

The study is made by using Cleveland heart data file.The dataset has 303 occurrences and 14 attributes. Eight categorical variables and six numerical variables are present. The dataset consists of 14 attributes, of which the target variable is the attribute "Heart Disease". The patients in this dataset range in age from 29 to 79. Gender values of 1 and 0 are used to identify male and female patients, respectively. There are four different types of chest pain that may be a sign of heart disease. Narrowed coronary arteries restrict blood circulation to the heart muscles, resulting at type 1 angina. Chest discomfort of type 1 angina is spurred on by mental or emotional distress. Chest pain that is not caused by angina may have many different causes and is not always related to genuine heart disease. Asymptomatic, the fourth category, might not be a sign of coronary disease. The classification purpose is to determine whether the patient will develop coronary heart disease in the next ten years (CHD). The attributes in the data file can be classified into: Demographic: Age and Sex. Information on medical History: Exercise, Cholesterol, Fasting Blood Sugar, Resting ECG, Chest Pain Type, Resting Blood Pressure, etc. If the Target attribute is 1: There is a risk of getting cardiovascular disease in the next 10 years. If the Target attribute is 0: There is no risk of cardiovascular disease.

When analyzing the data after preprocessing, we have plot certain graphs that provide us information about the distributions and variance of the attributes. During the data analysis, it is found that the variables are not highly correlated to each other as shown in Fig. 4a. Further Uni and Bi variant analysis is done on the data to determine the majority age group of the patients. This is shown as a histogram in Fig. 4b.

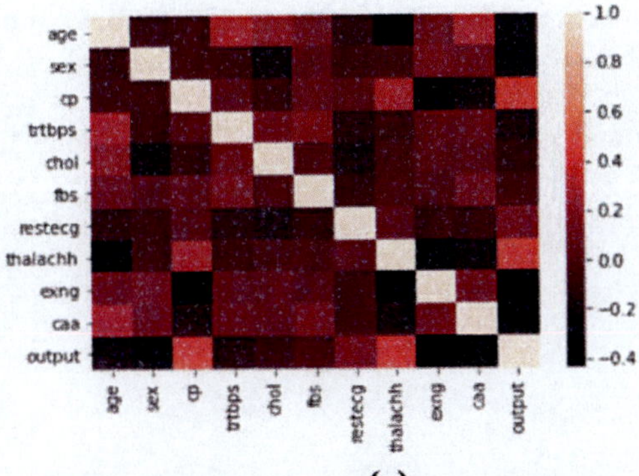

(a)

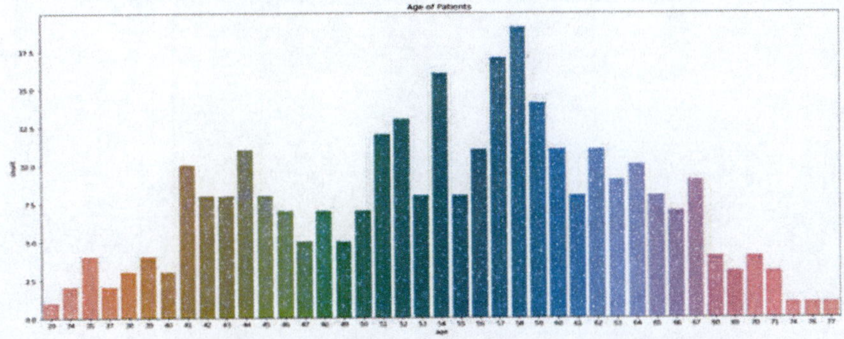

(b)

Fig. 4 **a.** Correlation Matrix, **b.** Uni and Bi variant Analysis

Table 1 a. Accuracies of different models in ML Table 1b. Accuracies of different models using Grid Search CV

Model	Accuracy (%)
Logistic Regression	81
KNN	80.5
SVM	81

Model	Accuracy
Logistic Regression	82.417582
SVM	79.120879
Random Forest	78.021978
K Nearest Neighbour	78.021978
Decision Tree	69.230769

The results of ML models are as shown in Table 1a. where Logistic Regression has the highest Accuracy followed by SVM, Random Forest, KNN and decision Tree. On tuning the hyper parameters of the data, and applying them on Logistic Regression, KNN and SVM, we obtain the accuracies to be 81%, 82.5% and 81% respectively as depicted in Table1b. As, the logistic regression model is the consistent one among all other models, we apply it on the data without hyper parameter tuning and we get the accuracy to be 82.4%. The Confusion matrix is obtained for Logistic Regression as shown in Fig 5.

The EvalML is run and the rank is provided by the Artificial Intelligence for suggesting the model is as shown in Table 2. EvalML produces an AUC of 91.23%.

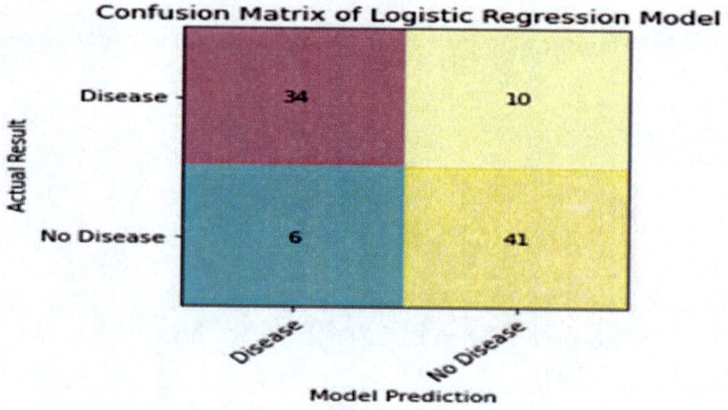

Fig. 5 Confusion Matrix for Logistic Regression

Table 2 Suggestions on Classifier provided by Auto

Sl No	ID	Pipeline_Name	Search Order	Validation Score	Mean cv score	Std deviation cv score	Percent better than base line	High variance cv	Parameters
1	5	Extra Trees Classifier w/ Label Encoder + Repl…	5	0.413358	0.413358	0.029595	97.366943	FALSE	{ Label Encoder“:{” positive_label’: None},
2	2	2 Random Forest Classifier w/ Label Encoder + Re…	2	0.420733	0.420733	0.032101	97.319964	FALSE	{ Label Encoder“:{”positive_ label’: None},
3	8	3 LightGBM Classifier w/ Label Encoder + Replace…	8	0.462099	0.462099	0.066745	97.056469	FALSE	{ Label Encoder“:{”positive_ label’: None},
4	4	1 Random Forest Classifier w/ Label	4	0.466918	0.466918	0.024541	97.025773	FALSE	{ Label Encoder“:{”positive_ label’: None},
5	9	Elastic Net Classifier w/ Label Encoder + Repl…	9	0.470037	0.470037	0.075389	97.005905	FALSE	{ Label Encoder“:{”positive_ label’: None},
6	7	XGBoost Classifier w/ Label Encoder + Replace …	7	0.488356	0.488356	0.042491	96.889214	FALSE	{ Label Encoder“:{”positive_ label’: None},
7	1	Logistic Regression Classifier w/ Label Encode…	1	0.489726	0.489726	0.0583	96.880485	FALSE	{ Label Encoder“:{”positive_ label’: None},

(continued)

Table 2 (continued)

Sl No	ID	Pipeline_Name	Search Order	Validation Score	Mean cv score	Std deviation cv score	Percent better than base line	High variance cv	Parameters
8	3	Logistic Regression Classifier w/ Label Encode…	3	0.526695	0.526695	0.022981	96.644997	FALSE	{ Label Encoder":{"positive_label': None},
9	10	CatBoost Classifier w/ Label Encoder + Replace…	10	0.638074	0.638074	0.006335	95.935522	FALSE	{ Label Encoder":{"positive_label': None},
10	6	Decision Tree Classifier w/ Label Encoder + Re…	6	6.429347	6.429347	2.401983	59.045611	TRUE	{ Label Encoder":{"positive_label': None},
11	0	Mode Baseline Binary Classification Pipeline	0	15.698798	15.698798	0.135402	0	FALSE	{ Label Encoder":{"positive_label': None},

5 Conclusion

A smart healthcare system is suggested that merges IoT and cloud technologies for real-time monitoring of heart failure patients. The framework collects information on cardiac patients using IoT-based devices. These signals are subsequently sent to the cloud unit for processing. Using signals, the machine learning model determines the patient's survivability. Improving through analyzing health-related symptoms, the prediction of heart disease patient mortality Machine learning algorithms will help save the lives of people suffering from sudden heart failure. A patient's survival rate may be increased by understanding the variables that contribute to acute heart failure. In this study, an efficient, effective, and a basic smart healthcare system is suggested for patient monitoring using failure of the heart. Findings are then sent to medical professionals so they can follow up with the patient. Our future goal is to investigate new machine learning models in the proposed intelligent framework. We examined the accuracy of various categorization algorithms and obtained the corresponding accuracies. It is obvious that Logistic Regression provides the highest accuracy, therefore ML plays an important role in analyzing heart disease. As a result, it is evident that Healthcare ML is a form of technological improvement that can solve a variety of problems.

References

1. Zhao W, Wang C, Nakahira Y: October. Medical application on internet of things. In: IET International Conference on Communication Technology and Application (ICCTA 2011), pp. 660–665 (2011). IET
2. Liu X, et al.: A hybrid classification system for heart disease diagnosis based on the RFRS method. Computational and Mathematical Methods in Medicine (2017)
3. Soni J, Ansari U, Sharma D, Soni S (2011) Predictive data mining for medical diagnosis: an overview of heart disease prediction. Int J Comput Appl 17(8):43–48
4. Thenmozhi K, Deepika P (2014) Heart disease prediction using classification with different decision tree techniques. Int J Eng Res General Sci 2(6):6–11
5. Chauhan S, Aeri BT (2015) The rising incidence of cardiovascular diseases in India: assessing its economic impact. J Prev Cardiol 4(4):735–740
6. Patil SB, Kumaraswamy YS (2009) Extraction of significant patterns from heart disease warehouses for heart attack prediction. IJCSNS 9(2):228–235
7. Alemdar H, Ersoy C (2010) Wireless sensor networks for healthcare: a survey. Comput Netw 54(15):2688–2710
8. Verma L, Srivastava S, Negi PC (2016) A hybrid data mining model to predict coronary artery disease cases using non-invasive clinical data. J Med Syst 40(7):1–7

An Overview of Security Issues in IoT-Based Smart Healthcare Systems

Shambhu Sharan, Amita Dev, and Poonam Bansal

Abstract Because of advancements in mobile computers and wireless networks, Internet of Things which is generally abbreviated as "IoT" is a hot topic in today's academic society. Patient care, on the other hand, is an essential component of healthcare procedures. The usage of IoT-based healthcare system might improve patient's life and the effectiveness of medical practitioner's services. The IoT is a dynamic and currently trending topic in various domains including smart buildings/ cities, automated patient monitoring, smart surveillance, and smart factories. As a result, it is offering novel research possibilities as well as business breakthroughs. Because of continual monitoring of patient's wellbeing, smart health is particularly significant and trending subject for scholars and scientists. The goal of smart healthcare is to give medical services to people at whatever time and from whatever location. The majority of intelligent healthcare surveillance devices are linked via the wireless communication means, which is very sensitive to attacks. Unfortunately, several vulnerabilities have been detected that potentially put such healthcare monitoring apps and services at risk. Router Attack, Select and Forwarding Attack, Replay Attack, Denial of Service (DoS) Attack, Fingerprint and Timing-based Snooping attack, and Sensor Attack are examples of such assaults. This review paper investigates the existing status of confidentiality and protection within healthcare system, the problems found in adopting security frameworks, and makes the argument for privacy and security measures.

Keywords Healthcare IoT · Internet of Things · Privacy · Security Threats · Security · Smart Health

S. Sharan (✉) · A. Dev · P. Bansal
Indira Gandhi Delhi Technical University for Women, New Delhi, India
e-mail: shambhusharan@igdtuw.ac.in

A. Dev
e-mail: vc@igdtuw.ac.in

P. Bansal
e-mail: poonambansal@igdtuw.ac.in

N. Hasteer et al. (eds.), *Decision Intelligence Solutions*, Lecture Notes in Electrical Engineering 1080,
https://doi.org/10.1007/978-981-99-5994-5_4

1 Introduction

The Internet of Things (IoT) currently is one of the well-established technology frameworks and a well-studied area. Sensing devices are now utilised practically all over, from daily use appliances to commercial surveillance systems [1]. The application of IoT and sensor-based sophisticated patient care system is fast rising. The Internet of Things tends to make our lives simpler, quite convenient, and quicker. For example, a framework delivers user-friendly speech recognition and alert functionality by utilising a smart device as the technology foundation [2, 3]. The IoT integration significantly changed overall standard of lifestyle in a variety of aspects, including relevant insights, productivity, and cost-effectiveness. IoT technologies are employed in healthcare to augment patient supervision, reduce expenses, and stimulate inventiveness in treating patients. Although industry 5.0 refers to the unification of IoT within production and consumer industries, medicine 5.0 and health 5.0 suddenly blown up in the realm of healthcare [4]. This has allowed new capabilities for online patient supervision, autonomous mobility aids, pharmaceutical administration, prior indication and effective and efficient healing programmes, asset management, as well as medical equipment upkeeping.

Monitoring patients remotely is one of the essential areas in IoT-based healthcare that helps in saving countless lives as well as wealth, but other functions are just as vital [5]. IoT-based devices can readily manage and control a variety of life-threatening conditions. Cardiovascular illness is a widespread disease that accounts for the majority of fatalities worldwide [6]. Mobile phone based health monitoring solutions are growing increasingly popular as the data communication technology transformation continues. These devices may capture authentic health data in real-time and provide recommendations to patients and the healthcare professionals. Allowing everyone to assess their health and recommending them to seek urgent care in case of an emergency can save a person's life. Wireless Body Sensor Networks (WBSN) are widely used as the key healthcare IoT solution throughout the medical industry [7]. It is commonly utilized in numerous health monitoring devices. Nevertheless, the potential application of IoT in healthcare is vast, and it might be used efficiently for advance detection, investigations, and efficacious therapies. The deployment of these screening systems has the potential to reduce medical costs for the country in the long term. Because of extensive mobile internet connectivity, combining cellular data with a healthcare system or platform based on an open-source Android architecture has become relatively simple. Electrocardiography (ECG) is becoming a commonly offered technology in current history. An ECG may correctly evaluate the heart's functioning by identifying minor alteration in voltage created by the heart tissue. Heartbeat rate and temperature of the body are two of the most crucial characteristics of the human body that contribute significantly to assessing a patient's health status. Medical professionals and the patients themselves may possibly make advantage of a smart device to continually keep track of respective heart rates, collect vital information, and undertake relevant actions to prevent catastrophic harm. The patient's heart rate is expressed by the frequency of heart beats every minute. It is sometimes described as the pulse rate of the human body. A healthy adult's typical pulse rate is between 60–100 bpm [8]. Another example is indeed the latest Apple's watch featuring Parkinson's disease symptoms detection systems [9]. WBAN sensors and actuators are attached to the patient's body dependent on the kind of ailment and caretaker data requirements. It allows medical practitioners to gather information autonomously and implement recommendation criteria to permit early involvement in the care of the patient. Diabetes is yet another extremely widespread disease all over the world. Diabetes affects around 42.2 crore individuals internationally, as per the World Health Organization (WHO) data, perhaps number is growing more and more [10].

The intelligent healthcare system domain is the realm that continually monitors the patient's health and diagnoses the patient in the event of an emergency. Figure 1 depicts the overall design of an intelligent healthcare monitoring system. Various sensing devices have been utilised in various sorts of health surveillance solutions. The sensors capture patient data and transfer it to the data pre-processing module. The pre-processing module analyses the data to earlier recorded online data to determine patient's health status. After establishing the patient's state, the system delivers feedback. Among the present works are several created methods for monitoring patient's health. Because of their wireless connection, these devices are prone to attackers. There are several risks and vulnerabilities posed by the attackers that might jeopardise such systems. Router attack, select & forwarding attack and replay attack are some of the routing-based attacks, whereas Denial of Service, fingerprint & timing-based snooping and sensor attack are examples of location-based assaults. We detail these cyberattacks including associated impact on healthcare tracking systems in this review paper. Furthermore, big data analytics, blockchain, machine learning, edge computing, biometrics, and nano-technologies are no longer limited to a single application or area. Rather, such innovative technologies might be incorporated into a variety of solutions, such as IoT security and privacy implementations. As a result, we analysed existing healthcare IoT security solutions, privacy-preserving frameworks, and procedures. Because the capabilities of IoT-based healthcare are extensive, we will concentrate on smart health tracking strategies and examine the uses of the aforementioned new techniques in the same perspective. Furthermore, we outline potential prospects for establishing safe and privacy-preserving new IoT-base healthcare applications.

2 Literature Review

Keshta defines the AI-driven IoT privacy and safety concerns & recommends how these concerns might be appropriately addressed [12]. The author's research employed an empirical research methodology, with qualitative data gathered using previous relevant and reliable information. According to the author's findings, the expansion of AI-driven IoT has brought a substantial amount of additional detectors and gadgets to the network, causing a variety of privacy and security issues

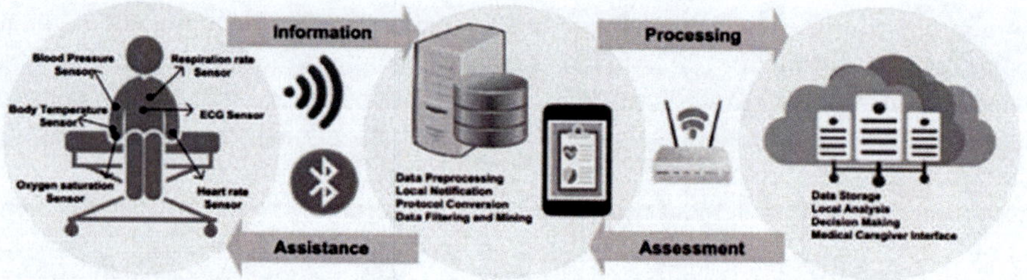

Fig. 1 Conceptual Framework of an IoT-Enabled Patient Monitoring System [11]

amongst individuals. His research suggests that the present necessity for AI-driven IoT should be to have well-defined architectural specifications including interfaces and data models that assure user's security and confidentiality.

When adopting cloud platform services, Kong et al. believe that customers are susceptible to progressively significant weaknesses in cybersecurity like information theft and privacy exposure [13]. The authors offer a security reputational system built on S-AlexNet CNN (Convolutional Neural Network) with DGT (Dynamic Game Theory) to tackle difficulties, and it is utilised for ensuring & protecting the legitimacy of healthcare information in IoT. First, adopting S-AlexNet CNN, the textual information of patient health data is categorised, followed by recommendation incentive approach as per DGT is provided. As a result, the reputational modelling of patient healthcare information security is created, & model assessment system is developed. Lastly, they conduct an exploratory investigation to validate the model's reliability and model index screening. Their experimental findings demonstrate that the system can handle the challenges of poor dependability of healthcare information screening index along with the lower precision of credit differentiation in a cloud setting. As a result, portable interface dependability is increased, and data protection and data confidentiality of portable web services were efficiently reinforced.

Butpheng et al. looked at scientific papers from the year 2017 to 2020 in order to investigate the use of intelligent approaches in health and how it has changed over the years, notably the incorporation of IoT devices with cloud-based technologies [14]. The web possesses the ability to safeguard individuals from damage and enable them to stay engaged in intelligent wellness related decision making as a storehouse for healthcare data and e-Health evaluation. High degrees of e-Health incorporation reduce possibility of discovering untrustworthy material using web. The authors reviewed several research approaches pertaining to protection and confidentiality in IoT context and mobile e-Health solutions, mainly with a focus on prospects, advantages & problems of implementing such systems. According to the authors, an intriguing emerging paradigm is the integration of IoT-based e-Health platforms with expert machines and technologies such as cloud services which enable smart goals and capabilities.

As per Thinakaran et al., while wireless connectivity and the rise of devices are ubiquitous, omnipresent, and seamless, the question of patient confidentiality continues to remain contentious [15]. This is especially visible in the health industry, where there has been an incredible increase in patient awareness about data privacy. The authors present a methodology for modelling the privacy needs for IoT-based healthcare applications in their study. They examined numerous security paradigms to determine the fundamental concepts necessary to create privacy-aware IoT health apps. Their suggested framework outlines critical privacy considerations that must be met in the creation of breakthrough IoT healthcare applications.

Data created by sensor-based gadgets, according to Hussain et al., require privacy, reliability, end-to-end (E2E) security, and authenticity for encrypted transmission over the secure channel [16]. Patient monitoring solutions related to IoT operate in a tiered fashion. Every layer of the tiered architecture possesses certain privacy & security problems that must be handled. The authors had done extensive study to address those security vulnerabilities in various IoT sectors. They also portray a security architecture for on-the-go healthcare surveillance solutions that uses two standard IoT protocols i.e., Message Query Telemetry Transports (MQTT) and Constrained Application Protocol (CoAP), to assure confidentiality, security, and authenticity of data. This security framework employs hypertext transfer protocols (HTTPs) to safeguard sensory information against cyber vulnerabilities since the same being continually transferred over the layers. As a matter of fact, it defends from the attack with a remarkably minimal risk-to-benefit ratio. The author's technique emphases on the way security architecture in Internet of Things centred healthcare systems is safeguarding using CoAP and HTTPs layers.

3 Discussion

The Internet of Things is altering human–machine interactions. A future era of technological and IoT advancements is being incorporated into the field of healthcare, which is assisting in saving crores of rupees annually. The four possible actions mentioned below must be taken in order to construct a smart patient monitoring system:

- comprehensive understanding of how to extract relevant data from smart entities linked to a network,
- data upkeeping and transferring
- system's sophisticated processing and decision making, and
- safeguarding against security flaws

For example, in a wearable device application to monitor patient vitals, numerous items are linked together, such as the watch being linked to a laptop and a smartphone, and the smartphone being linked to additional smart devices such as a blood pressure monitoring device, a heart rate monitor, and so on. If an attacker attacks the laptop, he will have easy access to the user's vitals.

3.1 Attacks on Smart Health System's Security

Healthcare gadgets that handle crucial confidential details and specific medical services data are anticipated. Similarly, such intelligent sensing devices may be linked to broader information applications for better accessibility at any time and from any location. To enable the optimal usage of IoT in the treatment and rehabilitation area, unique aspects of IoT, such as security requirements, vulnerability assessment, and defensive measures, must be recognised and examined from the medical care viewpoint. In various healthcare security related risks, cybercriminals attempt to:

- steal patient's health information,
- modify existing information or deny system services.

In IoT-based healthcare systems, there are primarily 2 main categories of attackers first being an internal attacker and second being an external attacker. The first type lives within the network & carry out hostile operations in the background. Because the attacker is present in the network, detecting it is somewhat simple. The external attackers reside outside of the network and engage in harmful behaviour. Since this types of attacker exists outside of the network, detecting it is comparatively challenging. It stealthily observes system functions before engaging in harmful activity.

3.2 Attack's Classification

In the e-Health context, there are 2 major sorts of cyberattacks commonly performed by the attackers: routing attacks and location-based attacks, as illustrated in Fig. 2. Router attacks, select and forwarding attacks, and replay attacks are some of the examples of routing-based attacks, whereas Denial of service attacks, fingerprint and timing-based snooping attacks, and sensor attacks are all examples of location-based attacks. In routing assaults, hackers primarily exploit the data route to transmit or discard packets of data. In a location-based assault, the hackers primarily target the destination or the endpoint node in order to disrupt the system's functionality. Expert systems are also particularly useful for detecting and treating different kinds of illness, maintaining private information of the patients, and providing personalised medical services. Data is likewise a difficult chore in this case.

Router Attack. Data routing is critical for healthcare related systems because it allows for distant information distribution and promotes network adaptability in large institutions. Nevertheless, routing presents a few challenges, owing to the extraordinary diversity of wireless communication systems. The adversary in this assault targets the data that is being sent across devices in a wireless sensor network. This is because the most important prerequisite in a mobile healthcare system is the secure transfer of health records at the receiver side, which might be a medical professional

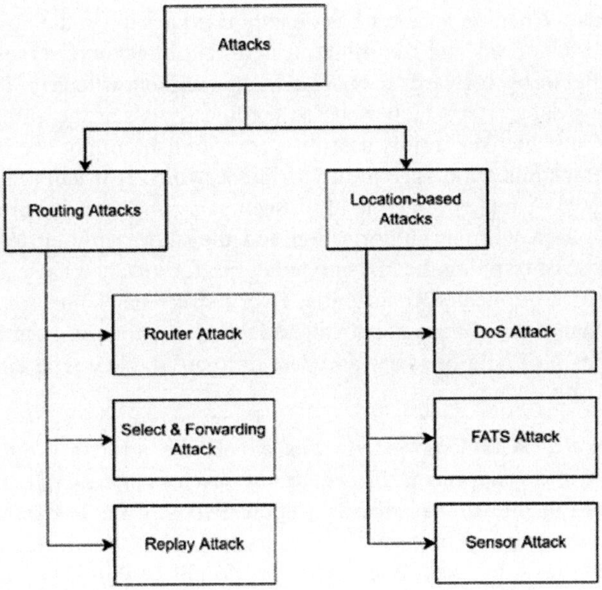

Fig. 2 Classification of attacks in smart health system

like doctor or be a health center like hospital. The directing of key and crucial information indicating the patient's health status is emphasized in this type of attacks. There are indeed a smaller number of applications which make use of multi-trust routing. Multi-trust routing essentially is critical in expanding system's incorporation area, hence providing adaptability at the price of complexities.

Select Forwarding Attack. Another sort of routing attack is the selective forwarding attack. In this type, the attacker, acting as a regular node in the routing, strategically denies packets from surrounding nodes. This attack is also referred to as community-oriented specialised forwarding. Whenever an attacker acquires accessibility to a sensing device in such kinds of assault, it loses packets of data besides sending them to other sensing device in order to raise suspicion. It will cause significant harm to the network since losing critical information in a monitoring system is undesirable. In case, sensing device is proximate to base station, then the assault does indeed have a severe influence on the system. As a result of packet drop caused by the select and forwarding attack, determining the cause of packet drop might be difficult. Insufficient data arriving the recipient side of this assault might be dangerous to any patient or intelligent healthcare system itself. These kinds of attacks might have higher risk than not having any data. Because of the reality, that, with inaccurate knowledge, individuals would be unable to perceive the wider understanding in relating to medical wellness. The modified patient data may be conveyed to the receiving side, resulting in incorrect patient care.

Replay Attack. When an attacker has unauthorised access to the system, a replay attack can be used to hack and control the system. The attacker observes the system's activity and afterwards forwards the packet to the recipient whenever the transmitter finishes sending data, after which it begins sending signals as the initial sender. The major goal of the attacker in this assault is to establish confidence in the system. The attacker transmits a message to the recipient, which is usually used during the access procedure. A replay attack is often seen as a cybersecurity breach in which certain data is kept without authorisation and then retransmitted to the recipient with the purpose of trapping the last into unlawful, for instance, false authentication or validation or a duplicated transaction. Every strike has some sort of effect on the system. Unauthorized access, data alteration, prohibition of monitoring system, changing the path of data delivery, and data drop are the key repercussions on the health monitoring system.

Denial of Service Attack (DoS). In this attack, the adversary overloads target network flow with mysterious traffic, rendering services unavailable to others since other nodes are unable to transmit data upon detecting the congested path. In a DoS attack, the adversary often takes advantage of NAV behaviour by tempering the signals in control packets. Because nodes do not double check or validate all of the signals in control packets, it is harder to identify such types of attack. In this attack, the information of patients can be accessed without authorization or data access privileges. The DoS attack also renders the system communication path busy, preventing any other information from passing to some other sensing device in the network. DoS attacks cause data transfer between node connections to be interrupted or unavailable. This form of assault jeopardises the accessibility of network or healthcare services, network operation, and sensor obligations. In a DoS attack, the adversary might easily modify patient's data, confuse recipients about health records pertaining to the patient, send fake health records, and add counterfeit health data, as well as an adversary can repeat existing transmissions to compromise message authenticity. All of these DoS attacks can result in fake therapy, a false patient condition, and a misleading distress call to any specified persons. Data manipulation can also result in patient mortality. The DoS attack primarily happens on every level of the network & employs various risks.

Fingerprint and Timing-Based Snooping (FATS). In case, message has been protected by an encryption, the wireless network may not be able to identify emerging security risks that obtain information by evaluating the data transfer from sensor to sensor. This physical layer risk simply demands the broadcast time and fingerprint of each transmission, wherein fingerprint is a collection of RF waveform properties unique to a single transmitter. This is known as a FATS attack. To counteract this technique, an adversary secretly observes all sensory data transmissions with dates and times and signatures. Considering this, the attacker employs a signature to associate each message with a distinct transmitter and use various times of deduction for each sensor site. If an attacker obtains such details, then he may also gain the ability to disturb the medical issues of the patients.

Sensor Attack. Sensors regularly leave or connect to the network owing to inadvertent malfunction of sensing units within the network as well as malevolent activity done by external attackers. Sensors in wireless networks may perish owing to a lack of power. In this instance, a skilled attacker may simply substitute the sensing unit with the genuine one, get access to the network, and carry out nefarious operations. As a result, if the patient data is not appropriately placed at numerous sensors, the attacker can modify the information as far as the hacker desires. In addition, due to inadequate of authentication schema, fake data might be entered or served as legitimate. For each round, the attacker is considered to be conscious about the origins of actual data and therefore might negotiate a group of sensors or nodes.

4 Conclusion

The IoT has arisen through several scientific topics that ultimately aids in making society smarter, including but not limited with smart buildings, smart vehicle, and smart healthcare. In smart patient monitoring system, sensing devices at a specified place transmit regular updates about the individual's health condition with the goal of improving the standard of health as well as securing the patient's life in an event of an emergent situation. Several systems have been built to provide constant intensive care of people's healthiness; however, it is getting more tough to sustain the continuous monitoring owing to security vulnerabilities posed by cyber criminals. In this paper, we compared the types of attacks based on their efficacy, strategy, and security needs as suggested security solutions for the system. These security risks have an influence on the system by denying system services, stealing data, updating data, changing the data path, and dropping data. As a result, specific security requirements for system design and development are required for safe healthcare monitoring systems. The prospect of this work will emphasis on how an intelligent healthcare surveillance system may avoid these types of assaults by mitigating enhanced security techniques to deliver continuous health records surveillance.

References

1. Javaid M, Haleem A, Rab S, Pratap Singh R, Suman R: Sensors for Daily Life: A Review (2021). https://doi.org/10.1016/j.sintl.2021.100121
2. Younis H, Hansen JHL: Challenges in real-time-embedded IoT command recognition. In: 7th IEEE World Forum on Internet of Things, WF-IoT 2021 (2021). https://doi.org/10.1109/WF-IoT51360.2021.9595903
3. Dev A, Bansal P: Robust features for noisy speech recognition using MFCC computation from magnitude spectrum of higher order autocorrelation coefficients. Int J Comput Appl 10(8):3638 (2010). https://doi.org/10.5120/1499-2016
4. Hussain F, et al.: A framework for malicious traffic detection in iot healthcare environment. Sensors 21(9):3025 (2021). https://doi.org/10.3390/s21093025

5. Malasinghe LP, Ramzan N, Dahal K: Remote patient monitoring: a comprehensive study. J Ambient Intell Humaniz Comput 10:5776 (2019). https://doi.org/10.1007/s12652-017-0598-x
6. WHO: Cardiovascular diseases. https://www.who.int/health-topics/cardiovascular-diseases#tab=tab_1. Accessed 02 Aug 2022
7. Akkaş MA, Sokullu R, Ertürk Çetin H: Healthcare and patient monitoring using IoT. Internet of Things (Netherlands) 11:100173 (2020). https://doi.org/10.1016/j.iot.2020.100173.
8. Edward R, Laskowski MD: What's a Normal Resting Heart Rate?. https://www.mayoclinic.org/healthy-lifestyle/fitness/expert-answers/heart-rate/faq-20057979. Accessed 5 Aug 2022
9. Farr C: The Apple Watch Just Got a Lot Better at Tracking Symptoms of Parkinson's Disease. https://www.cnbc.com/2018/06/09/apple-watch-adds-tech-to-track-parkinsons-disease.html. Accessed 4 Aug 2022
10. WHO: Diabetes. https://www.who.int/health-topics/diabetes. Accessed 2 Aug 2022
11. Sahu ML, Atulkar M, Ahirwal MK, Ahamad A (2022) Vital sign monitoring system for healthcare through IoT based personal service application. Wirel Pers Commun 122:129–156. https://doi.org/10.1007/s11277-021-08892-4
12. Keshta I: AI-driven IoT for smart health care: security and privacy issues. Informatics Med. Unlocked 30:100903 (2022). https://doi.org/10.1016/j.imu.2022.100903
13. Kong F, Zhou Y, Xia B, Pan L, Zhu L: A security reputation model for IoT health data using s-alexnet and dynamic game theory in cloud computing environment. IEEE Access 7:161822161830 (2019). https://doi.org/10.1109/ACCESS.2019.2950731
14. Butpheng C, Yeh KH, Xiong H: Security and privacy in IoT-cloud-based e-health systems-a comprehensive review. Symmetry (Basel) 12(7):1191 (2020). https://doi.org/10.3390/sym12071191
15. Thinakaran K, Dhillon JS, Gunasekaran SS, Chen LF: A conceptual privacy framework for privacy-aware iot health applications. Proceedings 6th International Conference Comput. Informatics, pp 175183 (2017).
16. Hussain A, et al.: Security framework for iot based real-time health applications. Electron 10(6):719 (2021). https://doi.org/10.3390/electronics10060719

Continuous Integration and Continuous Deployment (CI/CD) Pipeline for the SaaS Documentation Delivery

Bishnu Shankar Satapathy, Siddhartha Sankar Satapathy, S. Ibotombi Singh, and Joya Chakraborty

Abstract User documentation is crucial in any product development lifecycle, Software or hardware. Without it, stakeholders struggle to comprehend the more complex functionalities—usually the most valuable critical components of any software. Providing comprehensive and easily accessible product documentation always helps avoid this worst-case scenario. It enhances the user experience and takes full advantage of the Software. Many organizations prefer the unstructured local authoring license that is easy to use for the stakeholders involved in developing the product documentation. However, setting up an automated pipeline for this is a complex approach. Based on our several years of industry experience, we have presented an automated documentation deployment workflow in this article. This workflow helps stakeholders involved in decision-making to lay out a plan for documentation scope for the SaaS delivery. Following a doc-as-a-code approach, our proposed method will help the organization to save time and focus more on managing the end-user technical documentation related to SaaS customization and the application release delivery.

Keywords User documentation · Doc-as-a-code · Automation · Software as a Service · Help as a Service

B. S. Satapathy (✉)
Center for Multidisciplinary Research, Tezpur University, Napaam, India
e-mail: bishnu_mdp21@tezu.ernet.in

S. S. Satapathy · S. I. Singh
Department of Computer Science and Engineering, Tezpur University, Napaam, India

J. Chakraborty
Department of MCJ, Tezpur University, Napaam, Assam 784028, India

N. Hasteer et al. (eds.), *Decision Intelligence Solutions*, Lecture Notes in Electrical Engineering 1080,
https://doi.org/10.1007/978-981-99-5994-5_5

1 Introduction

In the recent timeframe, with the evolution of more and more Software as a Service (SaaS) based solutions, the application is becoming more intuitive with the help of an advanced UX user interface framework. Also, the solution deployment approach evolved and became quicker. A primary feature-rich solution end-to-end development cycle until deployment used to take six to twelve months. The SaaS-based system reduced even lesser than a month to complete the process. Now the question is, can the traditional method of product document delivery, which is more of a monolithic system, match the pace of SaaS delivery? The time taken for publishing and deploying the doc code manually always leads to last-moment firefighting to complete the tasks. The complexity somehow eases when any solution providers use the content management system. But it increases if any organization still uses standalone writing tools where multiple writers work in a suite of products (Theo [1].

2 Background of Software Solution Delivery Workflow

The idea of producing and maintaining technical documentation for applications is known as "help as a service" or, in short, HaaS. Many specialists sought to develop a solution that would make it easier to offer product documentation and satisfy the needs of software solutions so that when the key is complete, the documentation could be created and distributed along with the application. Doc as Code strategy has been developed recently to simplify the work by moving away from the monolithic documentation approach and delivering thin user documentation (Thomchick Richard, [11]. The objective is to enable stakeholders to integrate technical documentation with their development tools and processes, allowing them to generate, publish, and maintain it. What technical writers and UX designers may utilize the process, use the tools to update product documentation, and use the tools used by developers to manage the automated delivery pipeline, discussed in the subsequent sections (Heetae [2].

2.1 Lack of Proper Product Technical Documentation for SaaS Customization

The technical writers/UX writers' team is an essential part of the delivery team when using the conventional delivery method. Technical writers and subject matter experts spent much time working together to create high-quality product documentation. The transition from the traditional on-premises distribution method to a SaaS solution left a sizable procedural gap in the delivery of documents. Considering the fast-paced

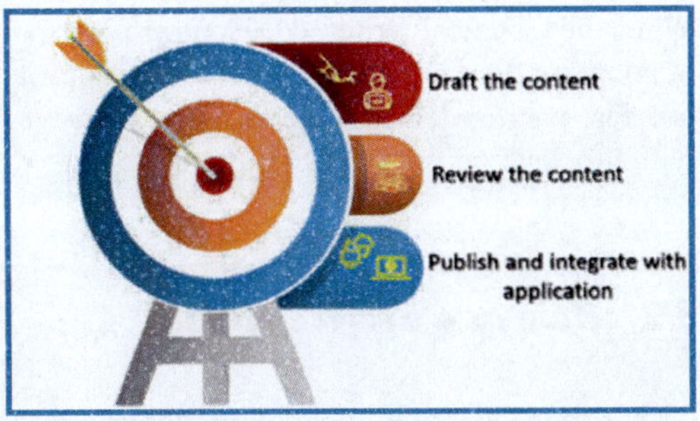

Fig. 1 Highlights the traditional documentation delivery process

solution delivery roadmap, we discovered that the development team had prioritized the development activities over the documentation process (Fig. 1). (Behutiye et al. [10]) From any incomplete product documentation, stakeholders may acquire some details. However, if referred, it will hardly be helpful during some critical issues. (Mary Poppendieck and Poppendieck [3], Rashina [9], Christian R. Prause and Durdik [4].

3 What is CI/CD Pipeline Approach?

The Continuous Integration and Continuous Deployment (CI/CD) pipeline help reduce the repetitive activities from the software development life cycle using automation tools. Mostly the pipeline runs in the background and during non-business hours, which allows the development team to focus on bringing more efficiency and productivity to the application instead of spending time on repetitive tasks. (D.M. Hutton et al. [7].

3.1 CI/CD Code and Manual Doc Approach

For collaborative authoring across the product team, many software organizations use open-source wiki-based tools like Confluence MediaWiki and GitHub/docs, which contains the documentation website code that helps push the content quickly to any external help URL [5],V. [6]. GitHub also provides the capability to maintain the application code base and the API documentation. However, the main challenge came when the team tried to implement a CI/CD pipeline for documentation using tools like Madcap Flare for authoring, and GitHub is in use only to maintain the source. Mainly organization uses such tools to maintain unstructured based product documentation. The skillset of technical writers/UX writers is more functional and lesser on the technical side. In small-scale industries, writers follow the manual approach or use an automation process in bits and pieces to avoid interrupting the development workflow.

When the technical writers/UX writers need to write documentation, it is just a matter of adopting the same development team approach, as the SaaS deployment approach is a fast pacing delivery mechanism. However, the major challenge is to integrate with the development C/CD pipeline.

3.2 *Help as a Service (HaaS)*

As per "docs-as-code," documentation is written and maintained with the same rigor as computer code. Technical documentation should adhere to a different standard than documentation created by developers. The main critical objectives of docs-as-code:

- Standardize the doc delivery strategy by outlining and polishing content eradicating the manual touch points by removing the repetitive tasks
- Collaborative authoring, docs are validated and deployed using the pipeline
- Use of a similar code base of the development team and deployment pipeline
- Usage of analytics and feedback to improve the documentation quality

The idea behind this approach is to treat documentation as an essential part of the software delivery process and for the writers to match the delivery timeline of the solution. Therefore, the source content should share the same code base developers use for coding-related activities. This approach is a fundamental shift that helps the software developer and the technical writer work in the same direction.

Advantages of Using Documents as Code. The application's code base and any required documentation should be the same. The availability of cloud platforms and a centralized code repository system have allowed application developers to concentrate on building code and correcting defects immediately rather than slowing down operations. Similar to how the agony of the traditional approach, which takes longer to resolve minor doc flaws, is felt by the stakeholders. In a conventional method, small content must go through a lengthy editorial procedure to correct a typo with the help development team.

Start with Docs-as-Code. Any organization can begin the initial step by establishing the documentation team's mission aligned with the organization. The documentation team evaluates the present workflow as soon as a company decides to move toward full SaaS delivery and finds ways to improve it. The Git-based platform is used for code management because it is open source and has several helpful DevOps tools.

Help as Code Workflow. Let's now probe into how the docs-as-code approach works in practice tools to adopt processes in place to implement it within any organization.

- **Automation**: The ability to use automation, which is implemented for source code, is another advantage of the docs-as-code approach. This approach makes it possible to introduce standard practices from the DevOps world, such as CI/CD.
- **Publishing:** Once the documentation has been drafted and staged on GitHub, the automated pipeline triggers and deploys the HTML outputs to the cloud server. This workflow provides the end users with a speedy deployment experience.

3.3 Measuring Success

Success criteria for an organization should be saving costs and time with a stable setup.

4 Research Methodology

The author has 15 years of experience in the software industry, in Technical Communication and Process Management, mainly audience analysis and working with stakeholders to manage the release reediness strategy. We have reviewed the current documentation delivery process and validated it based on the study selection criteria and skilled stakeholders from various SaaS project teams. Around 30 stakeholders were interviewed for this study. In addition, a quality assessment for the selected studies was performed by the authors.

4.1 Requirement Gathering and Analysis

During the requirement-gathering phase, we discussed with the stakeholders and business owners to understand the current process and objective they are planning to achieve. We also discussed the business challenges and bottlenecks and identified the potential business opportunities fed into the process as requirements.

- **Planning phase**. Gather details on the SaaS product documentation roadmap.
- **Prototype Designing activities.** The development stage covers designing the new automated doc delivery workflow.
- **Testing activities.** It involves pipeline performance and document validation.
- **Continuous support/monitoring and maintenance phase.** It involves monitoring and operations and other maintenance and support activities.
- **Reliability, Sustainability, Traceability, and Efficiency.** These attributes are essential in determining doc delivery and software quality.

5 CI/CD Documentation by Example

The solution's objective is to simplify the help deployment and be cost-effective, agile, and more productive in Cloud Environment. Also, this solution should provide customers with a seamless help deployment experience. To design the solution, the team needs to evaluate by decoupling all the application-specific documents so that multiple writers can collaborate. During deployment, the strategy should be configuring a pipeline to deploy changes in the subsequent releases for cloud customers.

5.1 Objectives and Key Results

Terms Used in This Document. The following table describes the application/terms used to explain the example of the CI/CD pipeline.

- **Product Documentation**: refers to the deliverable shipped with applications.
- **GitHub**: It's a web-based version control repository service owned by Microsoft.
- **Build**: Clone the files from the source to organize, control, and manage builds.
- **MadCap Flare**: An authoring and publishing tool.
- **Output:** The HTML output generated from the documentation tool
- **Jenkins:** It is an open-source tool that automates the manual part of the software development process with continuous integration and delivery.
- **Cloud Server**: A cloud server is a binary repository.

To build an end-to-end documentation workflow, identify the tasks mentioned in Image X. This will help plan/design the automation activities (Table 1).

5.2 Build Automation Workflow

As explained in Fig. 2, the tasks are executed through an automated workflow.

- **Step 1: Doc Source Update (GitHub)**: The writer clones and updates the repository to their local system, and once the content is finalized, push it to GitHub.
- **Step 2: Jenkins**: The admin triggers the build through the Jenkins dashboard.
- **Step 3: Build Server**: The utility performs the following tasks in the background:
- Clones/refreshes the Madcap Flare projects in the doc server
- Publishes the deliverables to a cloud server
- **Step 4: Cloud Server**: Microservice runs in the backend and refreshes the updated help outputs to make visible the latest output for the customers.

Table 1 Automation tasks

Area	Activities	Action Items	Description
Pre-publishing	Identify the documentation tasks to set up an environment	Evaluate the script designing scope for cloning the docs from the source	A script to clone/refresh the GitHub repositories to local repositories
	• Single sourcing • Authentication messages	Identify the files and establish the parent–child relationship across repositories, if any	A script to pull all the latest files from parent repositories to the child. Design messages for failure and other warnings
Publishing	Auto-scheduling the publishing job	Match doc publishing time with application build timing	Configure the scheduling properties to support batch publish
Infrastructure	Jenkins	Configure a doc pipeline with all scripts	Set up an end-to-end automation workflow

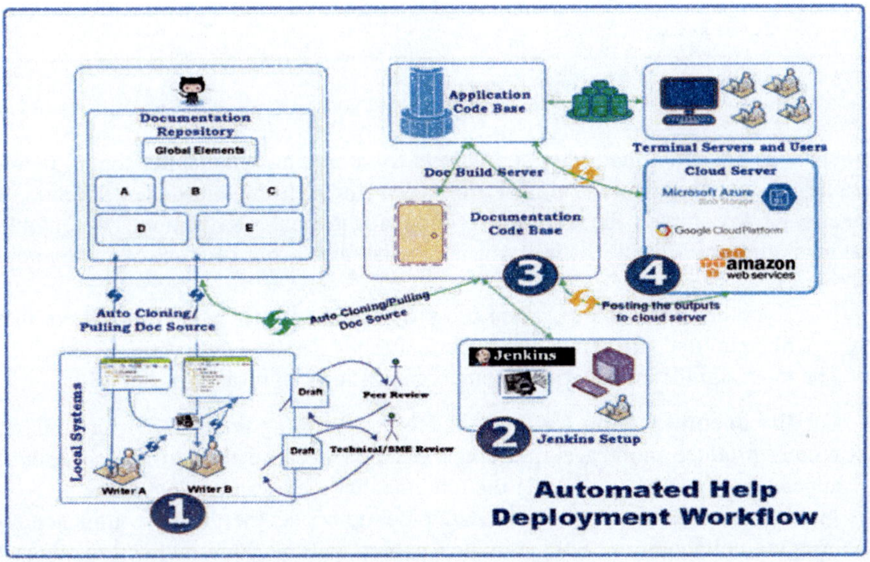

Fig. 2 Depicts the automated documentation deployment workflow

5.3 *Configure Build Automation Workflow*

To set up the CI/CD pipeline for product documentation, you may need to plan to design the following scripts. These suggestions are provided based on the input received during the design discussion. The main objective is to make the workflow scalable and modular to cater to the various application suits.

- **Clone**: You can use the shell scripting language to design the script. The script needs to clone the repositories from GitHub. When creating the jobs, consider other requirements like establishing the single sourcing, auto-triggering the build, and error reporting in the log. (Christian [8]
- **Publish**: Design the publishing script to stage/ refresh files on the cloud server.

6 An Experimental Study Using the CI/CD Pipeline

A study was done using the prototype of the CI/CD pipeline.

- Team Size: 10
- Number of active versions considered: 4
- Application suite (Eight modules) and 80 doc (50 large and 50 small) deliverables for each version, the volume ranges from 100 to 2000 pages
- Release frequency: By weekly

6.1 Result of the Testing Exercise

It has been observed that when ten writers tried cloning /publishing the 80 deliverables, it took ~ 80 h to complete the tasks. Through the automated pipeline, it takes 6–8 h to complete the whole job without any manual intervention. The tangible saving is almost 90–100%. This saving is calculated from one instance of deployment (Table 2).

In a hypothetical scenario, when the projected figure for a year considers that there will be around ~ 100 delivery occurrences for four active versions (Table 3).

The most significant benefits for any organization can be as follows:

- **Quality documentation for SaaS delivery**: Following the faster writing phase, writers produce more accurate documentation because the writing phase will appear shortly after developing the functionality they want to document.
- **Increased collaboration**: The CI/CD pipeline boosts the team's confidence by bringing collaboration. Subject matter experts will be more inclined to suggest changes and improvements by being very familiar with them.

Table 2 Result of the testing exercise

	Number of docs	Efforts
Manual	100	~ 80 h
Automated	100	~ 6–8 h during non-business hours

Table 3 Table captions should be placed above the tables

Number of docs	# Supported versions	# Supported versions	Number of deployment instances	Effort in minutes
Automated	100	4	24	~ 800 h during non-business hours
Manual	100	4	24	~ 5000 to 7,500 h

- **Complete version control**: This can track every change to the documentation quickly, and in case of an error, it can be fixed on the fly. Additionally, versioning helps to validate that the documentation is up-to-date.
- **Decoupling of huge volume content**: The stakeholders will only be responsible for the accuracy of the content, while another team will be in charge of the layout.

7 Conclusions

Adopting the Docs-as-code approach brings several benefits, both for the users of the documentation, who are often customers outside the organization. It is principal to increase the quantity and quality of documentation in SaaS delivery. While this remains challenging, practitioners will benefit from applying the identified practices and tools to mitigate the stated challenges. The team may make mistakes when implementing docs-as-code because of competency level. However, the team can be mature to use the concept over time as they have more time to focus on learning. This article may help any organization's stakeholders involved in decision-making and struggling to lay out a plan for documentation scope for the SaaS delivery. Following a doc as a code approach and the proposed workflow, an organization can manage the appropriate end-user SaaS technical documentation with a fast-pacing application release delivery.

References

Theunissen T, van Heesch U, Avgeriou P: A Mapping Study on Documentation in Continuous Software Development, Information, and Software Technology, 142:106733 (2022). ISSN 0950–5849,https://doi.org/10.1016/j.infsof.2021.106733.

Cho H, Lee S, Kang S: Classifying issue reports according to feature descriptions in a user manual based on a deep learning model. Inf Softw Technol 142:106743 (2022). https://doi.org/10.1016/j.infsof.2021.106743

Poppendieck M, Poppendieck T: Lean software development: an agile toolkit. Computer (ISSN: 0018–9162) 36 (8) (2003). http://dx.doi.org/https://doi.org/10.1109/MC.2003.1220585

Prause CR, Durdik Z: Architectural design, and documentation: Waste in agile development? In: 2012 International Conference on Software and System Process, ICSSP 2012 - Proceedings, in: ICSSP'12, IEEE Press, Piscataway,NJ, USA, pp 130–134 (2012). https://doi.org/10.1109/ICSSP.2012.6225956

Wattanakriengkrai S, Chinthanet B, Hata H, Kula RG, Treude C, Guo J, Matsumoto K (2022) Github repositories with links to academic papers: Public access, traceability, and evolution. J Syst Softw 183:11111

Jackson V, van der Hoek A, Prikladnicki R: Collaboration Tool Choices and Use in Remote Software Teams: Emerging Results from an Ongoing Study. 2022 IEEE/ACM 15th International Workshop on Cooperative and Human Aspects of Software Engineering (CHASE), pp 76–80 (2022). https://doi.org/10.1145/3528579.3529171

Hutton DM: Clean Code: A Handbook of Agile Software Craftsmanship. In: Martin RC (ed.) Clean Code: A Handbook of Agile Software Craftsmanshi, 27, (99) Prentice-Hall, ISBN: 9–780–13235–088–4 , In: Kybernetes 38.6 (June 2009), pp. 1035–1035 (2009). https://doi.org/10.1108/03684920910973252

Manteuffel C, Tofan D, Koziolek H, Goldschmidt T, Avgeriou P: Industrial implementation of a documentation framework for architectural decisions. in: Proceedings - Working IEEE/IFIP Conference on Software Architecture 2014, WICSA 2014, IEEE, pp 225–234 (2014). https://doi.org/10.1109/WICSA.2014.32

Hoda R, Noble J, Marshall S: How much is just enough? Some documentation patterns on agile projects. In: Proceedings of the 15th European Conference on Pattern Languages of Programs, ACM, pp 1–13 (2010). https://doi.org/10.1145/2328909.2328926

Behutiye W, Karhapää P, Costal D, Oivo M, Franch X: Non-functional requirements documentation in agile software development: challenges and solution proposal. In: International Conference on Product-Focused Software Process Improvement, Springer, pp 515–522 (2017). https://doi.org/10.1007/978-3-319-69926-4_41

Thomchick R (2018) Improving access to API documentation for developers with docs-as-code-as-a-service. Proceedings of the Assoc Inf Science Technol. 55:908–910. https://doi.org/10.1002/pra2.2018.14505501171

Secure & Trusted Framework for Cloud Services Recommendation-A Systematic Review

Urvashi Rahul Saxena, Parth Sharma, Gaurav Gupta, and Mihir Sahai

Abstract A dependable technology that gives us access to the with a confidential and secure environment that can be adaptable to the needs of consumers. With the help of cloud data storage, it is now feasible to store a lot of data at a low cost and with real-time access. The only those with access rights to the information kept in the cloud may access it. Cryptography can be used to provide more security to the data. Cloud with lots of benefits consists of various vulnerabilities due to its nature because of which breach of data is possible, it can be handled with TMS (Trust Management System) i.e. the secure and trusted framework by using CSTA (Cloud Security Trusted Authority) which judges the user activity on the basis of its activity. Furthermore, it is presumed that the proposed architecture lacks faith in the cloud service provider. Cloud consumer feedback is an important source for assessing trust management in the cloud framework. The CSTA is applied on the cloud where data is uploaded by the consumer and hence the judgment will be relied on the feedback of the user such that it does not provide fake recommendations. Thus in the cloud environment, it will be providing us to maintain the correctness of trust from the feedback of the consumers.

Keywords Feedback · Cloud Service · Security · Threats · Framework · Cryptography

1 Introduction

The last couple of decades with the ubiquitous use of the internet has sparked a revolution in technology that has affected every aspect of computing [1]. Interactive mobile devices are rapidly replacing computer systems, which were once created as separate systems. This has led to the creation of a form of computing that is all- pervasive/present worldwide. It has helped us form a densely networked world

U. R. Saxena · P. Sharma · G. Gupta (✉) · M. Sahai
JSS Academy of Technical Education, Noida, U.P. 201301, India
e-mail: gauravgupta200222@gmail.com

N. Hasteer et al. (eds.), *Decision Intelligence Solutions*, Lecture Notes in Electrical Engineering 1080,
https://doi.org/10.1007/978-981-99-5994-5_6

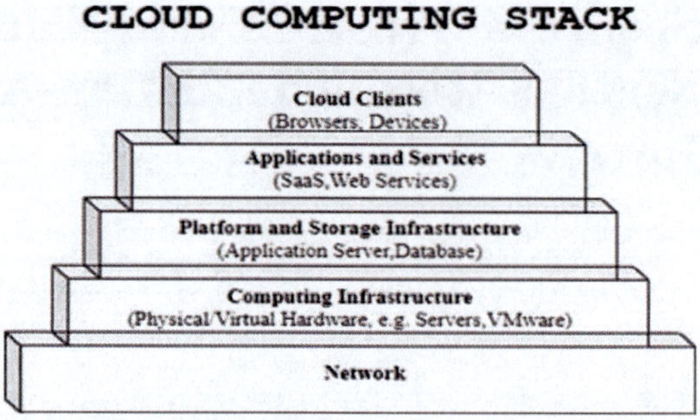

Fig. 1 Cloud Infrastructure Layers

of computers interacting and sharing with each other. With the advancement and everyday dynamism of technologies, the need for storage of data is also in demand day by day [2]. Cloud is the most prominent answer for storage of data as it is the most proficient way of real-time updating of data on cloud storage.

As we saw in the times of COVID-19 in the last couple of years, the COVID pandemic has significantly increased the use of online resources such that the need for the storage of data has grown and cloud services are the most efficient way for the storage of data. This has led to the rapid usage of cloud services. The major benefit of it is the ease of access to data for the user from anywhere and at any time. Still, the cost of maintenance and storage of the data is also high along with many other complexities. Figure 1 illustrates Cloud Infrastructure Layers, it consists mainly of three layers but can be expanded into five layers.

With all of these advancements in the field of cloud, the vulnerability has also increased tremendously for the breach of consumers' data. Threats to cloud computing services are always present, including repudiation and trust attacks like those described in [2]. The main reason for these assaults is the very dynamic, dispersed, and opaque nature of cloud computing services. That is also understandable given that it is made available to the general public, which makes it difficult for users and cloud service providers to uphold and manage confidence among cloud systems. These threats can be well understood by the fact that there are a huge number of malicious users who try to manipulate the feedback by frequently submitting fake reviews of their experiences on cloud services.

This lack of a structured infrastructure leads to an environment where user data can be compromised through various attacks. With these major points kept in the mind, we will develop a trusted and secure framework where we will be uploading the data over the cloud storage platform by moving it all through advanced encryption as the first step in the process of securing data [3]. The study revolves around how we can efficiently secure data over our cloud storage such that even if there is any breach that may happen it just is not able to manipulate the data. The whole model or STF (Secure and Trusted Framework) will rely on trust management to manage the feedback from the user by the use of CSTA (Cloud Security Trusted Authority) [4] and recommend cloud services to users by computing direct trust and indirect trust of users. Thus the review paper tries to do analysis of the available cloud frameworks with various security parameters available in the market and looks to find out the flaws, also it proposes a work which will be a secure and trusted for the cloud security need.

2 Related Work

The confidence for cloud services was accessed and managed through a variety of methods. The input from the users of the cloud resources was updated by Noor, Sheng, and Alfazi in [5, 9]. The main objective of the strategies is to determine the sporadic and recurrent reputational attacks that repeatedly attempt to compromise the Cloud computing security and privacy. As stated in [5], service providers' assessments of their clients are a trustworthy source of information that may be used to assess the dependability of cloud service consumers. Varalakshmi, Judgi, and Balagi stressed the importance of service providers' feedback in their study, which was published in [6], in order to increase consumers' trust in cloud-based services. Despite the existence of several repudiation attack models and trust management choices, a number of breaches are still possible, including collusion and Sybil attacks that allow for manipulation [1, 7, 8].

The security threats on the cloud service can be understood by the fact that every single user is able to access the data from multiple accounts, with which the wrong feedback suggestions possibilities are very much high. All of this complicates the cloud services. The study by Noor, Sheng, and Alfazi [13] in 2013 gave important details on specific assaults like on–off attacks. Instead, they examined the use of rare attack detection models in recognizing sporadic and repetitive reputation attacks like on–off attacks. The trust management review was also worked on in 2014 by Noor, Sheng, and Bouguettaya in [10, 14], but they did not address the remedies for the three attacks that were stated. A trust model was put forth by Tong, Liang, Lu, and Jin in 2015 in [11] and takes Score similarity and collusion size are taken into consideration, however the influence of scoring time or the inclusion of protections against all reputational attacks are not.

Decentralized trust management was provided by Blaze et al. in [14] using Policymaker. Its trust approach protected apps from potential harmful attacks by relying mostly on credential verification. But it didn't recommend choosing a reputed service. Kagal et al. [18, 19] say that large systems do not do well if we apply them to large-scale models with centralized security solutions. In place of this, they provide us with an alternate measure that evolves around providing the identity to the consumers.

Shand et al. in [12] proposed a mechanism that had an assessment model of risk for sharing resources. Although the model calculated suggestions very well, it struggled with lengthy chains of recommendations. A trust-based access control paradigm was developed by Ya-Jun et al. in [13] and depends on trust negotiation to establish initial trust for authenticating unknown users. Since it is a supplement to role-based access control, it is subject to its inherent problems.

Deno and Sun presented Pervasive computing's probabilistic trust management in [14] incorporates the trust value in the form of a probability and then enables a device to successfully connect with its neighbour. Given that all scenarios have the same probability, this model has the flaw of being unable to discriminate between having 100 favourable results from 200 contacts, with one positive result for every two exchanges. A methodology that details the challenges with cloud security and confidentiality is presented in the paper in [3]. Attacks and threats, security and privacy needs, and worries and dangers make up the three parts of the framework that is being given. Additionally, it gives us a safe paradigm that facilitates the resolution of challenging issues. However, it does not cover all threats. Therefore, it does not offer a cloud security approach for dealing with such attacks.

The research in [15] provides an example of a framework for assessing authenticity and secrecy in a variety of methods, by offering monitoring in real-time to ensure data integrity. The study in [16] examines a number of issues affecting user data security and privacy in a cloud environment. The report focuses on various of scalable approaches. In the research [17] analyses the number of clous security standard and advices. It suggests reporting, monitoring, auditing and security events to find the actual problem.

3 SWOT Analysis of Secure and Trusted Framework (STF) for Cloud Services Recommendations

In this section we illustrate the functionality of secure and trusted framework (STF) followed by its SWOT analysis. STF is a framework for securing the communication between different entities in a network in a trusted way of authenticating and authorizing each other.

3.1 Strengths of STF

Following represents various strengths of the STF(Secure and Trusted Framework) with their descriptions.

I. Less Vulnerable: The Third Party framework pattern upgrades the security of a company, specifically against cloud network attacks that may appear under a different mock-up of security as the data is stored in encrypted format in the cloud storage.
II. Strong policies for user access and authentication: Third-party framework demands greater user control within the network, leading to more secure accounts. User roles are categorized to give them the right to utilize accounts and encrypted data as necessary for their job responsibilities.

III. Smart segmentation of data: Separating a company's cloud, securing the important abstract property from unauthorized users. reducing the attack area by keeping vulnerable systems pretty secure with encryptions. There shouldn't be any lateral network migration of threats.
IV. Increased data protection: Using various encryption techniques, data security is maintained during both storage and transmission. The transmission of messages and automated backups are encrypted and hashed. Users authorized and listed in the access control list have access to the data.
V. Feedback and trust Decision Engine: Using a range of facts about the feedback histories saved in the central repository and the trust values of a specific client, it investigates and acknowledges the significance of trust in the data owners, roles, and entity values.

3.2 Weakness of STF

The Zero Trust method complicates security procedures due to all of these additional security advantages. Below are some weaknesses:

I. Time and effort to set up: Reorganizing policies in the existing cloud services might be difficult. Better to switch over after creating a brand-new network from begin. Starting from scratch is necessary for incompatible legacy cloud networks with a safe and reliable framework design.
II. Increased management of varied users: Reorganizing existing network policies while they continue to operate during the transition may be difficult. It would be better to switch over after completely redesigning the network.
III. More complicated application management: In Addition to cloud management, the system administrator also needs to manage framework requirements.

3.3 Opportunities of STF

The STF methodology makes no explicit mention of reaching total efficacy. STF underlines that user identity and feedback verification must come first when using cloud services. There must be a strong plan for identity governance and management. STF offers an abundance of opportunities and responsibilities, as the name suggests.

I. Multi-factor authentication (M.F.A.): Multi-factor authentication (MFA) provides an extra layer of security by requiring users to provide a number of different forms of identification (factors) in order to confirm their identity and obtain permission to use a network or multi-cloud environment.
II. Cloud Audit: On your cloud network, do an audit of the identities, access limitations and access policies that are currently in place. Understanding where your data and applications are located, as well as access policies and access controls,

are crucial as you start to develop the security and access protocols for your network.

III. Micro-segmentation: Whole STF is divided into multiple segments. They are the Central Repository, Feedback & Trust Decision Engine, Cloud Data Encryption Application (CDEA), etc.

IV. Assured Security: Using an identity and access management system that can automate, govern, regulate, check, and monitor how access is utilised inside the network, provision access based on user roles, and authenticate these users' identities before providing them access to your network and apps.

3.4 Threats of STF

When a new solution to a challenging issue develops and emerges through time, there are always certain risks. Implementing a Secure and Trusted Framework in a company requires more than a mental adjustment. It will be necessary to have a complete awareness of the roles played by the various departments of the business, as well as what the future requirements for each of those areas would entail. The greatest risk exists here. According to STF, building a Secure and Trusted Framework from the ground up is frequently simpler than redesigning an existing cloud Service because the present existing cloud services must continue to function during the alteration period. IT and security teams should always create a plan that outlines the desired end infrastructure as well as a step-by-step procedure for getting there.

3.5 This Review Paper Will Be Trying to Answer Some of the Research Questions(RQs)

RQ1. How is CSTA (Cloud Security Trust Authority) how is it going to handle its new, delicate duty of protecting when the data are not client data kept on its property?

A1.Data is supposedly held at a the data centre of a cloud service provider, but CSTA is now responsible for keeping such facilities secure. This is easily fixed by having the encrypted data transmitted to CSTA for decryption, processing, and re-encryption before being sent back or stored once more in the data centre of the cloud service provider, as seen in Fig. 1.

RQ2. How trust will be evaluated?

A2. The trust will be evaluated on the basis of the behaviour and feedback attributes of the user. The Trust Computation Module or TCM computes the trust value for Cloud Service Providers (CSP). It consists of the following components: (Figs. 2 and 3)

Trust Manager (TM): The trust value for the specific CSP is calculated by the trust manager (TM) using the behavioural and feedback characteristics.

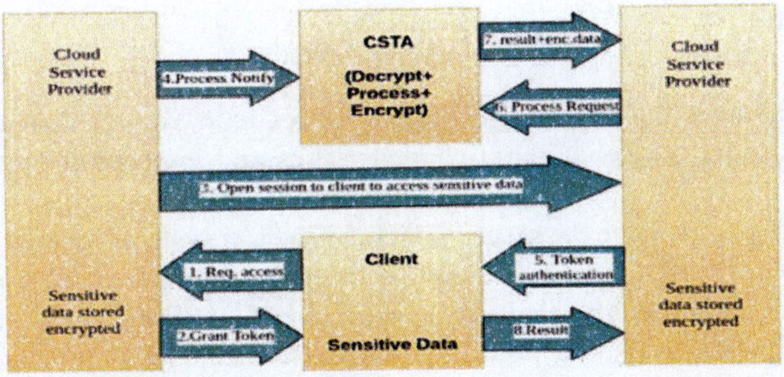

Fig. 2 Cloud security trusted authority scenario

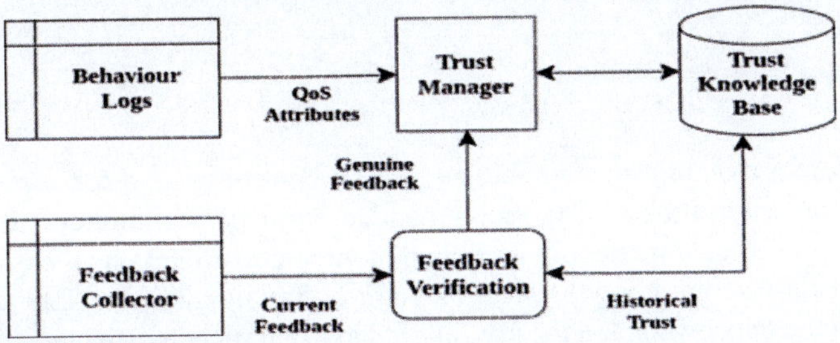

Fig. 3 Trust Evaluation Illustration

The Feedback Collector (FC) oversees regularly updating and maintaining the CSP's feedback properties.

Feedback Verification (FV): It is in charge of determining whether the feedback is authentic for calculating trust..

Behaviour Logs (BLs): It keeps track of all behavioural characteristics and updates them on a regular basis.

Trust Knowledge Base (TKB): The knowledge base known as the Trust Knowledge Base (TKB) is where calculated trust values are stored and retrieved as needed. Additionally, all CSPs update their trust levels at predetermined times.

RQ3. How will our data be encrypted?

A3. Homomorphic encryption can be used for private cloud computing and storage. This makes it possible for data to be encrypted while being treated in commercialized cloud environments. A fully homomorphic encryption scheme permits an infinite number of additions or multiplications of the ciphertext without compromising the integrity of the result. **Homomorphic Encryption** and Secure Multi-Party Computation are two techniques that the cloud team of experts is innovating in Privacy Preserving Machine Learning to help cloud services and its customers meet their security and compliance goals while enabling them to benefit from the flexibility, scalability, performance, and ease.

4 Conclusion

This study assesses the importance of trust for a Cloud Service Provider (CSP). It takes into account the behavioural characteristics and user comments on the CSP to calculate the cumulative trust value. This study concentrates on creating a trust model system to guarantee the security of cloud storage systems in order to safeguard cloud services against reputational assaults. Trust algorithms are used to protect cloud services from reputational assaults by successfully identifying malicious as well as unsightly behaviours. To stop cloud assaults, each trust algorithm includes a variety of scenarios. Control models and trust models can be integrated to address these issues.

5 Future Work

We'll present a dynamic trusted and suggestion control system for accessing such a widespread environment. The recommended strategy will address situations in which a requester has previously used the resource and an unknown entity is asked to utilize it despite not knowing who they are or having any prior interaction with them. Both the data owner and the user might have confidence in the model. The idea of maximum possible trust will be used to control vulnerable-disposed behaviour. Security and the level of encryption for the safety of the data will be further beefed up, seeing the past behaviour of the user. Further, the access control list will be managed according to the activity of the consumer behaviour and their feedback.

References

1. Zhou L, Varadharajan V, Hitchens M: Integrating trust with cryptographic role-based access control for secure cloud data storage. In: 12th IEEE International Conference on Trust, Security and Privacy in Computing and Communications (2013)
2. Mehraj S, Banday MT: Establishing a zero trust strategy in cloud computing environment. In: International Conference on Computer Communication and Informatics (2020)
3. Youssef AE. Alageel M: A framework for secure cloud computing. Int J Comput Sci Issues (IJCSI) 9(4):487–500 (2012)
4. Dawoud MM, Ebrahim GA, Youssef SA: In the introduction of A Cloud Computing Security Framework Based on Cloud Security Trusted Authority (2020)
5. Noor TH, Sheng QZ, Alfazi A: Reputation attacks detection for effective trust assessment among cloud services. In: 12th IEEE International Conference on Trust, Security and Privacy in Computing and Communications, pp 469–76
6. Varalakshmi P, Judgi T, Balaji D: Trust management model based on malicious filtered feedback in cloud. In: International Conference on Data Science Analytics and Applications, pp 178–87
7. Bhatt S, Sandhu R, Patwa F: An access control framework for cloudenabled wearable internet of things. In: IEEE 3rd International Conference on Collaboration and Internet Computing (CIC), pp. 213–233 (2017)

8. Deng W, Zhou Z: A flexible RBAC model based on trust in open system. Third Global Congress on Intelligent Systems (2012)
9. Noor TH, Sheng QZ, Alfazi A: Detecting occasional reputation attacks on cloud services. In: International Conference on Web Engineering, pp 416–23
10. Noor TH, Sheng QZ, Bouguettaya A: Trust Management in Cloud Services. Springer (2014)
11. Tong W, Liang J, Lu L, Jin X: Intrusion detection scheme based node trust value in WSNs. In: Systems Engineering and Electronics, pp 1644–1649 (2015)
12. Shand B, Dimmock N, Bacon J: Trust for ubiquitous, transparent collaboration. In: 1st IEEE International Conference on Pervasive Computing and Communications, Texas, USA, pp 153–160 (2003)
13. Ya-Jun G, Fan H, Ping-Guo Z, Rong L: An access control model for ubiquitous computing application. In: 2nd International Conference on Mobile Technology, Applications and Systems, China, pp 128–133 (2005)
14. Deno MK, Sun T: Probabilistic trust management in pervasive computing. In: IEEE/IFIP International Conference on Embedded and Ubiquitous Computing, China, pp 610–615 (2008)
15. Wang C, Liu C, Liu B, Dong Y: DIV: Dynamic integrity validation framework for detecting compromises on virtual machine-based cloud services in real-time. China Communications **11**(8):15–27 (2014)
16. Tari Z: Security and privacy in cloud computing. IEEE Cloud Computing 1(1):54–57 (2014)
17. Lang U, Schreiner R (2011) Analysis of recommended cloud security controls to validate open PMF "policy as a service." Elsevier, Information Security Technical Report 16(2011):131–141
18. Kagal L, Finin T, Joshi A (2001) Trust-based security in pervasive computing environments. IEEE Comput 34(12):154–157. https://doi.org/10.1109/2.970591
19. Kagal L, Undercoffer J, Perich F, Joshi A, Finin T: Vigil: enforcing security in ubiquitous environments. In: Grace Hopper Celebration of Women in Computing 2002 (2002)

Time-Series Based Prediction of Air Quality Index Using Various Machine Learning Models

Ishita Pundir, Nitisha Aggarwal, and Sanjeev Singh

Abstract Accurate prediction of the Air Quality Index (AQI) is of paramount importance as the negative health impact of poor air quality on humans has been widely established. The paper proposes an efficient model for AQI prediction based on the transformer algorithm. This model is compared with the widely used RNN-LSTM and regression model and outperforms both. Real-time data on pollutant concentration and meteorological data for the Anand Vihar, Delhi, India monitoring station from November 1, 2017, till August 6, 2022, has been used for analysis. From the data available for 11 pollutants, the AQI is determined as per the Central Pollution Control Board's (CPCB's) formula. The paper aims to predict the AQI along with the levels of PM2.5, CO, and PM10 pollutants. RMSE and MAE for PM2.5 evaluated using transformer are 17.74 and 11.15, respectively, best amongst all the models. Prior and accurate knowledge about Air Quality Index levels is directly relevant for policy-makers and the population at large so that the administrators may implement corrective measures. As the concept of smart cities is rapidly taking shape, the residents of such cities have the right to know about the air quality they are likely to breathe in the near future so that smart responses may be planned well in advance.

Keywords AQI · Time series analysis · Deep learning · LSTM · Transformer

I. Pundir (✉)
St Stephens College, University of Delhi, New Delhi, India
e-mail: ishitapundir0205@gmail.com

N. Aggarwal · S. Singh
Institute of Informatics and Communication, University of Delhi South Campus, New Delhi, India
e-mail: nitisha@south.du.ac.in

S. Singh
e-mail: sanjeev@south.du.ac.in

N. Hasteer et al. (eds.), *Decision Intelligence Solutions*, Lecture Notes in Electrical Engineering 1080,
https://doi.org/10.1007/978-981-99-5994-5_7

1 Introduction

The AQI is a critical index as it measures how unhealthy the ambient air is and can be used to predict how polluted it may become in the near future. Changes in AQI can significantly impact public health and the global climate. Effective methods of predicting air quality are critical, however, due to significant variations in pollutant levels and large amounts of data, AQI prediction in a metropolitan city is highly uncertain and challenging, yet very essential [1].

Air quality forecasting methods, particularly traditional ones, reflect the statistical connections using processes such as ARIMA, VAR and other regression methods. These models are able to achieve satisfactory accuracy in the AQI forecasts but require restrictive assumptions like stationarity in time series to be met before implementation. Therefore, they still have space for improvement in accuracy and practical approaches that still need to be explored for modeling AQI. Recently, AI-based forecasting models are attracting more attention achieving better performance and effectively dealing with large-scale data volumes, nonlinearities and interactive relationships amongst determinants in AQI modeling. Because of their versatility in modelling complex patterns underlying the data, neural network models such as convolutional neural network (CNN), deep belief network (DBN), and long short-term memory (LSTM) are the most widely used machine learning (ML) methods.

This study aims to i) Develop an LSTM architecture that can forecast short and medium-term AQI along with pollutant concentrations with enhanced accuracy. ii) Apply the attention-based transformer model to the time series data to compare its prediction accuracy with the earlier model. iii) To identify the optimum lags of the input variables and target variables that can predict future values of the target variables with maximum accuracy. iv) Identify the proposed models' relative merits, applicability, and effectiveness in predicting the AQI over a varying forecast horizon.

The contribution of this paper includes (i) a deep learning (DL) model for accurate short and mid-term prediction of AQI. (ii) identification of optimum lags of the past data required for future prediction, along with determining how far into the future can the proposed model give reliable predictions.

2 Related Work

The big data analytics approach has recently been widely used due to advancements in computational capacities and networks that can sense the environment and generate data using sensors. Gaganajot et al. investigate ML-based techniques to forecast quality of air on big data [2]. They reviewed and compared recent research on monitoring and evaluation of real-time air quality using artificial intelligence techniques for smart cities and shed light on some of the future research needs and current challenges. The study by Maleki et al. and Kelly et al., using an Artificial

Neural Network (ANN) method, estimated the hourly air contaminant concentration in terms of Air Quality Health Index (AQHI) and AQI parameters, which usually have health, agricultural, social, economical and political consequences [3, 4]. The study by Liu et al. focused on forecasting the levels of PM2.5 and suggesting measures to citizens about timely response mechanisms that can be put in place to offset the negative impact of excessive smog [5]. Another study used Lag-FLSTM, an LSTM model with a fully connected network based on Bayesian Optimization for prediction of quality of air using multivariate analysis [6]. The model attained 23.86% reduced RMSE than other contemporary techniques and could optimize the various meteorological features and pollutants that influence the prediction of PM2.5. In the Indian context, a study focused on determining the levels of air pollutants and AQI in the cities of Delhi and Agra to inform people about pollution trends well in advance [7]. Another study by Krishan et al. presented a forecasting model for PM2.5, NOx, and CO concentrations by applying the deep learning method, LSTM, at a location in National Capital Territory (NCT) - Delhi, considering the factors such as vehicular emissions, meteorological conditions and traffic data [8]. Rao et al. used LSTM to predict AQI and capture the dependencies in various pollutants. A real-time dataset consisting of 12 pollutants from *Visakhapatnam* city was taken and the modeling of time series data of each pollutant along with meteorological variables was performed for forecasting hourly concentrations [9].

3 Methodology

The methodology includes identifying the geographic area, data description and pre-processing, splitting of data, training and optimization of the model and finally, assessing the performance of models, as depicted in Fig. 1. The geographic study area is Anand Vihar (Latitude: 28.650218°, Longitude: 77.302706°), in Delhi, India, which borders the states of Uttar Pradesh and Haryana and one of the busiest road-railway junctions in the NCR. It is also one of the most polluted points in the world. As per our information, this study is the first attempt to forecast AQI utilizing AI for this study area.

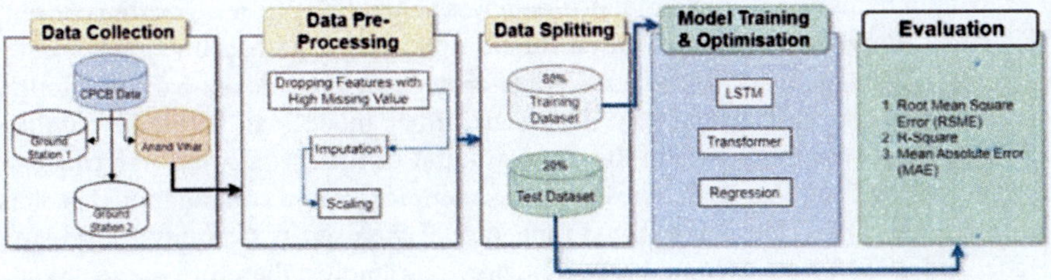

Fig. 1 Diagrammatic Representation of the Methodology

3.1 Data Description and Pre-processing

Data has been sourced from the CPCB's website. It's a time-series data obtained by combining observations from different sources and pollution monitoring points in real-time. Hourly data of the parameters (mentioned below) from November 1, 2017, till August 6, 2022 (41,748 labels) has been used for the analysis. Features having more than 80% missing values like temperature, rainfall, and total rainfall have been dropped from the experiments. All other missing values have been imputed using the interpolation method used widely for time-series data.

Since the scale and units of variables are different, the variables have been scaled in the range of [0, 1] before further processing. Using historical data, the AQI has been calculated for the Anand Vihar monitoring station for each hour over the aforesaid time-period according to the CPCB definition. The CPCB measures and reports data on eleven pollutants, namely PM2.5, PM10, CO, SO2, NO, NH3, NO2, Toluene, Benzene, Ozone and NOx and six features Relative Humidity (RH), Solar Radiation (SR), Pressure (BP), Wind Speed (WS), Ambient Temperature (AT), and Wind Direction (WD). The Sub-indices for each of the major pollutants at a monitoring station are calculated using their mean level over the past 24 h (8 h for CO) and the concentration range beyond which the level of the pollutant is harmful. AQI is the highest sub-index of any of the pollutants for that location. It is not necessary that each pollutant be recorded at all locations. Therefore, AQI at the monitoring station can be calculated only if data can be measured for at least three pollutants, one amongst them should essentially be PM2.5 or PM10. In this study, the initial 80% of the records is used for training and optimization and the last 20% records is for testing and evaluating the model performance.

3.2 Model Optimization and Training

The present study aims to suggest a framework to accurately forecast the levels of air pollutants and the AQI for the next 1, 4, 8, or 12 h using the lagged values of the target and the input features. This study has optimized models on the different lags (4 and 12 h) and future horizons (1, 4, 8, and 12 h). Several functions could relate the lagged input features with the target in time series depending on the algorithm chosen. This paper employs Regression-based Random Forest as the baseline model, deep learning (LSTM), and attention-based Transformer models to develop the prediction framework using the Darts library. The transformer model can parallelize, unlike LSTM and process more data in the same amount of time. It also has the capacity to capture long-range dependencies with less complexity and computational cost as compared to LSTM. Hence, the Transformer model was used in this study. Regression Model and LSTM were used in most of the previous studies; therefore, these models were used to compare the result obtained by the Transformer model. The number of epochs was 50 for both LSTM and the transformer.

3.3 *Performance Evaluation Metrics*

For measuring models' performance, three metrics were used, namely the (i) RMSE, (ii) MAE, and the coefficient of determination (R^2) for measuring the overall explanatory power of the model.

4 Experimental Result and Discussion

The summary statistics of the various pollutants and chosen meteorological parameters are depicted in Table 1.

The present study involves multivariate time series prediction where the behavior of the pollutants depends upon the meteorological input features mentioned above. Some of these features depict seasonality, as evident in Fig. 2. The AQI calculated and categorized at the Anand Vihar monitoring station is either moderate or very poor for most hours, Fig. 3. The AQI also depicted immense variation, ranging between 44 and 856 and a standard deviation of 155 (Table 1).

The performance of the deep learning models, namely LSTM and transformer, along with the baseline random forest-based regression model, is tabulated in Tables 2, 3, 4 and 5. Past four and twelve-hour data is used for prediction. Similar algorithms have been run to assess the performance of the three models for the forecast horizon of 4 h (Table 3), 8 h (Table 4) and 12 h (Table 5), respectively.

As evident from the results above, the LSTM model performs well for the next hour prediction when we use a shorter input length of 4-h for some pollutants. When the input length is 12-h, the transformer seems to give a lower MAE and RMSE value. Also, the transformer shows better results for longer forecast horizons. In other words, the transformer and LSTM models compete for short-term prediction. For long-term prediction, the transformer model is much better. The performance of all three models deteriorates as we try to predict further into the future time horizon (Tables 4 and 5).

Table 1 Summary statistics of major pollutants and meteorological input features

	PM2.5	PM10	CO	Ozone	AQI	AT	BP	WS	WD	SR	RH
Min	0.20	1.00	0.00	0.10	44.00	1.30	727.08	0.20	9.00	3.00	0.23
Q1	53.00	142.25	1.38	14.28	166.00	20.03	971.70	0.30	108.50	9.50	42.45
Q2	93.50	243.38	1.90	23.17	278.00	27.67	972.92	0.47	212.80	43.45	57.62
Q3	182.00	381.00	2.70	43.90	384.00	33.20	981.12	1.20	261.83	389.52	71.85
Max	985.00	1000.0	19.70	199.50	856.00	46.70	999.90	8.36	340.50	744.17	99.70
Mean	138.42	286.15	2.24	33.90	292.67	26.61	935.74	0.87	189.99	179.98	56.61
Std.dev	127.78	193.38	1.44	30.27	155.59	8.45	90.11	0.83	77.08	184.76	20.89

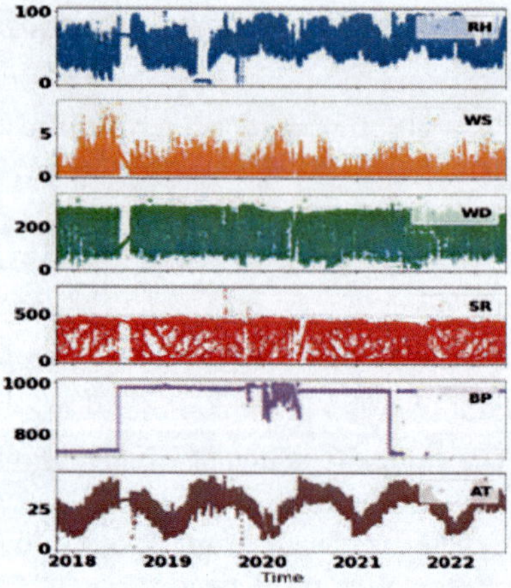

Fig. 2 Behavior of input meteorological features over time

Fig. 3 Air Quality categories over the chosen time horizon

Air Quality Category	count
Moderate	14041
Very Poor	9242
Severe	8907
Poor	7135
Satisfactory	2383
Good	25

Table 2 Model evaluation for a forecast horizon for the next hour [hrs - Input Length of Time]

hrs	PM2.5			PM10			CO			AQI		
	RMSE	MAE	R^2	RMSE	MAE	R^2	RMSE	MAE	R^2	RMSE	MAE	R^2
LSTM												
4	19.15	12.18	0.82	21.28	14.00	0.78	32.30	19.76	0.47	19.44	11.80	0.38
12	25.25	14.82	0.63	22.46	15.56	0.78	13.80	8.25	0.88	20.53	13.82	0.73
Transformer												
4	20.79	12.28	0.33	24.84	15.31	052	17.88	9.45	0.81	17.10	10.07	0.73
12	17.74	11.15	0.78	22.06	14.42	0.79	13.80	8.25	0.88	15.46	10.71	0.79
Regression												
4	28.09	21.17	0.78	33.14	25.00	0.78	21.38	16.68	0.72	28.13	21.47	0.78
12	25.57	18.04	0.77	30.45	22.13	0.79	17.40	12.78	0.73	26.46	19.39	0.79

Table 3 Four hours ahead forecast horizon

hrs	PM2.5			PM10			CO			AQI		
	RMSE	MAE	R^2	RMSE	MAE	R^2	RMSE	MAE	R^2	RMSE	MAE	R^2
LSTM												
4	31.94	20.26	0.70	37.19	25.57	0.65	23.38	15.31	0.44	27.90	18.77	0.44
12	25.41	13.52	0.74	26.34	16.28	0.65	20.14	12.86	0.75	23.01	14.51	0.60
Transformer												
4	30.95	18.75	0.27	37.07	26.01	0.68	22.79	14.94	0.54	36.88	21.90	0.72
12	21.04	11.85	0.79	24.51	18.01	0.64	19.69	11.51	0.83	21.24	13.34	0.74
Regression												
4	36.53	28.18	0.61	44.47	34.56	0.61	26.77	22.19	0.56	34.66	26.23	0.63
12	29.34	20.65	0.69	29.45	21.56	0.80	20.32	14.88	0.62	29.06	21.15	0.70

Table 4 Eight hours ahead forecast horizon

hrs	PM2.5			PM10			CO			AQI		
	RMSE	MAE	R^2	RMSE	MAE	R^2	RMSE	MAE	R^2	RMSE	MAE	R^2
LSTM												
4	32.32	21.83	0.59	49.35	32.57	0.45	29.17	18.13	0.43	32.25	22.04	0.57
12	29.94	19.36	0.58	39.44	26.74	0.56	24.67	17.05	0.21	28.22	18.07	0.53
Transformer												
4	33.66	22.03	0.50	40.79	28.35	0.61	24.67	16.21	0.44	30.50	20.30	0.45
12	28.16	17.57	0.63	37.37	25.39	0.50	19.81	12.34	0.51	25.92	16.60	0.68
Regression												
4	36.92	28.76	0.57	47.77	37.01	0.54	26.89	22.39	0.49	39.99	31.69	0.63
12	33.43	24.52	0.61	42.81	31.50	0.60	21.81	16.22	0.55	34.13	25.56	0.64

Table 5 Twelve hours ahead forecast horizon

hrs	PM2.5			PM10			CO			AQI		
	RMSE	MAE	R^2	RMSE	MAE	R^2	RMSE	MAE	R^2	RMSE	MAE	R^2
LSTM												
4	37.05	26.03	0.51	47.03	34.74	0.49	34.65	22.26	0.21	35.32	25.09	0.45
12	29.42	18.63	0.60	40.26	27.42	0.53	21.34	14.01	0.34	30.82	20.88	0.54
Transformer												
4	37.23	25.52	0.0.41	44.84	31.55	0.53	34.86	22.89	0.53	33.45	23.36	0.35
12	30.35	19.38	0.56	38.99	25.81	0.56	21.41	13.46	0.39	27.47	17.68	0.58
Regression												
4	42.67	34.08	0.47	54.24	42.73	0.42	32.57	27.87	0.38	45.13	37.13	0.43
12	34.63	25.58	0.55	45.05	33.54	0.55	23.45	17.89	0.48	37.31	28.45	0.56

The optimum lag for the AQI prediction for all pollutants is seen to be 12 h, as most pollutants give lower RMSE and MAE values for it. RMSE value of the three models for the next hour and next 4-h prediction is given in Fig. 4 (a) and (b). A comparison of the actual AQI value with the corresponding levels predicted by the transformer model for different forecast horizons is shown in Fig. 5 (a)–(d).

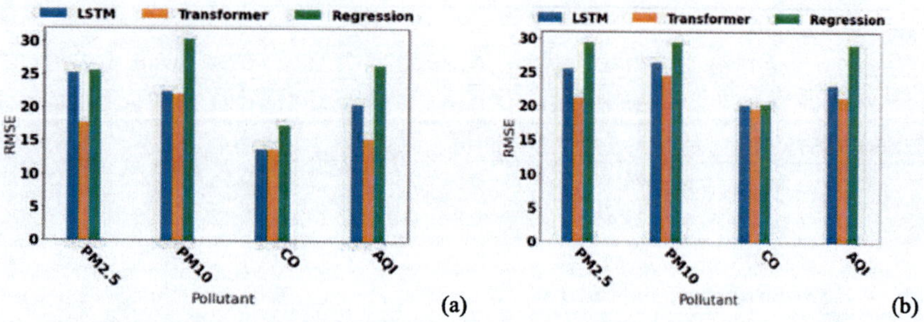

Fig. 4 Comparison of the RMSE value of the models for a) the next one-hour prediction and b) the next four-hour prediction

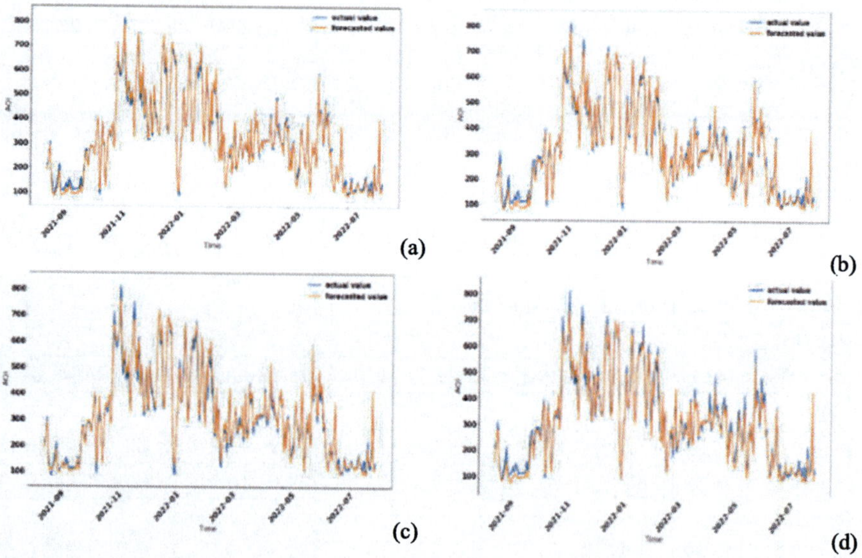

Fig. 5 Predicted and Actual AQI using transformer model with twelve-hour input length a) for next hour prediction b) next four-hour prediction c) next eight-hour prediction d) next twelve-hour prediction

5 Conclusion

The paper concludes that for a shorter forecast horizon, the LSTM and the transformer models are comparable. However, for longer forecast horizons, the transformer model performs well in terms of reduced MAE and RMSE values and higher R^2. The results align with those of Shi et al. (2022), where transformers outperformed the LSTM and GRU models [10]. The authors obtained a minimum RMSE of 23.57 and 38.13 for single and multiple-step prediction, respectively. The study [9] by Rao et al. (2019) established a supremacy of the RNN-LSTM framework over the traditional machine learning models SVR-linear and non-linear with the performance metrics RMSE ranging between 10.36–31.23 and R^2 between 0.26 to 0.78 for main 5 pollutants namely PM2.5, PM10, NO, NO_2, NO_x. Janarthanan et al. (2021) find LSTM to outperform SVR with an average RMSE of 10.56 and R^2 value of 0.57 [1]. Chaudhary et al. (2018) used LSTM to predict the pollutant concentrations at multiple points and obtained a comparison of input parameters on RMSE [7]. LSTM is found to do well in most studies for short-term prediction, but efficiency declines when predictions are made for the distant future wherever such predictions have been attempted. However, as per the information, no research work are available to compare the dataset used in this study. As transformers do not analyze their input sequentially, they can resolve the vanishing gradient problem that renders the RNNs, including LSTM, ineffective in long-term prediction. Moreover, transformers are believed to be faster than RNN-based models as they consider the entire input range together and are far less complex as the number of parameters involved is lesser. For this reason, we have seen the wide-scale application of Transformers to datasets with long historical information or in time series analysis. The transformer model scores over the widely used LSTM algorithm and can be successfully applied for Air Quality prediction, where well-in-advance prediction can facilitate timely action.

Acknowledgements The authors thank Dr. Geetika Jain Saxena, Associate Professor at the University of Delhi, for her valuable technical inputs.

References

1. Janarthanan R, Partheeban P, Somasundaram K, Elamparithi PN (2021) A deep learning approach for prediction of air quality index in a metropolitan city. Sustain Cities Soc 67:102720
2. Kang GK, Gao JZ, Chiao S, Lu S, Xie G (2018) Air quality prediction: big data and machine learning approaches. Int J Environ Sci Dev 9(1):8–16
3. Maleki H, Sorooshian A, Goudarzi G, Baboli Z, Birgani YT, Rahmati M (2019) Air pollution prediction by using an artificial neural network model. Clean Technol Environ Policy 21:1341–1352
4. Kelly JT et al (2019) A system for developing and projecting PM2.5 spatial fields to correspond to just meeting national ambient air quality standards. Atmos Environ 2:100019
5. Liu B et al (2019) A sequence-to-sequence air quality predictor based on the N-step recurrent prediction. IEEE Access 7:43331–43345

6. Ma J, Ding Y, Cheng JCP, Jiang F, Gan VJL, Zherui X (2020) A lag-FLSTM deep learning network based on Bayesian optimisation for multi-sequential-variant PM2.5 prediction. Sustain Cities Soc 60:102237
7. Chaudhary V, Deshbhratar A, Kumar V, Paul D (2018) Time series based LSTM model to predict air Pollutant's concentration for prominent cities in India. UDM
8. Krishan M, Jha S, Das J, Singh A, Goyal MK, Sekar C (2019) Air quality modelling using long short-term memory (LSTM) over NCT-Delhi, India. Air Qual Atmos Health 12(8):899–908
9. Rao KS, Devi GL, Ramesh N (2019) Air Quality Prediction in Visakhapatnam with LSTM based Recurrent Neural Networks. Int J Intell Syst App 2:18–24
10. Shi J, Jain M, Narasimhan G (2022) Time series forecasting using various deep learning models. arXiv:2204.11115 [cs.LG]

The Development of Internet of Things Skills to Enhance Youth Employability in Developing Countries: A Systematic Literature Review

Sijabuliso Khupe and Marita Turpin

Abstract Many developing countries are facing the severe problem of youth unemployment. Due to the changes in technology and the adoption of the Internet of Things (IoT) in developing countries, it is essential for the youth to identify which industries they can become involved in, what IoT skills to equip themselves with and how to use these skills to make them employable and adaptable to these changes. Through a systematic literature review, this paper identifies potential industries in developing countries where the youth have the most IoT related opportunities for work, as well as the IoT skills they would need to make themselves employable. Fifty academic papers from eight databases were analysed using thematic analysis to identify industry opportunities and required IoT skills for the youth. The results indicate that the technology and farming industries are where most IoT opportunities for the youth lie. The important skills identified were the use of sensors, data management and programming, as they form the backbone of the IoT and would allow the youth to be adaptable and employable in various industries. Lastly, suggestions are made of areas where governments can focus to facilitate youth skills development in IoT.

Keywords Fourth Industrial Revolution (4IR) · Internet of Things (IoT) skills · Youth employability · Developing countries · Systematic literature review (SLR) · Smart Agriculture

1 Introduction

The Internet of Things (IoT) is transforming how we live and work and is part of the modern technological ecosystem. The IoT has impacted education and skills development worldwide, affecting many demographics and industries [1]. Due to shortages of resources, particularly in developing countries compared to developed

S. Khupe · M. Turpin (✉)
Department of Informatics, University of Pretoria, Lynnwood Road, Pretoria, South Africa
e-mail: marita.turpin@up.ac.za

N. Hasteer et al. (eds.), *Decision Intelligence Solutions*, Lecture Notes in Electrical Engineering 1080,
https://doi.org/10.1007/978-981-99-5994-5_8

countries, the IoT has become a beacon for solutions to empower people and improve efficiency [2].

There is no standard definition for IoT. The Institute of Electrical and Electronics Engineers (IEEE) defines the IoT as a system of interconnected sensors, actuators, and smart objects that interact over networks to make them intelligent and programmable and enable them to communicate with humans and each other [3, 4].

A significant disparity exists in how developed and developing countries engage with technological advancements. While most technological advancements come from the Global North, developing countries primarily benefit from the productivity and efficiency of the technologies developed elsewhere [5]. Developing countries aim to grow their economies with efficient energy and sustainable initiatives, by adopting technological trends such as IoT [2]. This article considers the importance for developing countries to equip their youths with relevant skills, such as IoT skills, to make them employable and navigate the Fourth Industrial Revolution (4IR). Through a systematic literature review (SLR), this article examines the opportunities in developing countries for youth to become employable through IoT. The aim is to identify the different job opportunities available, the sectors for IoT-related jobs, and the training required to make them employable. While there is existing literature on these different aspects, it is not structured in a way that gives guidance to youth in a synthesized manner.

The following research questions have guided this study:

- *What job opportunities are available for youth with IoT skills in developing countries?*
- *What skills can make the youth employable in the IoT domain in developing countries?*
- *How can these skills be best developed in youth in developing countries?*

This paper is structured as follows. The background and motivation for the study will first be presented, where after the research method will be introduced. Following this will be a discussion of the findings from the data analysis. The paper ends with a conclusion containing reflections and suggestions for further studies.

2 Background and Motivation for the Study

The 4IR represents new ways people interact with technology. It brings disruption in terms of change in production, management, and governance systems in every country, including developing countries [6]. Technology is now immersed in how individuals carry out their everyday as well as complex tasks [7]. At the forefront of the 4IR is the IoT [4]. It has opened numerous opportunities and possibilities in the educational, agricultural, industrial, and medical sectors [4].

A developing country is defined as a country with a relatively low standard of living accompanied by low incomes and severe and limiting infrastructure impeding the sustainment of development, making it vulnerable to economic and environmental shocks and effects [5]. Developing countries can learn from how IoT was implemented in developed countries and use it to significant effect, such as to improve unemployment. High youth unemployment impacts the levels of investment in education and skill development in the Global South [8]. The International Labour Organisation (ILO) states that during the past 20 years, the global youth unemployment rate has been constant between the 12–14% range [8]. This is greatly overshadowed by countries in the Global South, such as Mozambique, whose youth unemployment is 20%, and South Africa, with 42% youth unemployment [8]. Youth unemployment is an unfortunate reality for many developing countries. The effects range from individuals becoming depressed and tending toward violence and crime [9], and substance abuse negatively affecting economic growth and development [10]. This paper will examine whether the IoT opportunities that 4IR has opened can be leveraged to enhance employment among the youth in developing countries. It will look at how youth can use their skills obtained from IoT to make them more employable or become economically active.

3 Research Method

This study follows a systematic literature review (SLR) method. A SLR assists in identifying, evaluating critically, and consolidating the findings and conclusions of high-level individual studies looking at one or more research questions [11]. In this study, an electronic online database search was conducted to find journal articles investigating IoT skills for youth in developing countries and how those skills can make them employable. The investigation was carried out in August 2022 on the following online databases: Science Direct, IEEExplore, Scopus, Emerald Insight, ERIC (EBSCOhost), EconLit (EBSCOhost), Taylor & Francis and SpringerLink. These databases were selected for their track record of having reliable journals and conference papers. They also provide a cross-reference to other journals and conference papers that are not easily accessible from databases that require special access or subscription. For this study, the following search string was used: *("IoT" OR "Internet of things") AND "Youth" AND ("Africa" OR "developing countr*") AND ("skills" OR "employment" OR "job opportunities").*

In total, from the eight online databases used, the starting pool of documents to consider for the study was 844. Due to the high number of documents discovered, steps were taken to narrow down the documents used for the SLR. The following inclusion and exclusion criteria were used. Inclusion criteria were: Publications that are in English; Publications where a research method was used, Publications that were peer-reviewed journal articles and conference papers; Publications limited to developing countries; Publications where the title, abstract, and keywords match and are relevant to the research question and objectives. Exclusion criteria were: Publications with only the abstract but not the full text available; Duplicate papers (the same papers discovered in different databases); Papers that do not talk about

IoT skills in developing countries; and papers that do not provide evaluations or information on how they got their results.

Of the 844 records identified, 765 records remained after duplicates were removed. After the abstracts were screened, 329 records were identified as potentially eligible. The screening process of applying the inclusion and exclusion criteria on the 329 full texts, resulted in 52 records that were selected. To ensure that all records were of sufficient quality, they were further screened by means of the following quality assessment questions: 1) Were the purposes and goals of the research outlined in a detailed manner?; 2) Were the research questions presented in a detailed way?; 3) Were the findings addressing the research questions fully?; 4) Are the findings credible?; and 5) Were conclusions, and future research questions reported suitable in line with the research?

As a result of the quality assessment, a further two records were excluded, resulting in 50 records that underwent a full-text analysis. A data extraction table was drafted to assist in identifying themes during the data analysis. While the dataset and data extraction table are not included in the paper due to space constraints, it is available upon request from the authors.

Data analysis was carried out by means of a thematic analysis. Thematic analysis is "a flexible and useful research tool, which can potentially provide a rich and detailed, yet complex account of data" [12]. It is used to group similar ideas and themes, offering a comprehensive overview of the data [13]. Following the thematic analysis method, the data analysis consisted of six stages that included getting to know the data by reading through the articles, generating codes, noting down themes that occur in the articles related to the research questions, defining and naming those themes and finally using the themes to inform the findings.

4 Discussion of Findings

The identified themes, supported by the data extraction table, assisted in responding to the research questions, as follows.

4.1 Job Opportunities Available for the Youth

The high unemployment rates of the youth in developing countries significantly impact the countries' development [8]. The SLR papers indicated several opportunities and challenges the youth face when looking for jobs in developing countries [14, 15]. Challenges included lacking the required or necessary skills for the job, lack of infrastructure in the country, little to no funding for projects and initiatives, and other socio and economic challenges [16]. Opportunities were noted for several industry sectors, namely: Technology, Farming, Education, Manufacturing, Supply

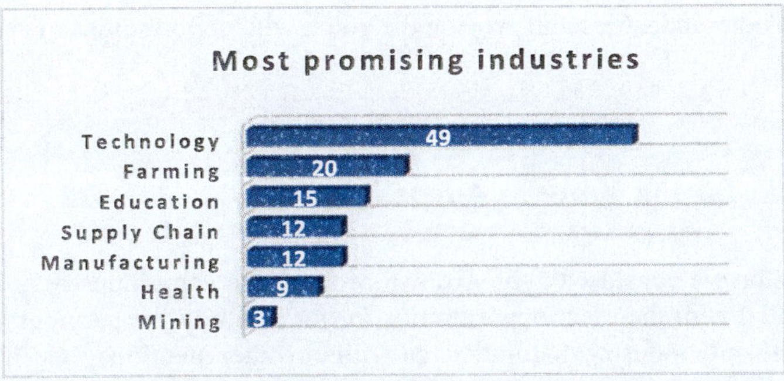

Fig. 1 Industry breakdown for available opportunities identified in papers

Chain, Health and Mining. Opportunities in these sectors were covered in the 50 papers as indicated in Fig. 1:

The technology industry was the one that occurred the most often, namely in 49 of the papers. The technology industry, in this case, is a combination of the mobile, ICT, cloud, communications, and networking sectors [15]. This industry supports most other industries in terms of application and implementation as it provides a channel for the other industries to communicate and interact.

The farming and agriculture industry was discussed in 20 of the papers. Most IoT used in farming and agriculture is linked to the technology industry. Technology is used to improve methods of farming by using sensors in implementing irrigation systems or monitoring the crops and livestock [17]. The use of sensors and networks and the need for data management and monitoring in this way reflects how IoT opens up opportunities for those who may not be fully skilled as some of the tasks may only involve installation and provides good experience and encourage hands-on learning [18].

The education sector ranked third in number of appearances. Education allows the youth to be well-equipped and develop skills that can be used when opportunities arise [19]. The growing use of IoT is changing how the education industry develops with more interactive platforms. It enables the youth to experience different technologies earlier than before and develop their skills to help them navigate the complex job market [20]. The manufacturing and supply chain industries each appeared in 12 of the 50 papers analysed. In the supply chain industry, the use of sensors and trackers could be used to keep track of products while they are being moved, through the use of radio frequency identification (RFID) [21]. Access to this information can enable organisations or businesses to plan accurately and estimate when their goods will arrive. The health industry was discussed in nine papers, and the mining industry in three papers.

The farming and agricultural industry, along with the education and technological industries, occurred the most often as industries where the youth may have the best IoT job opportunities in developing countries. Many developing countries focus on farming and agriculture [17]. The use of IoT can assist the technology industry to sustain the other industries and provide the youth with opportunities to get involved [22].

4.2 *Skills for the Youth to Adopt*

In this section we consider the breakdown of the technology industry as it was the one identified with the most opportunities for the youth in the previous section. It also was the only industry that interacted with the other identified industries due to the breakdown of the skills identified. Figure 2 shows the skills identified and their occurrences. Equipping youth with the appropriate IoT skills will enable them to contribute to developing countries as they can earn and thus improve their quality of life [14, 16].

Sensors received the most frequent mention: 33 or roughly two thirds of the papers identified working with sensors as essential. The use of sensors in most industries, including farming, suggests that the youth be taught how to use or install them, and this will go a long way to ensure that they can empower themselves and start to be innovative [23]. The use of sensors in IoT ties in with the amount of data collected, processed and stored for use [24]. This is reflected by data management skills discussed in 19 of the papers. Data management skills involve being able to read and record the correct values, for example, temperature or humidity from a sensor in farming, processing the data into information, and using the information correctly [25]. IoT involves many connected devices gathering, processing, or sharing data to achieve a goal [26]. As most IoT opportunities involve using sensors, having data management skills can ensure that youth have opportunities and be able to learn and practice these methods, as not all of them require expensive education to learn [27].

Networking skills were discussed in 13 papers. Networking skills reflect on the infrastructure required to use IoT. Networking is a big part of the connection between sensors and data management, allowing data to travel through set structures to where it can be stored or used [25]. Network infrastructure will require to be maintained and developed further over time [25]. Internet skills were discussed in eight papers

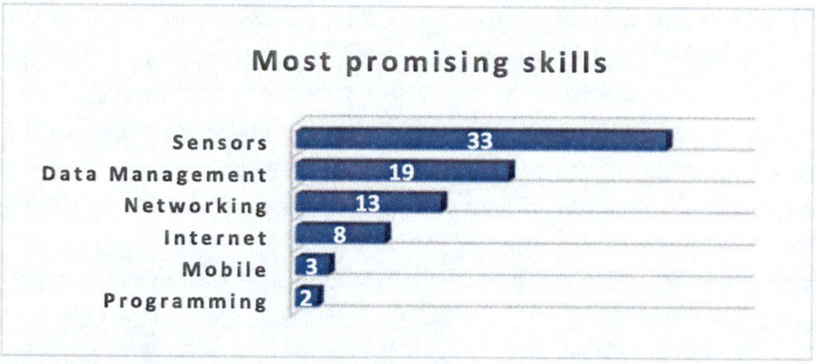

Fig. 2 IoT skills occurrences in the SLR papers

and mobile skills in three. For youth in developing countries to be innovative and continue to transform the environment with IoT, they need to understand and learn the power of the internet and mobile phones [15]. Productivity can be increased through the use of mobile phones and the internet and the time that can be reduced from travelling as they can now coordinate from their place and be able to sell their crops if they are farming, for example [28]. The opportunities are endless as they can also use the internet to look for other job opportunities [29].

Lastly, when programming was discussed, it considered how the youth can equip themselves with programming to develop systems that can be used in IoT [27]. The educational industry featured heavily in these papers. Programming skills and the knowledge of programming languages, particularly python, was identified as critical as it is easy to learn, widely used and considered the language of choice for IoT [27].

4.3 Recommendations for Developing IoT Skills in Youth

This section looks at how developing countries can enable IoT youth skills development to assist in addressing youth unemployment. For this section, the themes identified in the data analysis are used as a guide to reflect on the findings.

Figure 3 shows the breakdown of the themes identified during the data extraction process. In the 50 papers analysed, smart farming appeared 12 times. This was closely followed by sustainability (11 appearances), youth empowerment (nine), innovation (eight), digitalisation (seven), economic growth (six), and job creation (four). The last theme identified was energy management, which appeared three times.

Developing countries need to ensure that they invest and focus on the right strategies for the continued development of youth skills [8]. Developing countries should follow the Sustainable Development Goals (SDG) to be able to meet economic goals, which will improve the sustainability of the projects and provide for socio-economic growth [30]. In return, the youth's participation in economic growth will assist to

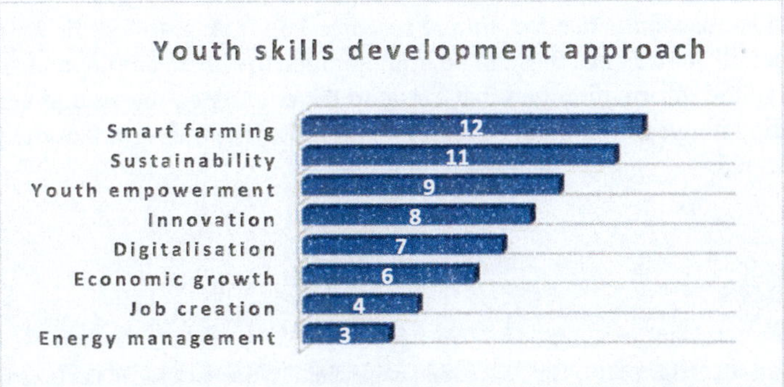

Fig. 3 Themes for youth skills development identified during data extraction

improve infrastructure and aid in the funding that may become available to assist in further development [31]. The need for energy cannot be emphasised enough for developing countries. Electricity drives the industries and power the sensors, networks and machines needed to enable IoT solutions. Focusing on sustainable energy generation and storage can drive growth sustainably [28]. A steady supply of power can assist sectors such as education in implementing new methods of teaching and skills development [32].

Overall, one may argue that to develop the youth's skills, proper infrastructure and structures need to be in place to ensure that there is access to the relevant options so that the youth can be exposed to opportunities that were identified [33].

5 Conclusion

The aim of this paper was to identify the IoT related job opportunities available to youth in developing countries as well as the IoT skills the youth can equip themselves with, to address youth unemployment in developing countries. Considering the evidence gathered and presented in this paper through a SLR and the thematic analysis of 50 academic papers, youth in developing countries can be optimistic about the job opportunities IoT provides.

The findings identified the technology industry and the agriculture industry as the areas with the most IoT related opportunities for youth in developing countries. Other industries with IoT opportunities are education, manufacturing, supply chain, health and mining. The skills required to do IoT related jobs, are: the use of sensors, data management, networking, internet, mobile and programming skills. These skills can be used in all the industries identified in the review. Enablers for youth skills development in IoT include a focus by governments on smart farming, sustainability, youth empowerment, innovation, digitalization, economic growth, job creation and energy management. There is a window of opportunity for the youth to equip themselves with IoT skills and use these skills to empower themselves. Countries that are adopting technology, must embrace the change and should start to prepare the youth with the adequate skills needed. Future research can look at IoT skills for the youth in more detail, at measures that can be implemented to ensure employment. This will help bridge the information gap that exists in developing countries and assist better coordination between scholars, educators, and industry, to ensure that the youth are well equipped.

References

1. Kayembe C, Nel D (2019) Challenges and opportunities for education in the Fourth Industrial Revolution. Afr J Public Aff 11:79

2. Routray SK, Hussein HM (2019) Narrowband IoT: an appropriate solution for developing countries. arXiv preprint arXiv:1903.04850 (2019)
3. Association IS (2015) Internet of Things (IoT) ecosystem study. IEEE Standards Association, The Institute of Electrical and Electronic Engineers, Inc. 1
4. Illahi AAC, Culaba A, Dadios EP (2019) Internet of Things in the Philippines: a review. In: 2019 IEEE 11th international conference on humanoid, nanotechnology, information technology, communication and control, environment, and management (HNICEM). IEEE
5. Dedrick J, Kraemer KL, Shih E (2013) Information technology and productivity in developed and developing countries. J Manag Inf Syst 30:97–122
6. Schwab K (2017) The fourth industrial revolution. Currency (2017)
7. Davis N (2016) What is the fourth industrial revolution. In: World economic forum
8. Fergusson R, Yeates N (2021) Global youth unemployment: history, governance and policy. Edward Elgar Publishing
9. Muhammad IK et al (2013) Effects of unemployed rural youth on rural development (ORIC-13-03). Afr J Agric Res 8:5562–5571
10. Obumneke E (2012) Youth unemployment and its socio-economic implications in Nigeria. J Soc Sci Public Policy 4:47–59
11. Siddaway A (2014) What is a systematic literature review and how do I do one. Univ Stirling 1:1–13
12. Braun V, Clarke V (2012) Thematic analysis. American Psychological Association
13. Maguire M, Delahunt B (2017) Doing a thematic analysis: a practical, step-by-step guide for learning and teaching scholars. All Ireland J High Educ 9
14. Kurt R (2019) Industry 4.0 in terms of industrial relations and its impacts on labour life. Procedia Comput Sci 158:590–601
15. Van Rensburg NJ, Telukdarie A, Dhamija P (2019) Society 4.0 applied in Africa: advancing the social impact of technology. Technol Soc 59:101125
16. Kummitha RKR, Crutzen N (2019) Smart cities and the citizen-driven internet of things: a qualitative inquiry into an emerging smart city. Technol Forecast Soc Chang 140:44–53
17. Nanda Kumar K, Vijayan Pillai A, Badri Narayanan MK (2021) Smart agriculture using IoT. Mater Today Proc (2021)
18. Musonda I, Okoro C (2021) Assessment of current and future critical skills in the South African construction industry. High Educ Skills Work-Based Learn 11:1055–1067
19. Abdelhamid MM, Elfakharany MM (2020) Improving urban park usability in developing countries: case study of Al-Shalalat Park in Alexandria. Alex Eng J 59:311–321
20. Cárdenas-Robledo LA, Peña-Ayala A (2018) Ubiquitous learning: a systematic review. Telematics Inform 35:1097–1132
21. Botha E, Malekian R, Ijiga OE (2019) IoT in agriculture: enhanced throughput in South African farming applications. In: 2019 IEEE 2nd wireless Africa conference (WAC). IEEE
22. Gupta S, Kumar V, Karam E (2020) New-age technologies-driven social innovation: what, how, where, and why? Ind Mark Manage 89:499–516
23. Oliveira J et al (2020) IoT sensing platform as a driver for digital farming in rural Africa. Sensors 20
24. Li Y, Zheng Y (2021) Regional agricultural industry economic development based on embedded system and Internet of Things. Microprocess Microsyst 82:103852
25. Benita F, Virupaksha D, Wilhelm E, Tunçer B (2021) A smart learning ecosystem design for delivering Data-driven Thinking in STEM education. Smart Learn Environ 8
26. Malhotra S, Kumar A, Dutta R (2021) Effect of integrating IoT courses at the freshman level on learning attitude and behaviour in the classroom. Educ Inf Technol 26:2607–2621
27. El Mrabet H, Ait Moussa A (2021) IoT-school guidance: a holistic approach to vocational self-awareness & career path. Educ Inf Technol 26:5439–5456
28. Anshari M, Almunawar MN (2022) Adopting open innovation for SMEs and industrial revolution 4.0. J Sci Technol Policy Manag 13:405–427
29. Morales-Avalos JR, Heredia-Escorza Y (2018) Igniting the innovation's competencies at engineering schools: IoT to the cloud labs network in Mexico. World J Educ 8:159–167

30. Sreenath S, Sudhakar K, Yusop AF (2021) Sustainability at airports: technologies and best practices from ASEAN countries. J Environ Manage 299:113639
31. Kamble SS, Gunasekaran A, Gawankar SA (2020) Achieving sustainable performance in a data-driven agriculture supply chain: a review for research and applications. Int J Prod Econ 219:179–194
32. Attallah B, Change Y-K, Il-Agure Z (2019) Developing IoT competencies in non-computer science students in higher education. In: 2019 sixth HCT information technology trends (ITT). IEEE
33. Lopez-Vargas A, Fuentes M, Vivar M (2020) Challenges and opportunities of the internet of things for global development to achieve the United Nations sustainable development goals. IEEE Access 8:37202–37213

Investigating Robotic Process Automation Adoption in South African Banking

Adheesh Budree and Mark Tew

Abstract Studies have shown that the traditional banking sector is under threat from digital banks and financial technology organisations. In response to this threat, leading banks have implemented Robotic Process Automation (RPA) powered by Artificial Intelligence to reduce costs and simplify operations. This has, however, proven to be challenging as in many cases the impact of automation technology implementations is perceived to affect the livelihoods of banking staff. Within the South African banking context, there is a particular sensitivity to factors that impede employment and labour unions are deeply involved in protecting workers. This study seeks to investigate the factors that drive RPA adoption in South African banks using the Technology-Organisation-Environment (TOE) framework, extended with Institution Theory, as a lens to structure an approach in organising RPA adoption factors in an extensive literature review on the phenomenon. Thematic analysis was used to analyse the interview data that was collected. Key findings were that the adoption of RPA in South African banks is driven by the expected benefits of RPA which are achieved when well-suited processes are targeted, an effective operating model for the program including business and IT personnel, with the right skills. A well-designed change program is critical for RPA adoption in banks.

Keywords Artificial intelligence · Robotic process automation · Banking · Fintech · Cost reduction · Business process reengineering

A. Budree (✉) · M. Tew
University of Cape Town, Cape Town, South Africa
e-mail: adheesh.budree@uct.ac.za

M. Tew
e-mail: TWXMAR002@myuct.ac.za

N. Hasteer et al. (eds.), *Decision Intelligence Solutions*, Lecture Notes in Electrical Engineering 1080,
https://doi.org/10.1007/978-981-99-5994-5_9

1 Introduction

Automation is not a new concept and is already prevalent in Automated Teller Machines (ATM) in the banking environment [1]. Traditional software automation is screen scraping and scripting which operates at a low level of automation [2]. Robotic Process Automation (RPA) is a new breed of software technology that makes use of software robots to carry out digital business processes that were previously performed by people. Technically, automation is a system that can be configured using 'if/then' logic to handle a task [3]. As industrial robots automate manual labour in manufacturing, RPA can automate manual labour with data and information [4].

Leading banks have started experimenting with RPA implementations, but that fact that few have managed to take advantage of the perceived benefits offered by the technology cannot be ignored by other leaders of banks [3, 5]. The benefits that are documented in manufacturing are more innovation, cost reduction and better decision-making, however research literature in other industries is lacking [6]. Many organisations remain sceptical about RPA and its benefits and some organisations that have implemented RPA are grappling with best practices to improve maturity [7].

This study seeks to investigate the factors that affect the adoption of RPA in the South African banking context. In the Global Competitive Report, South African banks are ranked 37th in the list of most sound banks in the world [8]. The banking industry contributes to 10.5% of South Africa's GDP which is significant contribution in terms of the percentage of the South African population that rely on banking for employment [9]. South African banks gain overall efficiencies through innovation in order to improve their cost-to-income ratio and improve their profitability [10].

RPA is an innovative technology that is believed to create efficiencies in organisations by automating business processes [11]. Banks are implementing RPA to take advantage of the associated benefits of RPA [3]. The labour costs in South Africa present less opportunity for banks to minimise staff costs than in more developed countries. It is predicted that just more than 40% of jobs in South Africa could be automated [8], and South African trade unions are already concerned about the impact on the labour force in local banks [12].

There is a lack of research cases on RPA in financial industries and further research needs to be added to literature to contribute for the purposes of research generalisation [3]. The social behaviour in the banking industry is unlike other industries and therefore it is challenging to generalise the factors that drive adoption [13]. Quantifying the benefits of automation is difficult and is moderated by the volume of financial transactions impacted by automation [14]. The research on the success and failure of Business Process Management (BPM) is conflicting with some claiming that the technology has not managed to deliver on the expected benefits [13]. In research articles relating to the implementation of RPA, the recent cases cite important lessons learned that when applied, impact the successful delivery of automation in organisations [11].

The industry contribution of this research is to help banks understand the factors that would improve adoption, the value of adopting RPA, and the possible effects of adoption by discussing the factors in the banking context and areas of application. The impact of automation has been documented in other studies, relating to other forms of automation and industries, to affect employees' livelihoods and this research aims to understand the impact of automation on bank employees in South Africa. The research also aims to contribute to the body knowledge, to present more cases of RPA adoption in the banking industry.

2 Literature Review

2.1 *Background*

RPA is the next level of automation where software robots (bots) are created as virtual works in order to fulfil tasks that would traditionally be carried out by human staff [15]. The most basic use case could be copying data from an email system and inserting it into a different application [3]. This has been described as "swivel chair" integration because RPA is able to log into one system and transfer data into another without any backend integration, imitating a human on a swivel chair accessing different systems [16].

RPA is not designed as a business application but rather to operate other business systems by sitting on top of existing infrastructure within an organisation [1]. In RPA terms, a robot is defined as a single software license [17]. The characteristics of a good RPA task candidate are repeatable, definable, and rules-based [1]. There are two defining characteristics that separate RPA from other automation software, including Business Process Management (BPM). Firstly, configuring RPA tasks does not require any programming skills and common providers use drag and drop functionality for configuration. Secondly, RPA does not need to integrate with any of the systems or applications that the automated process interfaces with, as all interaction is done through the front-end impersonating a user [15].

In terms of sophistication, RPA can be distinguished from other cognitive technology or AI by the type of data it is able to process, the processes, and the outcomes [18]. RPA can only process structured data based on rules-based processes with a single outcome, while AI is able to process unstructured data, processes inference-based processes, and produce a set of likely answers [11]. Cognitive technology can also be categorised based on intelligence into rules-based automation, knowledge-based automation, and Artificial Intelligence (AI) [20]. Other AI technologies include Natural Language Processing (NLP), Chatbot's, speech and image recognition, and machine learning [21]. The integration of other AI technologies and RPA increases the effectiveness of a solution since structured and unstructured data can be handled [4]. Intelligent Process Automation (IPA) is recently being referred to when combining AI and RPA to solve a more complex automated task [22]. The cost of cognitive robotics is much higher than RPA and should only be applied to processes of high value [5].

The banking industry has relatively high operating costs compared to other industries, which inherently results in a focus on savings and efficiencies that will drive down costs [10]. Competition between banks is another driver of efficiency through innovation and focus on costs [10]. Banks measure their efficiency by using a cost-to-income ratio [9] and digital technology, including automation, creates operational efficiency and rich data allowing banks to make swift decisions. Automation, however, would not replace personal customer relationships which are key to a market such as South Africa [16].

Some banks have already started implementing RPA and the challenge is for the technology to be properly embedded in the environment to gain a sustained advantage [3]. There are many reported business use cases in banking for RPA with opportunities for efficiency in terms of time and effort to be gained [23]. The efficiencies created by RPA will reduce the requirement for human workers that previously carried out tasks following automation [8]. However, organisations should not only look to efficiencies through staff reduction as there are more compelling factors that support the adoption of RPA [19] A major fear cited in literature is the loss of jobs brought on by the advent of AI [24]. However, banks that have implemented RPA have been able to reallocate staff to more interesting work and have augmented teams with bots rather than replacing staff [3].

A systematic approach was taken to develop the literature review following limited quality references in the literature. At the start, keyword searches for "Robotic Process Automation", similar technologies and variations of the phrase were used to acquire as many quality articles from scholarly databases. The quality articles were then forward and backwards searched, and in some cases, the authors contacted for full versions of their reports [25–27].

2.2 *Case Studies of RPA Implementations Specifications*

The London School of Economics (LSE) Outsourcing Unit has been researching RPA for the past few years and their research is commonly cited in the literature. The research is based on case studies in organisations in different industries. LSE's research has evolved as the information on RPA has matured in the literature. LSE has contributed two papers to shared services and IT as capabilities that would be impacted by RPA. Three of the papers have been based on cases at financial institutions [2, 6, 11, 16, 18–20].

In other cases, particular research has been conducted in financial institutions [3, 7, 23, 28]. A case was researched at an asset management organisation and RPA was used to help on a particular fund administration. An argument is made that automation in manufacturing is mature but is still new to white-collar processes [7]. A case study of a Norwegian bank was researched, and the literature is structured in the same way that an RPA project would be implemented [3]. Compliance automation research in an Irish bank found that repetitive banking roles could be effectively automated [23]. Research on the impact of RPA in India describes how the technology has impacted the need for human workers, particularly less-skilled ones [25].

3 Research Model

Different IT adoption models are relevant to organisation-level adoption and at an individual level. For this study, adoption of RPA is at a bank-level and therefore the adoption of theoretical models relevant to organisation levels need to be reviewed. A commonality in The Unified Theory of Acceptance Use of Technology (UTAUT) and the deriving theoretical models is that the technology focus is aimed at the individual level of adoption and not from an organisation perspective. Most organisation-level adoption models are derived from the TOE framework and Innovation of Diffusion Theory/ Diffusion of Innovations (IDT/DOI) model [26].

The TOE framework was selected because it is a well-used framework that has a firm theoretical basis which has been used in IS innovation research extensively. The TOE framework includes the environmental context which is not included in DOI and therefore better explains innovation adoption, particularly within organisations, and is more complete [26].

Based on the literature reviewed the following adaption shown in Fig. 1 of the constructs of the TOE framework was developed for the research.

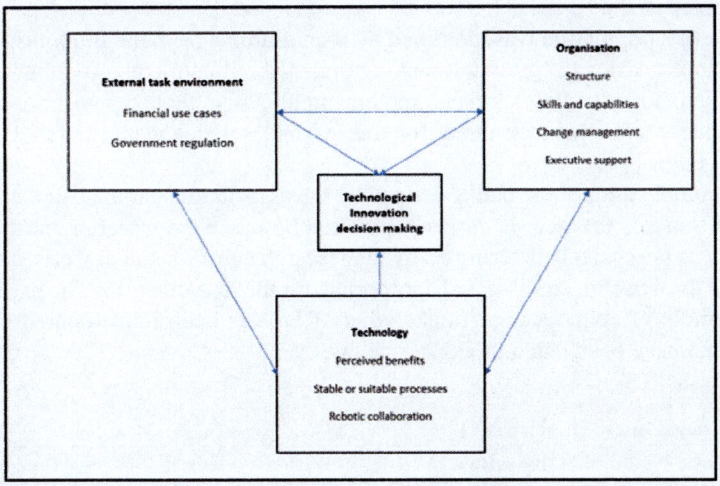

Fig. 1 Adapted Technology-Organisation-Environment framework for the adoption of RPA in South African banks

3.1 Propositions

The propositions for the research are based on the literature reviewed and considering the TOE frameworks are:

- P1. The factors affecting the adoption and usage of RPA in South African banks will comprise technology, organisation, and environmental factors.
- P2. Perceived benefits, stable or suitable processes and robotic collaboration are technological factors that affect the implementation and adoption of RPA.
- P3. Structure, skills, and capabilities change management and executive support are organisational factors that affect the implementation and adoption of RPA.
- P4. Financial use cases and government regulations will have an impact on the adoption of RPA in South African banks

3.2 Research Approach

A qualitative approach was most appropriate for this research because it aims to discover the context and background of the factors that affect the adoption of RPA in South African banks. Qualitative data allows the subject to be explored in the closest to the real setting resulting in rich and complete data [27]. Semi-structured interview questions guided by the TOE framework and Institution Theory formed the basis of the interviews to gain knowledge but the participants were encouraged to express themselves, which led to gaining deeper context of the impact when RPA was adopted in the bank [27].

The target population was identified by organisations from the Banking Association of South Africa (http://www.banking.org.za/about-us/member-banks) and then further defined by banks that have implemented RPA in their organisations. Major RPA vendors reference their clients on their websites which were used to determine the final targets [28].

The combination of the banks defined by the association and the listed clients by the RPA vendors focused the target population based on the research question. The sample was purposively determined by the research question and objectives and was intended to be useful, credible and appropriate for the research [29]. The participants are a sample of employees at Banks where RPA has been implemented in South Africa and have been listed in Table 1 below.

Table 1 Target List of Participants

Role	Expertise	Context
Bank executives	Business unit process owners	Domain
Business analysts	Business process mapping to technology	Technology
Programme managers	RPA programme delivery and change management	Technology

Executives representing the business need for RPA were identified for their perspectives of the environmental context of RPA adoption, in terms of a better understanding of why banks were making the decision to implement RPA. Business Analysts were deemed important for their perspectives on the process element of RPA and programme managers for the delivery perspective of the technology into the banks. These three groups of participants would represent the technology, organisation, and environment contexts of the organisation.

Homogenous sampling focusses on a subgroup where the sample is comparable which allows more depth to be explored and variances made more obvious and this is relevant since the sample is confined to South African banks [27]. A social media platform, Linked-in (http://www.linkedin.com), was used to search for participants based on bank and role search criteria. The interview was piloted with a purposive sample for the reliability of the interview questions and the time needed for the interview [30].

Open-ended questions were prepared by considering the TOE framework and used in semi-structured interviews. The questions were derived from the literature where other research has been conducted on similar technology adoption in the banking industry. The TOE framework defines categorisation relevant to the adoption perspectives and has been used to organise the data collected from the questions in the interview process.

The interviews allowed the participant to share additional context by following an unstructured approach for part of the interview that would not occur in structured interviews [27]. Sub-questions have been added to some of the questions to prompt exploration of deeper context based on the main question asked; the questions have been presented in Appendix A. Two questions were asked to determine the maturity of the RPA implementation at the participants' employers which was used to contextualise the information from the participant.

4 Data Analysis

4.1 Data Analysis Technique

The six steps of thematic analysis were used to guide the approach to establish themes and group data by themes [31]. These include familiarising yourself with the data, generating initial codes, searching for themes, reviewing themes and relationships, defining and naming themes, and producing the report [31].

4.2 Participant Demographics

Interviews were conducted with 12 experienced people working in South African banks. The roles of the participants ranged from business and IT executives responsible for function areas to RPA programme delivery heads to business analysts involved in RPA programmes.

The executives provided a more holistic view of the business impact on the bank compared to participants involved at a programme level who focussed more on the implementation of the technology. All the participants worked in banks that have more than a thousand employees and most worked in banks with more than ten thousand employees. The participants were asked at which stage their RPA implementation was at in order to determine the implementation maturity.

The options were: starting, work in progress, redefining or scaling. Ten out of twelve of the participants were involved in organisations that were either redefining or scaling the RPA capability in the bank and none of the participants cited their implementations to be a work in progress. The area of the business the participants work within was noted and although these include both business and IT stakeholders, the majority were leaders based in the IT department. The average years of experience of the sample group was seventeen years and nine months. The most experienced participant had 38 years of experience and the least experienced participant had five years of experience. All known participants held qualifications at a tertiary level and three of the participants were educated at Master's level. Most participants were males which was not intentional.

4.3 Thematic Analysis

The research analysis was conducted using thematic analysis. The six phases of thematic analysis resulted in codes being created and then aggregated into themes [32].

Codes that were similar in description were collapsed into single codes and renamed to provide a more general description. The codes were grouped, defined and themed using the TOE framework as a guide [32]. The research did not set out to prove, extend or disprove the framework but rather to structure and organise the research data using the framework as a guide (Table 2).

The organisational perspective was the most referenced perspective and is the dominant theme in the research. Perceived benefits were the sub-theme most referenced and was a common discussion point given that RPA is an emerging technology.

The sample size was determined by the criterion saturation. This is the point at which no new data is likely to be obtained based on recurring themes [33]. Theme saturation point occurred after the seventh interview onwards. No new themes emerged from the following interviews. Qualitative research samples are purposed

Table 2 Main themes and sub-themes defined by sources and references

Theme/Sub-theme	Sources	References
Technology Perspective	11	185
Perceived benefits	11	152
Challenges	10	24
Indirect benefits	2	7
Organisational Perspective	11	232
Change management	11	48
Executive support	8	19
Skills and capabilities	10	42
Structure	11	62
Trust	8	24
Use of technology	10	37
Environment Perspective	11	70
Government	9	32
Industry influence	8	14
Media	1	1
Social relationships	8	23
Future use	8	16
Maturity	10	24

for depth and detail of the phenomenon and not based on the quantification of the results [34]. At this point, the data collection process ended.

The sixth stage of thematic analysis is producing the report. The report includes a selection of the compelling extract examples that are then related back to the research question and literature [32].

5 Discussion

5.1 Technology

Perceived benefits were highlighted in the literature and were grouped by expected business and customer benefits. These business benefits included reduced admin [35], reduced costs [1, 36], simple administration, and easy integration [37]. For customer benefits, the literature highlighted faster release times [38], consistent quality [15], high availability [37], and increased scale to deal with requests efficiently [39]. These perceived benefits were consistent with what the participants shared as the benefits that they are expecting or are experiencing in the banks.

The findings for reduced administration were found to be a benefit of the participants in the interviews. It was revealed that the infrastructure build is complex and needs IT security and IT to buy in to the program. This was partially found in the literature where IT is needed to be engaged for successful implementation [40] but is not as simple as some researchers stated. The participants shared that once that was achieved, then automating processes was relatively straight forward which is what was found in the literature [35].

It was found through most of the interviews that the reduction in costs by implementing RPA was not attributed to a headcount metric but that the efficiencies that RPA brings have a positive impact on the cost-to-income ratio in the banks. This is congruent with what some of the research posits [40] and incongruent with researchers who state that the direct benefit of reduced costs was aligned to headcount reduction [1].

The benefit of consistent quality and the fact that RPA digitally records its actions for compliance purposes was found both in the literature and also in the interviews [15]. Using bots for twenty-four hours a day availability was not commonly found in the interviews, one of the participants shared that there were too many dependencies for RPA in their implementation to keep the bot busy overnight.

There is corroboration in the literature review that stable or suitable processes impacted the effectiveness of RPA. The success of automation relies on the processes being automated to be well defined [40]. This is prevalent in the feedback received from the participants. Participants also cited that they used process engineers and software programs as part of the process to potentially fix a broken process and then automate it or in some cases, the need for automation was satisfied with a fixed process. Application support was also referenced by the majority of the participants. This is related to the theme of organisational structure because of the dependency on the business owning and managing RPA [40]. IT security emerged as a technology challenge and was isolated from the organisational trust theme because of the impact security has on the technology perspective. The literature refers generically to the importance of alignment with IT [19] but does not reference the importance in a banking context that was shared by the participants.

5.2 *Organisation*

The literature referenced that successful RPA programmes were dependent on the business owning and managing the process and aligning with IT [40]. This was evident in the responses from the participants where most referred to using a centre of excellence (COE) structure that included IT and business stakeholders. There was also some variance in the interview data in terms of whether the COE was dominated by IT or by the business and some evidence supported that depending on the maturity of the COE would depend on how involved IT is.

The less mature RPA was in the organisation, the more IT dominated the COE. This could account for the conflicting research in the literature where some authors place a higher level of ownership on IT [15]. The reference to a formal COE was not evident in the literature and was cited by the participants that this formal approach was recommended by Blue Prism which was developed in conjunction with a bank and also dominated the interviewed participants' RPA implementations. The bank that had not adopted the COE approach had misaligned ownership of RPA in the bank. It is common in the literature that a well-defined change program is needed for RPA [11] and this was confirmed by the interview data.

The scepticism found in the interviews was attributed to the fear of the consequences of automating work rather more than what was found in the literature pertaining to the technology not delivering on its expectations [11]. That said, executives were won over by demonstrating the capabilities from what was shared in the interviews. This approach also built trust in RPA and ensured that leaders were supportive of the program which is referenced in the literature to be a key adoption factor [3].

The interviews also confirmed that unique change programs are needed for RPA [41] and participants gave examples of using videos, town halls, and forums to develop trust in the program which was not found in the literature. Some researchers state that RPA can be configured by business users [5] but the participants mostly supported the view that some advanced skills in RPA are required [14].

It is consistently found in the literature and the interview data that some consulting and RPA vendors augment teams initially until banks team can take over. The use of RPA technology was found to be consistent with the literature, mainly in back and middle-office applications [23]. The interviews found more specific use cases where bots are being used to open accounts, make payments, check debit order validity, make vehicle settlement calculations and letters as well as other operational tasks.

5.3 *Environment*

The literature covered the societal impacts both for and against automation from an RPA perspective [1] particularly referencing the possibility of jobs being lost. The participants shared that there are not any regulations governing the use of bots to carry out tasks in banks, and that banks are treating bots as workers to segregate areas of responsibility/functionality. The literature reviewed did not cover this from a banking regulation perspective but there was a participant who suspected that this may come into play when AI starts making decisions. Unions are engaged with automation programs and are being collaborated with to ensure alignment in the approach. In the interviews, it was found that participants feel banks have a responsibility to their workforce and are ensuring that they remain relevant though upskill programs and this corroborated with what is in the literature [3].

Industry influence is aligned with the literature, RPA is being implemented in the hope that efficiencies can be gained [3]. Some participants also said that RPA is viewed as a gateway to a more holistic digital strategy which was not directly referred in the literature but can be inferred by the opportunity for banks to better integrate their systems [37]. This is elaborated on by participants saying that emerging fintech's have less technical debt and can take advantage of digital opportunities more easily, driving the need for RPA in legacy banks. Although there was a reference to media reporting South African banks using robots and automation was found during the interviews but was not a significant influence on RPA adoption across all the banks.

6 Conclusion

This study looked at the factors that drive adoption in South African banks and intended to answer the research questions posed as part of the objectives of the research. Many factors influence the adoption of RPA and although research in RPA is limited, similar automation technologies have been previously researched. South African banks have been implementing RPA and the factors affecting adoption have been varied. Automation and the impact on peoples work lives has been prevalent in research and media.

The major findings of the research are that South African banks are adopting RPA to derive the expected benefits. This is consistent with organisations in other industries using similar automation technology. South African banks experienced similar challenges automating ill-defined processes, supporting automated processes, and aligning with IT security teams. South African banks have experienced similar organisational challenges than what has been recorded in the research. A unique finding is that South African banks need to work closely with trade unions to ensure that the approach to automation is responsible by considering the impact on employment rates in the banks. South African banks are augmenting teams with bots and allowing staff to work on more engaging work like other industries and contexts.).

South African banks wanting to successfully embed RPA into their organisations need to ensure that they are aligned with the expectations from the unions. Banks have their own responsibility to ensure that staff are equipped to remain relevant in a digitally inspired organisation so that they can continue to positively contribute to the bank. Processes need to also be carefully selected to ensure that they are fit for automation to avoid impacting the credibility of the program. Process engineers and software can be used to analyse the requirements.

6.1 Recommended Further Research

Future research can include a comparison of several accounts of RPA adoption cases. This would allow the testing of adoption models and determine the similarity and significance of the adoption factors. There would be value in conducting the research as RPA and AI mature in organisations and understand the impact on staff in banks following pervasive RPA adoption.

References

1. Dilla R, Jaynes H (2015) Introduction to Robotic Process Automation: A Primer. Carnegie Mellon University. The Institute for Robotic Process Automation, Pittsburgh
2. Willcocks LP, Lacity M, Craig A (2015) Robotic process automation at Xchanging. Outsourcing Unit Work Res Pap Ser 15(3):1–26
3. Stople A, Steinsund H, Iden J, Bygstad B (2017) Lightweight IT and the IT function: experiences from robotic process automation in a Norwegian bank. In: NOKOBIT 2017, vol 25, no 1
4. Kirchmer M (2017) Robotic process automation – pragmatic solution or dangerous, illusion? In: Business transformation and operational excellence summit insights
5. Lamberton C, Brigo D, Hoy D (2017) Impact of robotics, RPA and AI on the insurance industry: challenges and opportunities. J Financ Perspect 4(1):8–20
6. Willcocks LP, Lacity M, Craig A (2017) Service automation: cognitive virtual agents at SEB bank. Outsourc Unit Work Res Pap Ser 17(1):1–29
7. Britton BL, Atkinson DG (2016) An investigation into the significant impacts of automation in asset management. J Econ Soc Dev (JESD) 4(1):2–14
8. World Economic Forum (2017) The future of jobs and skills in Africa. Geneva: world economic forum
9. Erasmus C, Makina D (2014) An empirical study of bank efficiency in South Africa using standard and alternative approaches to data envelopment analysis (DEA). J Econ Behav Stud 6(4):310–317
10. Du Toit E, Cuba YZ (2018) Cost and profit efficiency of listed South African banks pre and post the financial crisis. Res Int Bus Financ 45:435–445
11. Lacity M (2017) Reimagining professional services with cognitive technologies at KPMG. UMSL Bus: 1–21
12. SASBO: (2018) Nedbank-job-cuts-due-to-robotics. Retrieved from SASBO: http://www.sasbo.org.za/nedbank-job-cuts-due-to-robotics/
13. Ringim KJ, Razalli MR, Hasnan N (2012) Critical success factors for business process management for small and medium banks in Nigeria. Bus Manag Rev 2(1):83–91
14. Trkman P (2013) Increasing process orientation with business process management: critical practices'. Int J Inf Manage 33(1):48–60
15. Zaharia-Radulescu AM, Pricop CL, Shuleski D, Ioan AC (2017) RPA and the future of the workforce. In: Proceedings of the international management conference, pp. 384–392. Bucharest: Faculty of Management, Academy of Economic Studies
16. Willcocks LP, Lacity M, Craig A (2015) The IT function and robotic process automation. Outsourc Unit Work Res Paper Ser 15(5):1–39
17. Banking Association of South Africa: (2020) The banking association of South Africa. Retrieved from Banking Matters, 10 January 2019. https://www.banking.org.za/reports/
18. Willcocks LP, Lacity M, Craig A (2015) Robotic process automation: mature capabilities in the energy sector. Outsourc Unit Work Res Paper Ser 15(6):1–19
19. Willcocks LP, Lacity M, Craig A (2015) Robotic process automation at Telefónica O2. Outsourc Unit Work Res Paper Ser 15(2):1–19
20. Willcocks LP, Lacity M (2015) Robotic process automation: the next transformation lever for shared services. Outsourc Unit Work Res Paper Ser 15(7):1–35
21. Grung-Olsen H (2017) A strategic look at robotic process automation. BPTrends, pp 1–5
22. Bollard A, Larrea E, Singla A, Sood R (2017) The next-generation operating model for the digital world. Digit McKinsey: 1–8
23. Cahill SM (2017) A multi stakeholder perspective on audit and automated compliance: bank of Ireland. University of Dublin: Masters dissertation
24. Naudé W (2021) Artificial intelligence: neither Utopian nor apocalyptic impacts soon. Econ Innov New Technol 30(1):1–23
25. Varghese FC (2017) The impact of automation in IT industry: evidences from India. Int J Eng Sci Comput 7(1):5000–5004

26. Oliveira T, Martins MF (2011) Literature review of information technology adoption models at firm level. Electron J Inf Syst Eval 14(1):110
27. Saunders M, Lewis P, Thornhill A (2016) Research methods for business students, 7th edn. Pearson Education Limited, Harlow, Essex, England
28. Le Clair C, Cullen A, King M (2017) The 12 providers that matter most and how they stack up. Forrester Wave™ Rob Process Autom Q1:1–17
29. Patton MQ (2002) Qualitative Research and Evaluation Methods. Sage, Thousand Oaks, CA
30. Neuman WL (1994) Social research methods: qualitative and quantitative approaches, 2nd edn. Allyn and Bacon, Boston
31. Fereday J, Muir-Cochrane E (2008) Demonstrating rigor using thematic analysis: a hybrid approach of inductive and deductive coding and theme development. Int J Qual Methods 5(1):80–92
32. Braun V, Clarke V (2006) Using thematic analysis in psychology. Qual Res Psychol 3(2):77–101
33. Thompson CB, Walker BL (1998) Basics of research (part 12): qualitative research. Air Med J 17(2):65–70
34. Miles M, Huberman M (1994) Qualitative data analysis: a sourcebook of new methods. Sage Publications, Beverly Hills, CA
35. Vedder R, Guynes CS (2016) The Challenge of botsourcing. Rev Bus Inf Syst (Online) 20(1):1–4
36. Davenport TH, Kirby J (2015) Just how smart are smart machines? MIT Sloan Manag Rev 57(3):21–22
37. Herbert IP, Dhayalan A, Scott A (2016) The future of professional work: will you be replaced, or will you be sitting next to a robot? Manag Serv J: 22–27
38. Davenport TH, Kirby J (2016) Just how smart are smart machines? MIT Sloan Manag Rev 57(3):21
39. Anagnoste S (2017) Robotic automation process-the next major revolution in terms of back office operations improvement. In: Proceedings of the international conference on business excellence. De Gruyter Open, Berlin:pp 676–686
40. Rutaganda L, Bergstrom R, Jayashekhar A, Jayasinghe D, Ahmed J (2017) Avoiding pitfalls and unlocking real business value with RPA. J Financ Transform 46(1):104–115
41. Krishnan G, Ravindran V (2017) IT service management automation and its impact to IT industry. In: Computational intelligence in data science (ICCIDS), 2017 international conference. IEEE, Chennai:pp 1–4

Calculation of Polarization Conversion Ratio (PCR) of Proposed Polarization Conversion Metamaterials (PCM) is Employed in Reduction of RCS Using AI Techniques for Stealth Technology

Ranjeet Prakash Rav, O. P. Singh, and A. K. Singh

Abstract An artificial intelligence (AI) technique approach used for achieving better Polarization Conversion Ratio (PCR) of the novel L shape cross layout of polarization Conversion Metamaterial (PCM). Important dimensions of the PCM unit cell have used to get the relationship of the input-outputs for AI model. This work shows the analysis and designs of cross-polarization, co-polarization and PCR Ratio for frequency band between 5 to 10 GHz. The AI model is proposed to compute the magnitude variation of S-parameters of the PCM for different combinations. The AI model can be introduced to be as exact as an EM simulator and its computation more effective in the PCM design. The simulation work is done by the HFSS 15.0 and MATLAB software.

Keywords Metamaterial · Polarization Conversion Ratio (PCR) · Polarization Conversion Metamaterial (PCM) · Artificial Neural Networks (ANN) · AI techniques

R. P. Rav · O. P. Singh (✉)
Department of Electronics and Communication Engineering, Amity School of Engineering and Technology (ASET), Amity University Uttar Pradesh, Lucknow Campus, Lucknow, India
e-mail: opsingh@amity.edu

R. P. Rav
e-mail: ranjeet.rav@s.amity.edu

A. K. Singh
Department of Electronics and Communication Engineering, Rajkiya Engineering College, U.P (A Government Engineering College), AKTU Uttar Pradesh Lucknow, Kannauj, India
e-mail: arun@reck.ac.in

N. Hasteer et al. (eds.), *Decision Intelligence Solutions*, Lecture Notes in Electrical Engineering 1080,
https://doi.org/10.1007/978-981-99-5994-5_10

1 Introduction

A metamaterial is artificially created from any smart material that is not found in the real world like natural materials. Metamaterials are as artificially designed structures which reflect rare electromagnetic effects as discussed by different researchers with the various possible applications such as shielding, cloaking, etc. Metamaterials geometry, size, shape, and arrangement makes them spruce properties which are capable of changing electromagnetic waves by enhancing, absorbing, blocking and reflecting. Some of the proposed designs are chessboard, checkerboard topology, pyramid and non-periodic structures etc. The study of simulation findings of the design over frequency band yields good absorption of EM waves. It has been noticed that the significant absorption of proposed designs in frequency band shows that it has potential applications in stealth technology. In the recent efforts, few researchers have proposed novel MMAs are ultrathin, less weight, and easy fabrication. In general, they made up of more than three layers, can be described as "top metallic patch layer-dielectric-bottom metallic layer". Various Kinds of dielectric material films are Arlon, Neltec, Rogers, FR4 etc., can be taken as the lossy layer of MMAs. Yajuan Han [1] et al. proposed that change in dispersion, geometric parameters, absorption property of the MA can be modified due to this low insertion loss in C band and simultaneously strong absorption in X band can be obtained. Yahong Liu et al. [2] suggested that dendritic type structure unit cell of absorbing metamaterial able to achieve high absorptivity for a wide incident angle. Zi-Jian Han [3] et al. proposed a metamaterial chessboard surface based on polarization dependent AMC for low RCS antenna design. Qi Zheng [4] et al. presented Checkerboard polarization conversion electromagnetic band-gap (EBG) structures having wideband circular polarization and low radar cross section (RCS) for application in satellite communications and stealth aircrafts. Yongtao Jia [5] et al. proposed L-shaped and a square patch type unit cell is coined as PRRS and due to this design a high polarization conversion ratio (PCR) is achieved. Ying Liu [6] et al. proposed Fishbone-shaped element PCM arranged in chessboard manner which can reduce the mono-static RCS of the object. Ying Liu [7] et al. presented and investigated that PCM can reduce radar RCS in the presence normal incident waves. Junyi Ren [9] et al. presented that a periodicity of artificial structures called PCM exhibit 80% polarization conversion of incident waves.

In this paper, we investigate the affect of dielectric materials and geometric parameters on PCR of metamaterials. The analysis is based on analytical models as changing the geometric parameter of metamaterials and obtained the different simulation results within the frequency band.

Recently, many applications of ANN have shown advantages, including language processing, voice recognition, computer vision and detection of face [10–12]. Our next step is to predict PCR design for the metamaterials with the help of ANNs, some of which are Narrow NNs, Medium NNs, Wide NNs, Bilayered NNs, and Trilayered NNs. An ANN is usually used to replace current methods of analysis because it can be trained and the imported knowledge can be generalized from the data and maintained

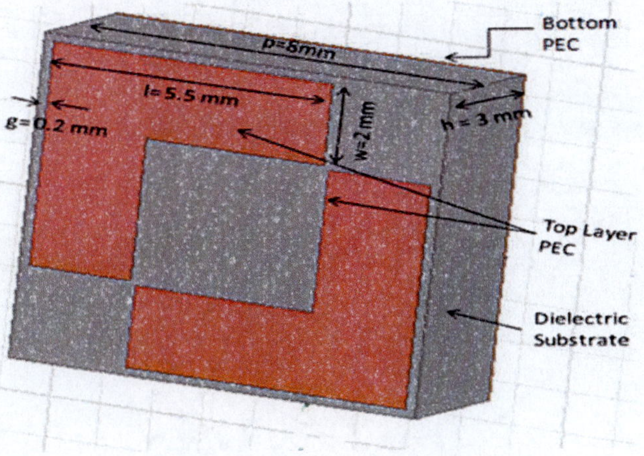

Fig. 1 PCM unit cell

the exactness of the original techniques [13, 14].A Linear Regression Neural Network (LRNN) based PCR prediction method is presented in this paper that can accurately determine co-polarization, cross-polarization, and PCR of MMAs. After the training process of the LRNN models is accomplished then observed that percentage of the coefficient of correlation (R^2) is 96% for target PCM.

2 PCM Unit Cell Design

Junyi Ren [9] et al. proposed that PCM unit cell made up of three layers: on the top is the patterned metallic L patch which is printed on $h = 3$ mm thick dielectric sheets having dielectric constant ($\in_r = 3.5$) and the metallic bottom layer. The dimension 'p' is 8 mm; the thickness 'h' of the dielectric material is 3 mm. The pattern on the top resistive film is an L shape patch is characterized with parameters [$l = 5.5$ mm, $w = 2$ mm] (Fig. 1).

3 Artificial Neural Networks

This study is based on Multilayer Perceptron architecture of ANN. The basic buildings blocks of the MLP are artificial neurons. In this block, inputs weighted sum added with the threshold weights are computed and move through the Activation Function. MLP (see Fig. 2), the block output in one layer is used as next layer inputs. It is usually necessary to train a network by using suitable algorithm to calculate its weights. The MLP consists of minimum number of two layers can be extended Up to

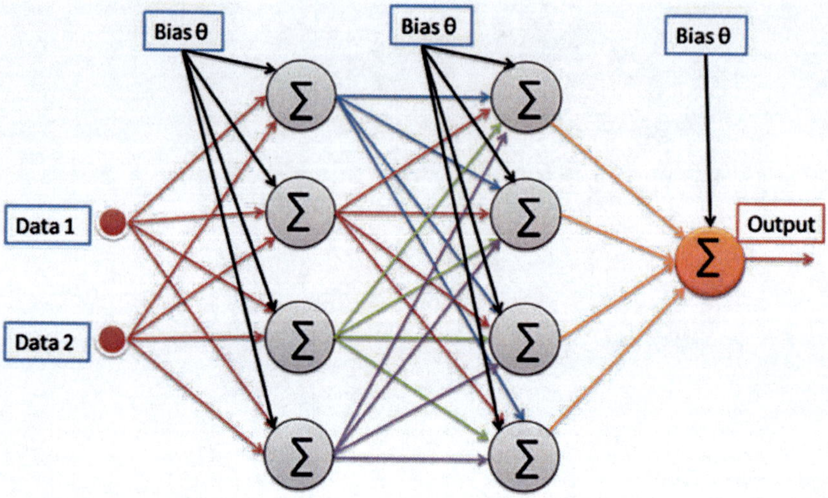

Fig. 2 ANN

N layers A layer is made up of numbers of neurons, and the range of neurons may be varied layer to layer [13, 14].The ReLU activation function is a vastly used activation functions in AI techniques. Its mathematical expression is as shown in equation [15, 16].

$$\begin{aligned} Relu(x) &= 0\ for\ x < 0; \\ x\ for\ x &\geq 0; \\ Relu(x) &= \max(0, x); \end{aligned} \tag{1}$$

In this study we observe the affect of few geometric parameters of the structure on the final response. Training data is imported to train an ANN and this trained model is used to predict the structure to a change in the parameter within a given area of interest. In this paper, LRNN Model [18] is used listed below (Table 1).

4 Data Collection

4.1 Training Data of PCM

This paper is presenting ANN prediction model for Polarization Conversion metamaterial. For Trained Model, training data is required which can be obtained by simulation of PCM unit cell on HFSS simulator. For good-trained model, we have to collect approx 420 data sets in Table 3. These data sets are prepared by 14 different materials shown in Table 2 with varying thickness of $h = 2.8$ mm to 3.2 mm. Total

Table 1 Linear Regression Neural Network

S.No	Neural Network Model	Narrow N.N	Medium N.N	Wide N.N	Bilayered N.N	Trilayered N.N
		1	2	3	4	5
1	No. of fully connected layer	1	1	1	2	3
2	First Layer Size	10	25	100	10	10
3	Second Layer Size	NA	NA	NA	10	10
4	Third Layer Size	NA	NA	NA	NA	10

420 combinations of optometric setup are made for frequency band 5 GHz to 10 GHz with sweep 1 GHz.

Polarization Conversion Ratio (PCR) is set to output variable which is depended on 9 input parameters are permittivity of dielectric material ($\in_r$), copper metal for bottom layer (cu) and top, length dimension of unit cell (p in mm), length of L shape (l in mm), width of L shape (w in mm), thickness of dielectric material (h in mm), gap between Unit cell border and L shape (g in mm), dielectric loss tangent of material and operational frequency (f_z in GHz).PCR is defined as mathematically as:

Table 2 Dielectric materials

S.No	Materials	$\in_r$	Dielectric loss tangent (tan δ)
1	Arlon AD300A (tm)	3.00	0.002
2	Rogers RO3203 (tm)	3.02	0.0016
3	Polyflon Copper-Clad ULTEM (tm)	3.05	0.006
4	GIL GML1000 (tm)	3.12	0.005
5	Arlon AD320A (tm)	3.20	0.0032
6	Rogers TMM 3 (tm)	3.27	0.002
7	Krempel Akaflex PCL (tm)	3.30	0.0052
8	Megtron6_1035_PP	3.35	0.002
9	Plexiglass	3.40	0.001
10	Megtron6_1078_PP	3.41	0.002
11	Megtron6_1035_Laminate	3.46	0.002
12	Neltec NH9348 (tm)	3.48	0.003
13	Dupont Type 100 HN Film (tm)	3.5	0.0026
14	Rogers RO4003 (tm)	3.55	0.0027

Table 3 Training Data

ϵ_r	cu(S/m)	p(mm)	l(mm)	w(mm)	h(mm)	g(mm)	Dielectric loss tangent	f_z (GHz)	Polarization Conversion Ratio
3	5.80×10^7	8	5.5	2.8	2	0.2	0.002	5	0.096486212
3.02	5.80×10^7	8	5.5	2.8	2	0.2	0.0016	5	0.113995926
3.02	5.80×10^7	8	5.5	2.8	2	0.2	0.0016	6	0.747859057
3.05	5.80×10^7	8	5.5	2.8	2	0.2	0.006	5	0.10092100
3.05	5.80×10^7	8	5.5	2.8	2	0.2	0.006	6	0.69219307
3.35	5.80×10^7	8	5.5	2.8	2	0.2	0.002	5	0.33427600
3.02	5.80×10^7	8	5.5	2.9	2	0.2	0.0016	8	0.71605100
3.02	5.80×10^7	8	5.5	2.9	2	0.2	0.0016	9	0.38482800
3.02	5.80×10^7	8	5.5	2.9	2	0.2	0.0016	10	0.20326700
3.35	5.80×10^7	8	5.5	2.8	2	0.2	0.002	6	0.96599200
3.48	5.80×10^7	8	5.5	3.2	2	0.2	0.003	9	0.19085169
3.48	5.80×10^7	8	5.5	3.2	2	0.2	0.003	10	0.10887654
3.55	5.80×10^7	8	5.5	3.2	2	0.2	0.0027	10	0.13102329

$$PCR = R_{xy}^2/\left(R_{xx}^2 + R_{xy}^2\right) \tag{2}$$

where $R_{xy} = crosspolarization$

$R_{xx} = co - polarization$

The data collection is done by HFSS simulation which is represented in Table 3.

4.2 ANN Trained Model

Using training data, trained five linear regression neural networks (LRNN) which predicts the development of Metamaterials. In Table 4, Design model parameter of LRNN is shown. Among the five neural network models, RSME, correlation coefficient $\left(R^2\right)$, MSE and MAE of the Medium neural network is better than others. In the next step, validation of test data of PCM with LRNN models is done.

Table 4 ANN model design

	Narrow NN	Medium NN	Wide NN	Bilayered NN	Trilayered NN
RMSE	0.0641	0.0598	0.0611	0.0638	0.0699
R^2	0.96	0.96	0.96	0.96	0.95
MSE	0.0041	0.0035	0.0037	0.0040	0.0048
MAE	0.0495	0.0456	0.0462	0.0478	0.0493

Table 5 Validation of Simulated and ANN

Test data									Desired	Predicted
ϵ_r	cu(S/m)	p(mm)	l(mm)	w(mm)	h(mm)	g(mm)	Dielectric loss tangent	f_z (GHz)	PCR_HFSS	LRMNN
3.05	5.80×10^7	8	5.5	2	3	0.2	0.006	5	0.22351	0.2227
3.05	5.80×10^7	8	5.5	2	3	0.2	0.006	6	0.92614	0.8764
3.05	5.80×10^7	8	5.5	2	3	0.2	0.006	7	0.99979	0.9347
3.05	5.80×10^7	8	5.5	2	3	0.2	0.006	8	0.94285	0.6338
3.05	5.80×10^7	8	5.5	2	3	0.2	0.006	9	0.68192	0.3448
3.05	5.80×10^7	8	5.5	2	3	0.2	0.006	10	0.41426	0.1978

5 Validation of Linear Regression Medium Neural Network (LRMNN)

In validation, Polyflon dielectric material is selected along with input parameter and simulated result of HFSS is PCR_HFSS and prediction results of LRMNN are mentioned in Table 5. From the Table 5, PCR_HFSS and LRMNN results are 96% correlated with minimum RMSE is 0.0598.

6 Result

On the basis of validation results explained in the last section, we have computed complete analysis of the PCR of PCM metamaterial using ANN. The results are presented for PCM unit cell composed of L shape patch structure and Polyflon Copper-Clad ULTEM (tm) (Permittivity ($\in r$) = 3.05, thickness (h) = 3 mm) dielectric material. The PCR with respect to frequency is plotted and shown in Fig. 3 in Red Line, which displays the reflection behavior of simulated HFSS results using Eq. 2.

Results were obtained by LRMNN model has fully connected layer is one and one hidden layer with 25 neurons (see Fig. 3) by blue line. Above curve showing that LRMNN is able to draw out and pick up the pattern of variation for the simulated PCR. The finding of this work should be considered in metamaterial designs. Comparison has been done to examine the accuracy between the simulated values and the desired ones.

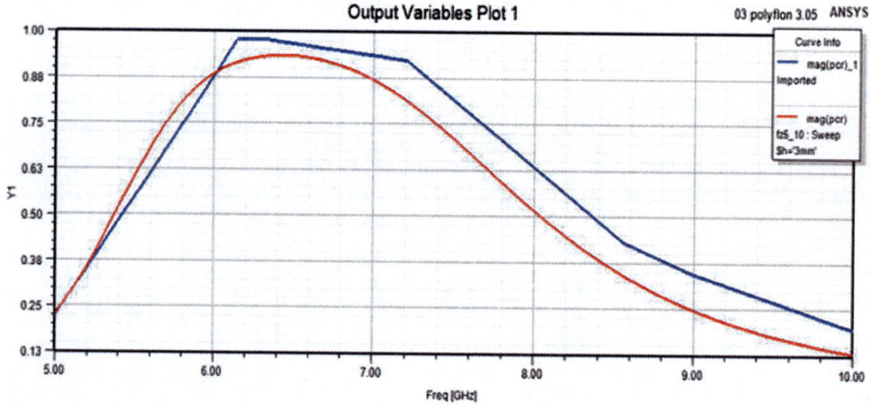

Fig. 3 Validation of PCR

7 Conclusion

In this paper, a Polarization Conversion Metamaterial (PCM) based on a proposed L shape patch has been analyzed using an AI technique to predict metamaterial PCM's PCR characteristics and found that this technique is taking less time as compared to normal technique. The effect of change of geometric parameters on the PCR of metamaterials has also been verified. Also it was investigated that the accuracy of LRMNN model is better than other works (see Fig. 3) with 96% of Coefficient of correlation.

References

1. Han Y, Gong S, Wang J, Li Y, Qu S, Zhang J (2020) Reducing RCS of patch antennas via dispersion engineering of metamaterial absorbers. IEEE Trans Antennas Propag 68(3):1419–1425
2. Liu Y, Zhao X (2014) Perfect absorber metamaterial for designing low-RCS patch antenna. IEEE Antenna Wirel Propag Lett 13:1473–1476
3. Han Z-J, Song W, Sheng X-Q (2017) Gain enhancement and RCS reduction for patch antenna by using polarization dependent EBG surface. IEEE Antenna Wirel Propag Lett 16:1631–1634
4. Zheng Q, Guo C, Vandenbosch GAE, Ding J (2020) Low-profile circularly polarized array with gain enhancement and rcs reduction using polarization conversion EBG structures. IEEE Trans Antennas Propag 68(3):2440–2445
5. Jia Y, Liu Y, Jay Guo Y, Li K, Gong S (2017) A dual-patch polarization rotation reflective surface and its application to ultra-wideband RCS reduction. IEEE Trans Antennas Propag. 65(6):.3291–3295
6. Liu Y, Li K, Jia Y, Hao Y, Gong S, Jay Guo Y (2016) Wideband RCS reduction of a slot array antenna using polarization conversion metasurfaces. IEEE Trans Antennas Propag 64(1):326–331
7. Liu Y, Hao Y, Li K, Gong S (2015) Radar cross section reduction of a microstrip antenna based on polarization conversion metamaterial. IEEE Antennas Wirel Propag Lett 15:80–83
8. Zhou Y, Cao X, Gao J, Li S, Zheng Y (2017) In-band RCS reduction and gain enhancement of a dual-band PRMS-antenna. IEEE Antenna Wirel Propag Lett 16:2716–2720
9. Ren J, Jiang W, Gong S (2018) Low RCS and broadband metamaterial-based low-profile antenna using PCM. IET Microwaves Antennas Propag 12(11):1793–1798
10. Deng L, Yu D (2014) Deep learning: methods and applications. Found Trends Signal Process. 7(3–4):197–387
11. Hatcher WG, Yu W (2018) A survey of deep learning: platforms, applications and emerging research trends. IEEE Access 6:24411–24432
12. Chen X-W, Lin X (2014) Big data deep learning: challenges and perspectives. IEEE Access 2:514–525
13. Hu YH, Hwang JN (2002) Handbook of neural network signal processing. CRC Press
14. Chistodoulou C, Georgiopoulos M (2001) Applications of networks in electromagnetics. Artech House
15. Qiumei Z, Dan T, Fenghua W (2017) Improved convolutional neural network based on fast exponentially linear unit activation function. IEEE Access XX:1
16. Hu X, Niu P, Wang J, Zhang X (2019) A dynamic rectified linear activation units. IEEE Access 7:180409–180416

17. Liu T, Yuming D (2020) U.S. pandemic prediction using regression and neural network models. In: 2020 International conference on intelligent computing and human-computer interaction (ICHCI)
18. Fan J, Lian T (2022) Youtube data analysis using linear regression and neural network. In: International conference on big data, information and computer network (BDICN)

A Mobile Application for Currency Denomination Identification for Visually Impaired

Kasarapu Ramani, Irala Suneetha, Nainaru Pushpalatha, K. Balaji Nanda Kumar Reddy, and P. Harish

Abstract Inspite of availability of various payment apps, currency notes are playing a prominent role as primary mode of exchange for many day to day transactions. For a person with good vision denomination identification is a straight forward task, whereas visually impaired it becomes a challenge. Each currency note is embossed with unique symbols, identification becomes much tougher to them to differentiate between currency notes. This paper mainly focuses on extracting distinctive features from an Indian Currency note using Oriented FAST and Rotated BRIEF (ORB) algorithm. This system also recognizes currency notes with different orientations. Apart from identifying currency denominations, it also calculates total amount available at hand. As now-a-days smart phone is carried by everyone, therefore a mobile application is developed for currency detection to aid visually impaired, which provides detected currency notes information through audio.

Keywords Currency note detection · ORB algorithm · Mobile application · Visually impaired people · Audio output · Feature point detection

1 Introduction

As per World Health Organization (WHO) report, there 284 billion people are visually impaired and 39 billion are blind [1, 10] in the world. Modern currency recognition systems find many applications in real-world including banknote currency counting with denomination, forged currency detection, cash counters at shopping malls, hospitals, and wherever money transactions are involved. For the most of the daily activities the people with blindness and visually impaired depend on others.

K. Ramani (✉)
School of Computing, Mohan Babu University, Tirupati, India
e-mail: ramanidileep@yahoo.com

I. Suneetha · N. Pushpalatha · K. B. N. K. Reddy · P. Harish
Department of Electronics and Communication Engineering, Annamacharya Institute of Technology and Sciences, Tirupati, India

N. Hasteer et al. (eds.), *Decision Intelligence Solutions*, Lecture Notes in Electrical Engineering 1080,
https://doi.org/10.1007/978-981-99-5994-5_11

The currency identification and counting denomination is another major challenge to them. A convenient mechanism for identifying currency notes is a dire need of visually impaired for their daily money related exchanges. A system which can recognize the currency, irrespective of its orientation, brightness, and quality is needed. Also, different Indian currency notes are available with different texture and sizes. Recent developments in smart phone platforms enable dynamic recognition of the currency notes with the help of mobile cameras and provide the output in audio form. The main goal of this paper is to develop a mobile app to recognize the given currency notes and calculate the total amount and enable the safety and deterministic financial transactions by visually impaired.

2 Related Work

In [2] the authors proposed CNN based method for currency detection, with 300 raw images and augmented images and obtained an accuracy of 88%. Though the model accuracy is good, the detection rate is 85% and identification of small features is difficult. In [3], authors proposed Artificial Neural Network for identifying currency color and texture with median level of accuracy. In [4] authors applied Scale Invariant Feature Transform (SIFT) for portable currency identification, but its accuracy is around 70%. The authors in [5] proposed a currency detection system for Arabia currency using Radial Basis Function Network and its performance is good for quality currency, but accuracy is decreasing with tilted and noisy notes. In [6] authors provided matching template method for different country currency detection. In [7, 8] the authors proposed image processing methods such as segmentation, equalization, Region of Interest for image template matching. It is used for recognize the Egyptian currency with accuracy of 89%. In [9] the authors used the Local Binary Pattern (LBP) technique to recognize the features of currency note with 92% accuracy. Though existing methods are achieving more than 90% of accuracy, an efficient computation method is required for feature point detection. The Oriented FAST (Features from Accelerated and Segments Test) and Rotated BRIEF (Binary Robust Independent Elementary Feature) is one such algorithm providing good accuracy with fast and efficient computation of feature points by considering orientation of the given images.

3 Proposed Method for Currency Identification

This paper presents a mobile application for identifying Indian currency notes using Oriented FAST and Rotated BRIEF (ORB) algorithm. This ORB algorithm is a combination of FAST and BREIF. The first one builds a pyramid to get multi-scale features. But, orientation of the given image and its descriptors are not considered by this method. To meet this requirement BRIEF is used to find out orientation key points. BRIEF rotates the descriptors according to the orientation of the key points. As a rotation matrix is computed to rotate BRIEF and to obtain rotated version of the given image. Therefore, it is better than Speed-Up Robust Features (SURF) and Scale Invariant Feature Transform (SIFT) methods. It has three steps: significant feature point generation, developing feature point descriptors and performing feature point matching. ORB is invariant to illumination, rotation and scaling. The proposed system's block diagram is as shown in Fig. 1. The flow of activities followed for the implementation are shown in Fig. 2.

Capture the test image through mobile camera and convert it into gray scale image. Before applying the ORB algorithm, the preprocessing is applied to eliminate noise and sharpening the image using Gaussian blurring equation as given in Eq. (1).

$$G(a, b) = \frac{1}{2\pi\sigma^2} e^{\frac{-(a^2+b^2)}{2\sigma^2}} \tag{1}$$

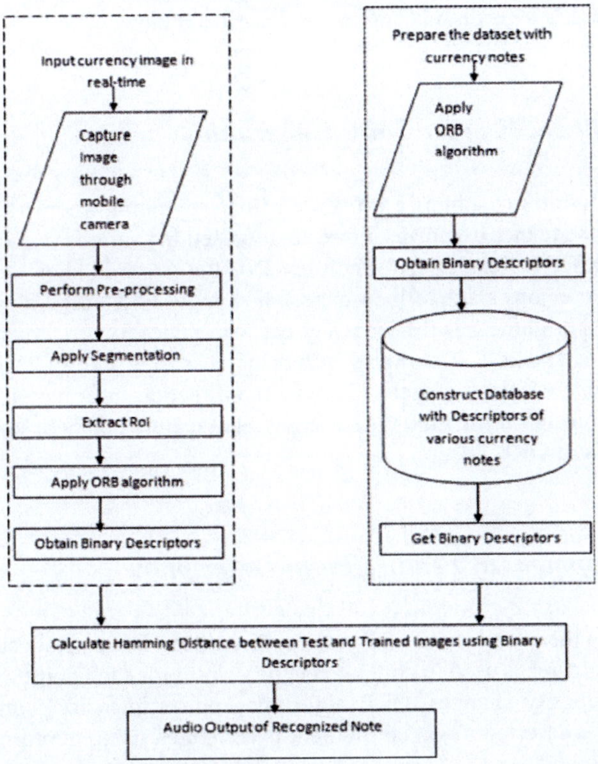

Fig. 1 Automated Currency Detection System

Fig. 2 Functional diagram of the Mobile Application

As currency note has many recognizable features such as watermark, intaglio, Fluorescence etc. To identify such significant feature Region of Interest is obtained through flood filling algorithm.

Then, the ORB algorithm is applied as described below:

3.1 Significant Feature Point Generation

Apply ORB algorithm to obtain a multi-scale image pyramid, where a single image is converted into sequence of images, those are representing various versions of original image with different resolutions. Each level of the pyramid shows downsampled version of its previous stage of the image. Then apply FAST algorithm to choose a pixel Pi and its brightness is Ib. Form a circle with radius 3 with the pixels. Select a brightness threshold as T. The pixels with brightness Ip + T or Ip-T are considered as feature points within the circle. To have the direction information apply Harris corner detection algorithm. Find the scale invariant feature points by applying image scale pyramids of ORB.

3.2 Development of Feature Point Descriptors

After obtaining these Oriented FAST feature points, Steer Binary Robust Independent Elementary Features is used to compute the descriptor for each point with direction. To identify intensity changes, ORB algorithm applies intensity centroid method, where the ORB descriptor's patch moment is computed using the formula given in Eq. (2):

$$m_{ab} = \sum\nolimits_{x,y} x^a\, y^b I(x, y) \tag{2}$$

After obtaining the moments using Eq. (1), the centroid is calculated using the following formula:

$$C_m = \left(\frac{m_{10}}{m_{00}}, \frac{m_{01}}{m_{00}}\right) \tag{3}$$

Then, the orientation of ORB patch is obtained using the following formula:

$$\theta = atan2(m_{01}, m_{10}) \tag{4}$$

Then, to obtain rotation invariance of patch, canonical rotation is applied.

The feature points obtained from FAST are given as input to BRIEF and binary feature vectors of size 128 to 512 bits are obtained in 1's and 0's bit strings. Then, BRIEF chooses a pair of pixels randomly around the neighborhood of the keypoint, known as a patch. To form this pair the Gaussian distribution is applied the keypoint as centre with spread of sigma. If the first pixel is brighter than the second one, then its intensity value is assigned as 1 otherwise 0. BRIEF applies this procedure 128 times around each keypoint to obtain 128- bit vector. Then all such vectors form the image.

3.3 *Feature Point Matching*

After handling scaling and rotation issues, the query currency note image features are compared with reference currency note image. To measure the similarity among the features, the Hamming distance is used to measure the distance between descriptors for matching with Brute Force Matcher.

3.4 *Mobile App*

A mobile App is developed to capture currency image in real time through mobile camera and determine its value with audio.

4 Results and Performance Evaluation

The mobile application is designed using Java in android studio environment. It captures the image in any orientation and provides detected currency output though audio as shown in Fig. 2.

The query currency image is dynamically captured through mobile camera, which will be compared with reference images based on ORB algorithm and identified currency is given as audio output. If the option taken is for generating total denomination, it reads each individual note, recognizes its denomination and provides grand total of available amount. In this work we extracted various unique properties of currency notes especially related to Indian and the algorithm developed for recognizing every unique feature using ORB algorithm. This paper demonstrates a template matching method by computing cross-correlation between the query currency image captured through mobile camera and the reference images available the dataset to ensure fast processing and robust identification system. Also, the currency notes with different orientations are recognized with the present system and obtained results are given in Fig. 3, 4, 5, 6 and Fig. 7, which is displayed on mobile screen and audio output is generated. It also finds the denomination and total amount available with the person. This research work presents an efficient and quick Indian Currency recognition system enabling the visually impaired individuals to identify uniquely the currency denomination and gross amount available. This entire model is implemented as a mobile application. The overall accuracy of proposed system is 88% and detection time is 0.682 s.

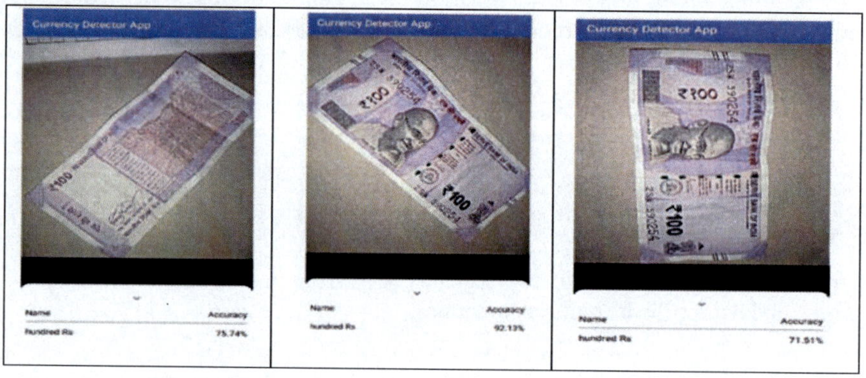

Fig. 3 Detected Hundred Rupees note with different Orientations

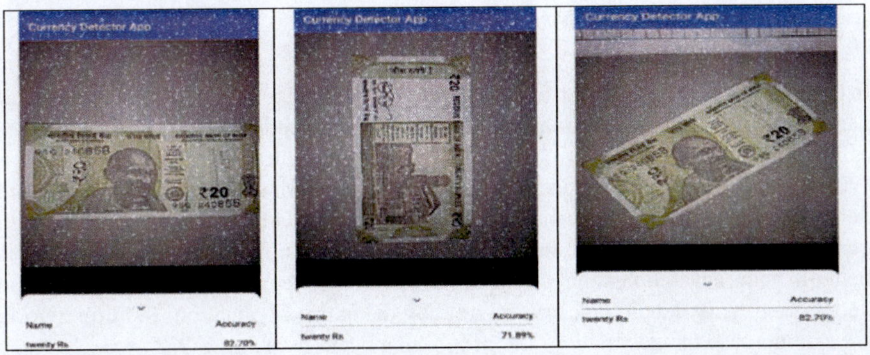

Fig. 4 Detected Twenty Rupees note with different Orientations

Fig. 5 Detected Five Hundred rupees note

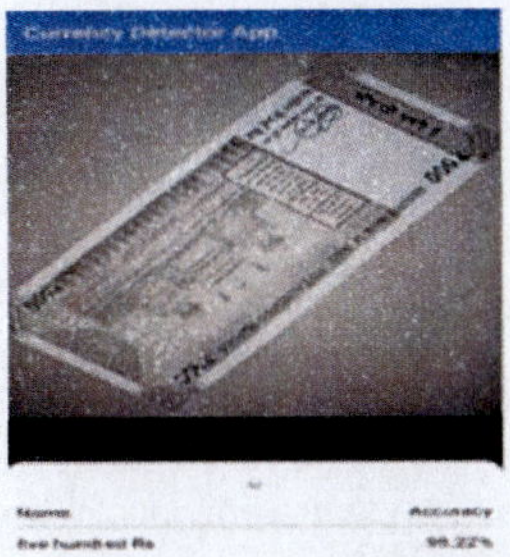

Fig. 6 Detected Fifty rupees note

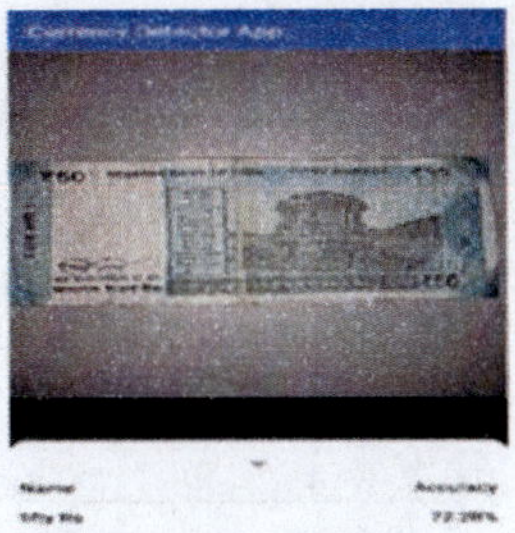

Fig. 7 Detected Ten rupees note

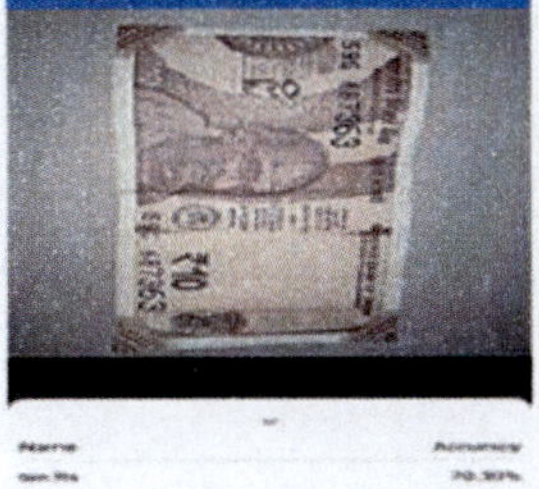

5 Conclusions and Future Work

In this work, a mobile application is presented to recognize Indian Currency denomination and as well as total amount available at hand with audio output using ORB algorithm, which enables visually challenged to handle cash transactions independently. We have trained the model using 7000 images in different orientations with different denominations. The accomplishments of this system are validated and compared with other existing systems, the obtained results demonstrate that the present method is better in accuracy and fast in recognition of currency. We can conclude that the blind people can easily detect the currency note by using this application. In Future work, we can increase the accuracy of the proposed system by using YOLO v5.

References

1. https://www.blindlook.com/blog/detail/the.-population-of-blind-people-in-the-world
2. Zhang Q, Yan WQ (2020) Currency detection and recognition based on deep learning. In: 15th IEEE international conference advanced video and signal based surveillance (AVSS), pp 1–6
3. Almisreb AA, Saleh MA (2018) Transfer learning utilization for banknote recognition: a comparative study based on Bosnian currency. Southeast Eur J Soft Comput 8(1):28–31
4. Abburu V, Gupta S, Rimitha SR, Mulimani M, Koolagudi (2017) SG Currency recognition system using image processing. In: 10th international conference on contemporary computing, IC3
5. Selvi Rajendran P (2019) Virtual bulletin board using man-machine interface (MMI) for authorized users. Indian J Sci Technol 12(34)
6. Hassanpour H, Masoumifarahabadi P (2009) Using hidden Markov models for paper currency recognition. Exp Syst Appl 36(6):10105–10111
7. Tong C, Lian Y, Qi J, Xie Z (2017) A novel classification algorithm for new and used banknotes. Mob Netw Appl. 22(3):395–404
8. Selvi Rajendran P (2018) Virtual information kiosk using augmented reality for easy shopping. Int J Pure Appl Math (IJPAM) 118(20):985–994
9. Selvi Rajendran P (2019) AREDAI: augmented reality based educational artificial intelligence system. Int J Recent Technol Eng (IJRTE) 8(1)
10. Selvi Rajendran PP, Anithaashri TP (2020) CNN based framework for identifying the Indian currency denomination for physically challenged people. IOP Conf Ser Mater Sci Eng 992. https://iopscience.iop.org/article/10.1088/1757-899X/992/1/012016/pdf

Deep Belief Network Algorithm-Based Intrusion Detection System in Internet of Things Environments

C. Geetha, A. Jasmine Gilda, and S. Neelakandan

Abstract Intelligent environments can improve the quality of human existence by reducing inconveniences and maximizing productivity. An Internet of Things (IoT) concept has recently established into a method for making smart atmospheres. Protecting users' personal information and preventing unauthorized access are paramount in any IoT-based smart ecosystem used in the real world. Smart environment applications face security risks from Internet of Things-based technologies. Accordingly, intrusion detection systems (IDSs) that are adapted to IoT settings are crucial for reducing the risk of attacks that exploit vulnerabilities in the IoT. Due to the inadequate processing and storage abilities of IoT technology and the specific protocols in use, traditional IDSs may not be applicable. In this article, we will discuss the most up-to-date intrusion detection systems (IDSs) for the IoT, with an emphasis on the appropriate approaches, features, and procedures. This research presents a Deep Belief Network (DBN) based Intrusion Detection System (IDS) employing the K-Nearest Neighbor technique for monitoring IoT networks for intrusions (KNN). Step one of this process involved using an Absolute Maximum Scaling (AMS) notion of normalization to the UNSW-NB15 dataset in order to cut down on data loss. This dataset, which consists of both recent attacks and normal network activity, covers nine distinct attack types. The next step was to use Principal Component Analysis (PCA) to minimize the dimensionality (PCA). Additionally, we have compared our method to other state-of-the-art studies. Experiments showed that DBN-KNN was superior to other intrusion detection systems in IoT technology.

Keywords Intrusion Detection Systems (IDS) · Deep Belief Network (DBN) · Absolute Maximum Scaling (AMS) · K-Nearest Neighbor (KNN) and Principal Component Analysis (PCA)

C. Geetha (✉) · A. J. Gilda · S. Neelakandan
Department of Computer Science and Engineering, R.M.K. Engineering College, Kavaraipettai, Tamil Nadu 601206, India
e-mail: cga.cse@rmkec.ac.in

N. Hasteer et al. (eds.), *Decision Intelligence Solutions*, Lecture Notes in Electrical Engineering 1080,
https://doi.org/10.1007/978-981-99-5994-5_12

1 Introduction

Our modern environment has become more sophisticated because to the growth of Internet-connected devices. The connected objects are able to monitor their environment, react appropriately, and share data with one another. Almost every industry today makes use of Internet of Things (IoT) technologies, including healthcare, municipal infrastructure, urban planning, agriculture, manufacturing, and transportation. Hence, the services enabled by the Internet of Things have significantly impacted people's lives. Over a hundred billion devices are predicted to be connected to the network by 2040 [1]. Participant-provided solutions, such as Internet of Things support, allow impaired people greater freedom and access to community activities.

IoT has improved people's lives, especially those of the elderly, by making it possible to monitor their health in real time. Appliances that can be worn, such as activity trackers, blood pressure monitors, and glucose meters, give patients easier access to individualized care. Daily usage of electricity to be reduced with the help of smart grids and meters. Through this supply-demand must be balanced. Incorporating smart agriculture into farming operations helps farmers pinpoint and quarantine disease-prone agricultural zones, forecast crop yield, and determine fertilizer needs.

As the Internet of Things becomes more widely used, security will provide a significant problem. Whenever the Internet of Things is exploited, not only data is destroyed, but also the Internet of Things' hardware. As a result, it is necessary to optimize and improve the performance of the intrusion detection for the IoT environments.

The information in this paper is presented in the following order: first, a literature study of intrusion detection systems in IoT environments based on DBN-KNN models is presented; second, the system that is being suggested is described; fourth, the findings and discussion are presented; and finally, a summary and conclusion are presented.

2 Related Work

Ding et al. [2]. shows the effectiveness of DBN in identifying malicious software. Each layer underwent 30 epochs of RBM pretraining, and then the network was fine-tuned with the use of the backpropagation algorithm and five-fold cross-validation. The classification results were put to the test using datasets consisting of 3,000 helpful and 3,000 harmful records. The suggested model was evaluated on a variety of training data (features), and it was found that 400 features yielded an accuracy of 96.1%.

There may be a large cost associated with keeping track of object neighbors, as shown by Yang et al. [3]. They depend on an important aspect of sliding windows—which indicates the "predictability" of the expiry objects and to identify which items are most likely to develop into outliers. To predict the design structure in successive windows, they only employ information about the items in the present window (the elements in the current window). All these projected design constructions can be conceptualized as a "approximate perspective" of each next iterated window, which will be used to spot outliers.

In [4], Kang and Kang proposed an in-car IDS that uses a DBN to process data. Although essentials of the DBN's architecture, such as the total amount of input layer units, were not provided, this IDS method is conceptually similar to previous DBN-based methods. CAN is used for in-car communications networks. The suggested detection system uses a total of 200,000 packets, 70% of which are used for training and 30% for testing. In order to train DBN with binary labels denoting normal or attack packets, features are retrieved from OCTANE's training CAN packets. The ROC curve was produced using the false negative and positive rates from the proposed IDS. DBN performed better than SVM and FFNN using the ROC curve. It was said to be accurate 97.8 percent of the time by the authors.

By combining word net with traditional methods, Fang et al. [5] established a new methodology that improves likeness rank. When compared to the standard review comparison and the adaptable KNN technique, the implemented grouping calculation is more precise. It's challenging to compile a client-facing report that uses semantic tagging, and even more so to discover a relevant example in the context of a growing online database. This means addressing the issues as they arise. In order to facilitate highlight extraction and pattern matching, Tao et al. [6] adopted an unsupervised class technique typical of machine learning. The limitations of previous efforts are numerous; for instance, they lack a flexible and robust system. It is not dealt with efficiently when situations arise where biased information is present.

S. Manimurugan [7] presented an intrusion detection system that, like the DBN algorithm, relies on the concept of hidden nodes. Tests on the CICIDS 2017 dataset were used to estimate that the proposed solution performing well in accuracy, recall, F1-score, precision and higher accuracy. The accuracy of this method is 97.93% for the usual class, 97.71% for botnets, 96.67% for brute strength, and 96.37% for DoS/DDoS.

3 Proposed System

Figure 1 shows that the initial dataset from unsw-nb15 is sent to the Deep Belief Network. The processed data is then transmitted to the K-neighbor algorithm, where it is checked one last time before being sent to the intrusion detection system to assess whether an intrusion has been committed. Otherwise, it will submit the processed data to be validated, but it will sound an alarm if it detects any kind of intrusion.

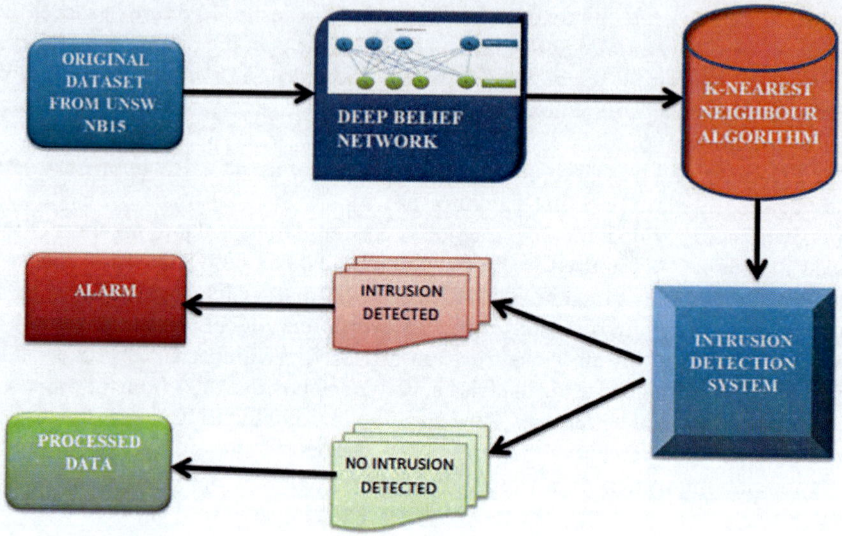

Fig. 1 Proposed structure for DBN_KNN

3.1 Deep Belief Network (DBN)

Deep Belief Network (DBNs) are reproductive methods. Restricted Boltzmann machines (RBMs) that are arranged one upon another in a DBN go through greedy layer-wise training to make sure they will function properly in an unsupervised environment. Deep Belief Networks (DBNs), which employ a different approach, are a number of RBM layers which are used for initial training and then it is being transformed into a forward network for weight fine-tuning. Lack of labelled data is likely to prevent widespread adoption of RBMs and auto-encoders because It will be fine-tuned on a little quantity of labelled data after being pretrained on unlabeled data.

The visible units in BMs represent observations, whereas the concealed units represent hidden aspects. The dissemination of an RBM's visible and concealed layers, which are based on the observable adaptable $v = (v_1, v_2, ...v_b)$ and the concealed variable $h = (h_1, h_2, ...h_a)$, is shown in Fig. 2.

$$p(v, h) = \frac{i}{z} \exp(-E(v, h)), \tag{1}$$

where

$$z = \sum_{v,h} \exp(-E(v, h)), \tag{2}$$

Fig. 2 DBN Architecture

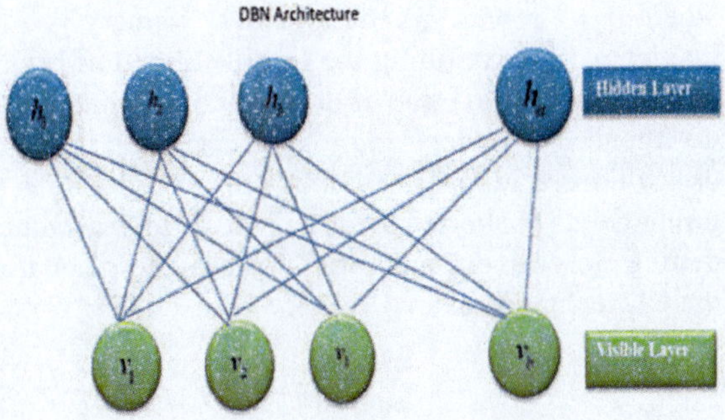

Are the normalization aspects for the divider purpose and E(v, h) energy purpose different as

$$E(v,h) = -\sum_{i=1}^{b} m_i v_i - \sum_{j=1}^{a} n_j h_j - \sum_{i=1}^{b}\sum_{j=‘}^{a} v_i w_{ij} h_j \tag{3}$$

where m_i and n_j are the preferences of the hidden and visible components, w_{ij} is the weight among hidden units, and v_i and h_j are the binary conditions of observable and hidden units. The probability of the previous layer is regarded as the marginal dissemination of p using the readily available observable information (v, h).

$$p(v) = \sum_h p(v,h) = \sum_h \frac{1}{z} \exp(-E(v,h)) \tag{4}$$

Approaches like those in the instance below can be used to estimate the probability of the concealed units.

$$p(h) = \sum_v p(v,h) = \sum_v \frac{1}{z} \exp(-E(v,h)) \tag{5}$$

As a consequence of the elements in the viewable and hidden layers being not linked, the ith observable unit's and j^{th} hidden unit's commencement probabilities are taken to be

$$p(v_i = 1|h) = \sigma\left[\sum_{j=1}^{a} h_j w_{ij} + m_i\right] \tag{6}$$

$$p(h_j = 1|v) = \sigma\left[\sum_{i=1}^{b} v_i w_{ij} + n_i\right] \tag{7}$$

where $\sigma(\bullet)$ represents the logistics' primary goal. The weights of the model are maximized by exploiting the log-likelihood of p(v) with deference to RBM limitations, albeit due to issue of delayed convergence, a technique known as contras-tive divergence (CD), devised by Hinton (2002), is frequently utilised. The CD method uses a limited number of Gibbs sample repetitions with observable elements set to training data to shorten the time it takes to determine the log-likelihood gradients by getting closer to expectations. The heaviness updating equation is transcribed using the CD technique as

$$\Delta w_{ij} = \eta \langle v_i h_j \rangle_{data} - \langle v_i h_j \rangle_{recon} \tag{8}$$

Reconstruction of previous layer to 1 is referred to as recon, and η is the learning rate.

3.2 K Nearest Neighbor

Distance metrics are utilized in the KNN technique for discovering class similarity and ranking articles. Classifications are assigned to objects when they are first created. The centre is correlated with the mean of the group's size distribution. An object is assigned a class at the outset if its similarity is maximally close to the class to which it is most closely related. Based on the results of the distance similarity technique, we were able to

$$\mu_j(N) = \sum_{i=1}^{k} \mu_j(N_j)\, SIMI(N, N)_i \tag{9}$$

where, $\mu_j(N_j) \in \{0,1\}$gives whether N_i communicates to $\mu_j(N_j) = 1$ or 0; SIMI(N, Ni) defines the detachment in among the assumed authenticating inquiries. Therefore the consequence of result can be; for $\mu_j(N) =$ extreme $\mu_j(N)$), and N $\in m_j$

The KNN algorithm, described by Eq. [9], cannot give the right answer for the mentioned question characterization, particularly if the outcome class is set as unlabeled. In order to enhance the precision provided by Eq. [10], this study presents the KNN with Fuzzy membership approach.

$$\mu_j(N) = \frac{\sum_{i=1}^{k} \mu_j(N_i)\, SIMI(N, N)_i \frac{1}{(1-SIMI(N,N1))^{2/(b-1)}}}{\sum_{i=0}^{n} \frac{1}{(1-SIMI(N,N1))^{2/(b-1)}}} \tag{10}$$

3.3 Intrusion Detection System

Researchers' key concern is technology's growth. IOT intrusion detection systems are mainly host-based and network-based. Interdisciplinary technology is widely used in IoT intrusion detection systems. The IoT intrusion detection system's efficiency and performance have been increased. Faced with network complexity and assault variety, scientists are creating a genetic approach to enhance neural network weights for intrusion detection in networks. The genetic method is used to discover the best BP neural network weight. Compared to standard network intrusion detection algorithms, this one has faster training sampling times and better identification and detection effects. After each of the procedures has been finished, this system does a check to determine whether or not there has been an unauthorized access attempt of any kind. In the event that there has not been an intrusion, it will deliver the processed data; otherwise, it will send out a signal.

4 Result and Discussion

4.1 Environmental Setup

In order to reflect the frequency of contemporary low-impact attacks more accurately, the new UNSW-NB15 dataset was developed (Nour et al., [27]). It was developed using the research tool IXIA Perfect Storm by the Australian Centre for Cyber Security (ACCS). The 49 attributes can be grouped according to a variety of criteria, including flow, necessity, substance, timing, broad categories, links, and labels. The details of the features are described in (UNSW dataset). The name of the attack category may be found in feature #48 (attack cat), whereas normal and attack are represented by the numbers 0 and 1 in feature #49 (Label). Unlike NSL-four, KDD's UNSW-nine NB15's attack families include fuzzers, investigation, backdoors, denial-of-service, exploits, generic, inspection, shellcode, and worms. Files containing the UNSW-NB15 data set's comma-separated values (CSV) can be downloaded here.

4.2 Validation Dataset (Sec)

Table 1 shows IDS-based Deep Belief Network Validation Data using K-Nearest Neighbor. Figure 3 compares the old and new models. Old models (GBDT-GWO, Random forest (RF), Linear Regression (LR), Butterfly Optimization Algorithm (BO), Support Vector Machine) have validation values of 4.034, 3.745, 3.204, 2.856, 2.345. The 100 dataset validation values for the current model (GBDT-GWO, Random Forest, Linear Regression, Butterfly Optimization Algorithm, and Support

Table 1 Validation Dataset for DBN-KNN

No of data from Dataset	GBDT-GWO	RF	LR	BOA	SVM	DBN-KNN
20	4.034	3.745	3.204	2.856	2.345	1.945
40	4.189	3.920	3.450	2.954	2.509	2.224
60	4.423	4.123	3.545	3.145	2.764	2.467
80	4.508	4.302	3.945	3.482	2.903	2.722
100	4.873	4.403	4.001	3.567	3.004	2.872

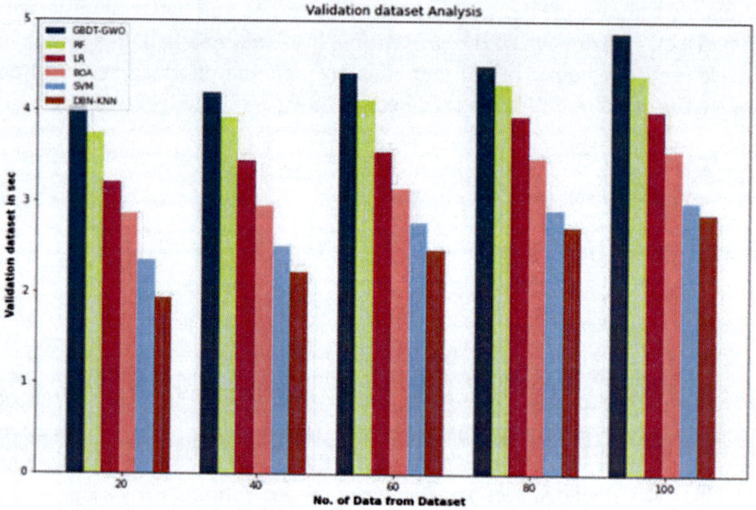

Fig. 3 Validation Dataset for DBN-KNN

Vector Machine) are 4.873, 4.403, 4.001, 3.567, 3.004. The 20 validation dataset's recommended DBN-KNN value is 1.945, whereas the 100 dataset's is 2.872.

4.3 Accuracy (%)

Table 2 illustrates Deep Belief Network with K-Nearest Neighbour method based on Intrusion detection system accuracy. Figure 4 compares the new and old models. Accuracy Analysis of 50 datasets for existing models (GBDT-GWO, Random forest(RF), Linear Regression(LR), Butterfly Optimization Algorithm(BOA), Support Vector Machine) are 56.89, 62.45, 68.14, 76.12, 84.90. Accuracy analysis of 250 datasets for existing models (GBDT-GWO, Random forest(RF), Linear Regression(LR), Butterfly Optimization Algorithm(BOA), Support Vector

Table 2 Accuracy Analysis for DBN-KNN

No of data from Dataset	GBDT-GWO	RF	LR	BAO	SVM	DBN-KNN
50	56.89	62.45	68.14	76.12	84.90	89.15
100	58.03	64.83	72.56	79.43	86.36	91.78
150	61.98	67.90	75.87	81.34	86.93	93.45
200	60.93	69.12	74.93	84.04	88.33	94.90
250	63.86	71.03	77.24	86.94	90.13	95.34

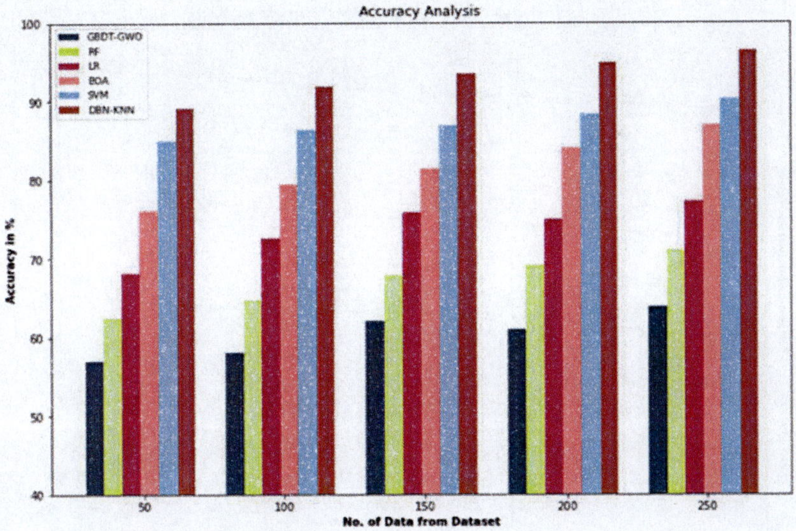

Fig. 4 Accuracy Analysis for DBN-KNN

Machine) are 63.86, 71.03, 77.24, 86.94, 90.13. But DBN-50 KNN's validation dataset is 89.15 and 250 is 95.34.

4.4 Mathew Correlation Coefficient (%)

In Table 3, we can see the IDS-based K-Nearest Neighbor Deep Belief Network's Mathew correlation coefficient. Figure 5 shows the state of the art compared to our model. Mathew correlation coefficients for the existing models are in the 0.332–0.546, 0.641–0.725, range. In comparison, the validation values for the existing models (GBDT-GWO, Random forest(RF), Linear Regression(LR), Butterfly Optimization Algorithm(BO), and Support Vector Machine) on the same 100-sample dataset are 0.432, 0.532, 0.639, 0.699, and 0.817, respectively. The F1-score for DBN-KNN is 0.823 on 20 datasets and 0.913 on 100 datasets.

Table 3 Mathew correlation coefficient for DBN-KNN

No of Dataset	GBDT-GWO	RF	LR	BAO	SVM	DBN-KNN
20	0.332	0.435	0.546	0.641	0.725	0.823
40	0.376	0.456	0.573	0.652	0.747	0.848
60	0.392	0.489	0.594	0.669	0.762	0.863
80	0.402	0.512	0.623	0.683	0.798	0.896
100	0.432	0.532	0.639	0.699	0.817	0.913

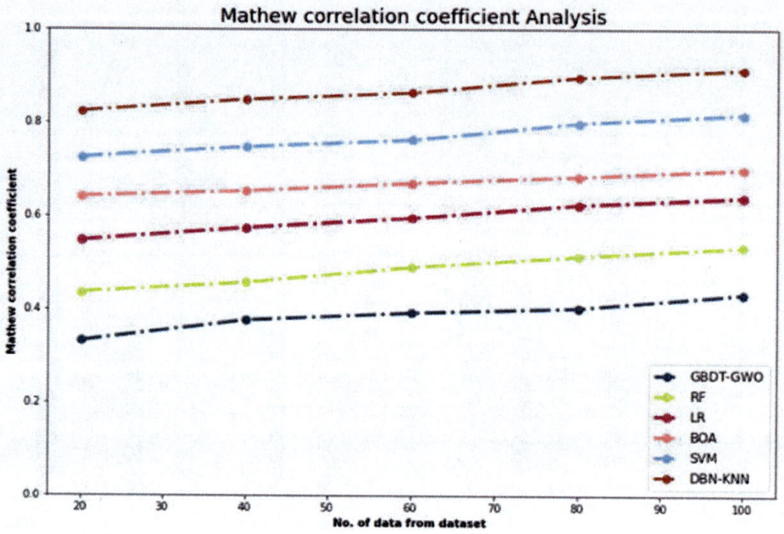

Fig. 5 Mathew correlation coefficient for DBN-KNN

5 Conclusion

The Internet of Things intrusion detection system's performance is optimised while it is being studied. The DBN-KNN network structure and the core parts of the system for un-authorized person detection for the Internet of Things are contrasted at the outset of this methodology. As a result, the environment of Internet of Things is now more secure. In comparison to previous neural network models like GBDT-GWO, RF LR, and others, the DBN-KNN network model that was constructed in this paper offers clear advantages. The detection rates of DOS, R2L, U2L, and probing attack types are improved, and the frequency of false alarms is decreased, among other benefits. An experimental simulation analysis was used to uncover these benefits. As a result, it will act as a guide and a reference for better services provided by smart homes, smart campuses, and smart supply chains in light of the continued development of the global Internet of Things and the introduction of 5G. Additionally, it will hasten the global deployment of the Internet of Things and support more productive industrial sector growth in the economy.

References

1. softbank-son-iot-1000-devices-2040@ www.techinasia.com," May 2021. https://www.techinasia.com/softbank-soniot-1000-devices-2040
2. Ding Y, Chen S, Xu J (2016). Application of deep belief networks for opcode based malware detection. In: Proceedings of IJCNN, pp 3901–3908
3. Yang D, Rundensteiner EA, Ward MO (2009) Neighbor-based pattern detection for windows over streaming data. In: Proceedings of the 12th international conference on extending database technology: advances in database technology (EDBT), Saint Petersburg, Russia, pp 529–540
4. Kang MJ, Kang JW (2016) Intrusion detection system using deep neural network for in-vehicle network security. PLoS ONE 11:1–17
5. Fang J, Guo L, Wang X, Yang N (2007) Ontology-based automatic classification and ranking for web documents. In: IEEE fourth international conference on fuzzy systems and knowledge discovery (FSKD), pp 7695–2874
6. Tao X, Li Y, Liu B, Shen Y (2012) Semantic labelling for document feature patterns using ontological subjects. In: IEEE/WIC/ACM international conferences on web intelligence and intelligent agent technology, pp 530–534. 978-07695-4880
7. Manimurugan S, Al-Mutairi S, Aborokbah MM, Chilamkurti N, Ganesan S, Patan R (2020) Effective attack detection in internet of medical things smart environment using a deep belief neural network. IEEE Access 8:77396–77404
8. Anand P, Singh Y, Selwal A, Alazab M, Tanwar S, Kumar N (2020) IoT vulnerability assessment for sustainable computing: threats, current solutions, and open challenges. IEEE Access 8:168825–168853
9. Tanwar S, Tyagi S, Kumar S (2018) The role of internet of things and smart grid for the development of a smart city. Intell Commun Comput Technol 19:23–33
10. Hinton GE (2012). A practical guide to training restricted Boltzmann machines. In: Neural networks: tricks of the trade. Lecture notes in computer science, vol 7700, pp 599–619
11. Ieracitano C, Adeel A, Morabito FC, Hussain A (2020) A novel statistical analysis and autoencoder driven intelligent intrusion detection approach. Neurocomputing 387:51–62
12. Pu J, Wang Y, Liu X, Zhang X (2019) STLP-OD: Spatial and temporal label propagation for traffic outlier detection. IEEE Access 7:63036–63044
13. Mohamed T, Otsuka T, Ito T (2018) Towards machine learning based IoT intrusion detection service. In: Recent trends and future technology in applied intelligence, IEA/AIE 2018, Lecture Notes in Computer Science, vol. 10868, pp 580–585. Springer, Cham
14. Sharath MN, Rajesh TM, Patil M (2022) Design of optimal metaheuristics based pixel selection with homomorphic encryption technique for video steganography. Int J Inf Technol 14:2265–2274. https://doi.org/10.1007/s41870-022-01005-9
15. Gupta S et al (2023) Mobility aware load balancing using Kho-Kho optimization algorithm for hybrid Li-Fi and Wi-Fi network. Wireless Netw. https://doi.org/10.1007/s11276-022-03225-0
16. Mardani A, Mohan P, Raj Mishra A, Ezhumalai P (2023) A fuzzy logic and DEEC protocol-based clustering routing method for wireless sensor networks. AIMS Math 8(4):8310–8331. https://doi.org/10.3934/math.2023419
17. Ezhumalai P, Prakash M (2022) A deep learning modified neural network (DLMNN) based proficient sentiment analysis technique on Twitter data. J Exp Theor Artif Intell. https://doi.org/10.1080/0952813X.2022.2093405
18. Mary Rexcy Asha S, Roberts MK (2022) Artificial humming bird with data science enabled stability prediction model for smart grids. Sustain Comput Inf Syst 36. https://doi.org/10.1016/j.suscom.2022.100821

19. Metwally AM, Gupta MS (2022) Metaheuristics with deep transfer learning enabled detection and classification model for industrial waste management. Chemosphere 136046. https://doi.org/10.1016/j.chemosphere.2022.136046
20. Sathishkumar VE et al (2022) Deep learning based sentiment analysis and offensive language identification on multilingual code-mixed data. Sci Rep 12:21557. https://doi.org/10.1038/s41598-022-26092-3
21. Sambath N, Ramanujam V, Ram M (2022) Deep learning enabled cross-lingual search with metaheuristic web-based query optimization model for multi-document summarization. Concurrency Comput Pract Exp. e7476. https://doi.org/10.1002/cpe.7476
22. Ying Z, Peisong L, Xinheng W (2019) Intrusion detection for IoT based on improved genetic algorithm and deep belief network. IEEE Access 7:31711–31722
23. Alotaibi Y, Alghamdi S, Khalafand OI, Nanda AK (2022) Improved metaheuristic-driven energy-aware cluster-based routing scheme for IoT-assisted wireless sensor networks. Sustainability 14:7712. https://doi.org/10.3390/su14137712
24. Sridevi M, Chandrasekaran S, Lingaiah B (2022) Deep learning approaches for cyberbullying detection and classification on social media. Comput Intell Neurosci 2022. https://doi.org/10.1155/2022/2163458
25. Fiore U, Palmieri F, Castiglione A, De Santis A (2013) Network anomaly detection with the restricted Boltzmann machine. Neurocomputing 122:13–23
26. Nix R, Zhang J (2017). Classification of android apps and malware using deep neural networks. In: Proceedings of IJCNN, pp 1871–1878
27. Nour M, Slay J (2015). UNSW-NB15: a comprehensive data set for network intrusion detection systems. In: Proceedings of IEEE MilCIS, pp 1–6. NSL dataset. https://www.unb.ca/cic/datasets/nsl.html

Crop Classification Based on Multispectral and Multitemporal Images Using CNN and GRU

C. Sagana, R. Manjula Devi, M. Thangatamilan, T. Charanraj, M. V. Cibikumar, G. Chandeep, and D. Mugilan

Abstract Food is very essential in our daily life and demand for food is increasing day by day. Crop classification can be used to identify the suitable crops to grow on the land in such a way that crop production can be increased. Accurate and mapping of land at correct time can be used in numerous applications such as management of sustainable agriculture and food security. With the growing technology, deep learning methods along with high resolution Sentinel-2 images can be used for mapping crops and crop classification. In this study, we have proposed a new model that works with Convolutional Neural Network as the base layer along with Gated Recurrent Unit to improve its performance and accuracy. The farming land images are collected from different dates of the growing season covering all growing stages of the crop. For acquiring multispectral data, bands from the Sentinel-2 images shall be used, and the features in each patch of the image will be extracted and used. Following this proposed ideology, by feeding both the multitemporal and the multispectral data to the model, several crop classes can be identified directly using the satellite images easily and efficiently. To measure the performance, we have compared the models by finding their accuracy. On concluding, our study found that using CNN model along with GRU by remote sensing produces higher accuracy than just CNN alone in classifying the crops using the satellite imagery.

Keywords Deep Learning · CNN (Convolutional Neural Network) · GRU (Gated recurrent unit)

C. Sagana (✉) · R. Manjula Devi · M. Thangatamilan · T. Charanraj · M. V. Cibikumar · G. Chandeep
Kongu Engineering College, Perundurai, Tamil Nadu, India
e-mail: sagana.c@gmail.com

D. Mugilan
K S Rangasamy College of Technology, Tiruchengode, India

N. Hasteer et al. (eds.), *Decision Intelligence Solutions*, Lecture Notes in Electrical Engineering 1080,
https://doi.org/10.1007/978-981-99-5994-5_13

1 Introduction

Crop classification is a well-organized way of classifying various plants that aids in recognizing and cataloguing the vast amount of data regarding several varieties of plants. For calculating agricultural yields, enhancing crop production management, and improving crop insurance, timely and precise crop type categorization is crucial.

In the recent years, crop classification using remote sensing images have been developed which could give more accurate results. For the classification purpose, using both the multispectral and multitemporal remote sensing data will provide results with better precision.

After classifying the crops using the remote sensing images, the classified data can be used for the following purposes.

1. Measuring wealth of land: Although wealth of a land can be measured in terms of crops grown area and total area manually, it is a difficult job to do the same with large areas of land like for a city or state. But with satellite images and the suggested model, the richness of area can be found easily.
2. Helping government to provide subsidies: With the vast scale of farming area that government needs to manage, it will be hard for them to know their agricultural wealth. This is a crucial process and government cannot provide support for farmers without proper information. With the model in place, providing the farming information, it will be useful for the government to choose the areas to provide subsidies with better precision.
3. Crop recommendations: There are a multitude of crops available which can make it confusing for a farmer to choose what to plant and harvest. Once the model is implemented across vast region, the data can be used to recommend farmers which crops to grow together and the best period to plant it.

With the method of detecting the radiation that is reflected and emitted from a spot from a distance, typically via a satellite or an aircraft, remote sensing allows us to identify and keep track of the physical properties of a location. Researchers can sense the ground by using cameras to collect remotely sensed images. In addition to identifying agricultural conditions and improving farming precision, remote sensing is used in agriculture to anticipate crop production, track crop damage and progress, identify crops, and analyze crop health.

Food is an essential part of our life. They are essential for every human and is required in massive amounts. With the current cultivation strategy, it is very hard to keep up with the growing food demands. Also, productivity of the farmer decreases due to improper knowledge about soil and water requirements. Some crops should be grown under certain climatic conditions but growing them under different weather conditions might lead to crop loss. Many lands have left unused and became barren as they are not fit for agriculture. Farmers have very little idea when it comes to deciding what crops to grow in their land. Using modern techniques like deep learning crops can be classified and the one which best suits the farming land can be determined. Traditionally either multispectral or spatial data are used for this process which do not account for other missing data. This produces inaccurate crop maps leading to imprecise crop maps and food estimates.

2 Existing Systems

Seyd Teymoor Seydi (2021) [1] released a revolutionary framework based on a dual attention module (DAM) and a deep convolutional neural network (CNN), employing time-series data from Sentinel-2 to benefit from both the spatial and spectral properties of Sentinel-2 datasets to recognize crops in a novel way. The outcomes showed that the suggested strategy has a Kappa coefficient of 98.54% and a high overall accuracy of 0.981.

Shunping Ji [3] (2018) offers a unique spatiotemporal remote sensing crop classification approach based on 3D (three-dimensional) convolutional neural networks (CNN). To increase labelling accuracy up to a necessary level with the highest effectiveness, he added an active learning method to the CNN model. He then compared 3D CNN with 2D CNN, and in the end concluded that 3D CNN fared better than other widely used approaches and was appropriate for describing the dynamics of crop growth.

Luan Pierre Pott (2021) [4] developed a model for classifying crops to estimate large-scale crop areas using a testing transfer learning model and Sentinel 2 and SRTM digital evolution data. Depending on the model's output, features are chosen and discarded using MDA and Spearman correlation of the variables. Sentinel-1, Sentinel-2, and Digital Elevation (SRTM) data fusion, with an overall accuracy of 0.95, was important for creating this intricate regional categorization. Like other research, adding SRTM data to the combination of SAR and optical data enhanced crop classification.

Lijun Wang (2022) [5] evaluated deep learning algorithms for classifying crops. The study found that, in comparison to the choice of backbone architecture, patch size had a stronger impact on model accuracy. The UNet++ architecture with Timm-RegNetY-320 backbone was found to have superior classification accuracy to other backbone options like EfficientNet-b7, VGG16, Timm-ResNet101e and ResNet50.

Mehmet Ozgur Turkoglu (2021) [6] attempted to map crops using a time series of satellite images. Using crop type labels with a hierarchical tree structure allowed him to map unusual crop types much more effectively. ConvRNN was used to encode the three-level labels, enabling the model to acquire joint feature embeddings of exceptional classes. The proposed hierarchical convRNN model outperformed numerous baselines by a minimum 9.9% points in F1-score when he compared it in the end.

Hongwei Zhao (2019) [7] utilized a combination of 3 deep learning models for identifying early crops which includes (GRU RNNs, LSTM RNNs, and 1D CNNs). sentinel-1A imagery time series which could help avoiding the need to train optimal architectures and hyper-parameters on the unique data of each time series. The result was found that the RNNs (GRU- and LSTM-based) performed better than 1D CNNs in the case of very brief time series durations, and the F-score achieved by three deep learning methods were more than 0.90.

Shuting Yang (2020) [8] used hybrid convolutional neural network-random forest (CNN-RF) networks and optimal feature selection (OFSM) networks to analyze optical multi-temporal remote sensing images to develop an inventive system for classifying crops that achieves a balance between processing speed and classification accuracy. A hybrid CNN-RF network model constructed with one-dimensional convolution (Conv1D) and RF was successful in categorizing the data and produced better outcomes, as measured by overall accuracy (94.27%) and K coefficient (0.917).

Aiym Orynbaikyzy (2021) [11] evaluated the capacity of random forest models to classify crops spatially. Through an examination of solely optical, only SAR, and optical-SAR feature combinations, the research investigates the spatial transferability of random forest models. It demonstrated that random forest models using optical-SAR combinations perform better than models using single sensor data in training sites and geographical spaces that are not visible to the model. It also demonstrated that SAR-based models exhibit the lowest accuracy losses when applied to an area outside of the training regions.

Xia Zhang (2016) [14] used an agricultural classification technique that combines the object-oriented classification (OOC) methodology with the generation and optimization of a vegetation feature band set (FBS). The findings show that the suggested strategy can greatly increase crop classification accuracy, decrease edge effects, and be a useful way for monitoring invasive species, classifying crop species, and other applications linked to precision agriculture.

Henning Skriver (2011) [15] proposed the classification of crops using multi-temporal short-revisit SAR data. The classification was done using dual and single polarizations and data that is fully polarimetric and concluded that multitemporal acquisitions are crucial, and the ideal method is cross polarized backscatter. In the absence of sufficient acquisitions, the polarimetric mode might outperform the dual and single polarization modes.

3 Proposed System Design

In the proposed system, cropland data images from Sentinel-2 are given as input. Then, the data is preprocessed, augmented, and standardized for the further process. A deep learning model consisting of CNN layer, group normalization layer and a GRU layer is built, and the model is trained and validated using the processed data. Using that trained model, crops can be classified efficiently.

Although CNN alone can be used to classify and predict images, it won't be enough in certain cases. Input data fed to the model won't be persistent in a traditional neural network as they won't retain the information. A Recurrent Neural Network can be used which has loops making it possible to get information that has been already processed. GRU is a newer generation of RNN with two gates,

Table 1 Crop size

S. No	Crop	No of fields
1	Paddy	262
2	Sugarcane	329
3	Cotton	287
4	Coconut	372
5	Chilies	219

- Update gate to select what data to add, remove, and maintain.
- Reset gate to choose how much of the past information to let go.

From the survey, it has been identified that using both the temporal and spatial features of the crop gives better results. Hence, in the proposed hybrid model, the CNN layer will be helpful in extracting the spatial features and the GRU layer will be helpful in extracting the temporal features corresponding to the CNN layer output.

4 Proposed Methodology

4.1 Dataset

The dataset is collected from the Sentinel-2 satellites during the growing season. It can be downloaded separately using the services provided by Copernicus Open Access Hub of the European Space Agency.

The dataset for this project was taken in Tirunelveli, Tamil Nadu region, during the months of June to August 2022 consisting of the crops namely: paddy, sugarcane, cotton, coconut, and chilies (Table 1).

The dataset selected consists of 12 band layers such as B01, B02, B03, B04, B05, B06, B07, B08, B8A, B09, B11, B12 and Cloud Probability Layer. First, a standard grid with a 10 m spatial resolution is used to map all bands.

4.2 Data Preprocessing

The process in which the raw data is converted into useful and efficient data is called data preprocessing. Here, the SNAP application is used for data preprocessing.

SNAP is an open-source sentinel toolbox developed by European Space Agency (ESA) can be used to explore the satellite data. It contains several tools and filters to process, modify, and visualize the data. The sentinel-2 dataset downloaded is opened with SNAP. The dataset is viewed with different filters and modifications to identify the area of interest. Once that is done the required area with all the bands are exported in TIFF format for training & testing the model.

Instead of working with large tile data, the data is cropped into small areas so that it would be easy to train the model. Every field in the images are cropped into small patches and then converted into numpy arrays needed for further calculation. For each patch, ground truth data is collected which is the collection of information at a particular location.

4.3 Data Augmentation

Data augmentation can be used to create new training data from the existing data. It can be used to improve the model's efficiency. It also helps to avoid overfitting problems in the dataset. Data augmentation process is done on the training data to the quality, size and features of the data, so that the model performance will be increased.

4.4 Data Standardization

Data standardization is done to produce clear and consistent data. It converts data to a common format that helps in processing and analyzing the data. It also improves the speed and stability of the model. To standardize the data and minimize the dissimilarities between the data, Eq. (1) is used in every band of the data collected.

$$x = \frac{(x - \mu)}{\sigma} \tag{1}$$

where, x is image data stored in each band, μ is mean of all bands and σ is standard deviation of all bands.

4.5 Train and Test Data Split

The split is configured so that 70% of the data are used for model training and 30% are used for model testing. It is designed in above manner so that efficiency of the model would be increased using the large set of training data.

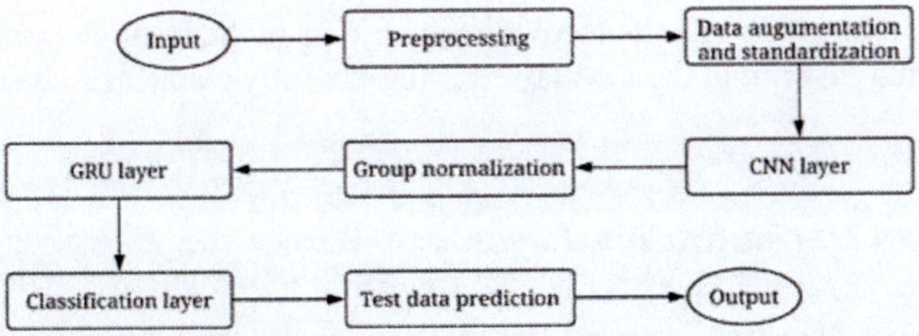

Fig. 1 Flow diagram of proposed work

5 Model Architecture

5.1 Flow Diagram

The model is designed by using 3 layers of 2D CNN followed by Group normalization layer and then next to 3 layers of GRU and then finally to classification (Fig. 1).

5.2 2D CNN Layer

A CNN model can be used to process images and identify the objects in it or also used to classify the images. These are considered as most effective architecture for image classification with high accuracy.

CNN can be classified based on the dimension of the convolution network that is used. The 2D convolution is a type of CNN model which uses two-dimensional convolutional kernel, and this model is used in this project. The whole advantage of using 2D-CNN is that it can extract the spatial features from the data using its kernel, which other networks are unable to do so that the multi-spatial features of the crop can be found from the image and can be used for further process (Fig. 2).

A 2D-CNN consists of two basic operations namely convolution and pooling. Several features are extracted from the preprocessed dataset during the convolution

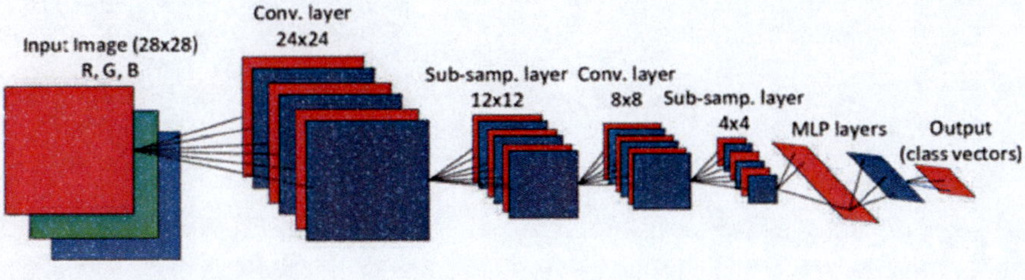

Fig. 2. 2D CNN architecture

phase. The pooling operation reduces the dimensions of the features extracted. Then, this data is fed forward to the next layers using the fully connected layer.

5.3 Group Normalization Layer

Group normalization layer is used to normalize the features extracted. This layer groups input into batches where each batch has a specific number of features and then applies normalization over them. The standard deviation and mean are calculated and then used in Eq. (2) to normalize the data given.

$$y = \left(\frac{x - E[x]}{\sqrt{Var[x] + \epsilon}}\right) * \gamma + \beta \tag{2}$$

Here, $x - E[x]$ is the no of channels in input, $(Var[x] + \epsilon)$ indicates the no of channels to separate the channels into and γ & β are learnable per channel parameter vectors.

5.4 GRU Layer

Gated Recurrent Units are improvised version of standard recurrent neural network which are more efficient (Fig. 3).

It is very similar to Long Short-Term Memory, but GRU uses gates to regulate the information flow. GRU has a simple architecture, and it is faster to train. It has different gates such as Update gate to store the value of knowledge that needs to be passed to next stage and Reset gate to store the amount of knowledge gained to forget and Current memory gate to reduce the impact of previous input on the current input.

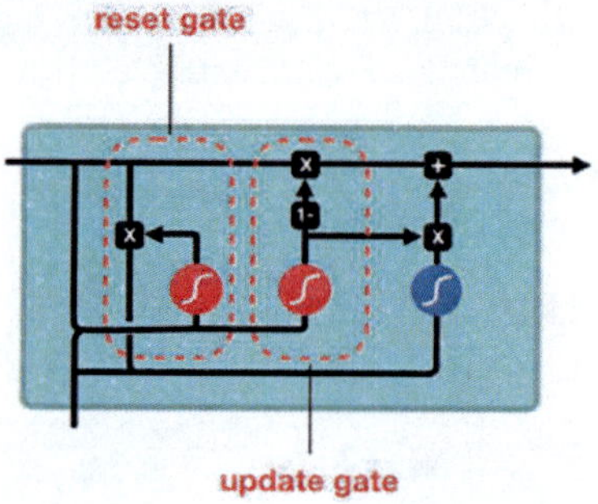

Fig. 3 GRU architecture

5.5 *Linear Layer*

The linear layer produces a single layer feed forward network with n inputs and m outputs. It applies a linear transformation on the incoming data using Eq. (3).

$$y = xA^T + b \tag{3}$$

where, x is input data, A^T is transpose of weight matrix, b is bias.

6 Results and Comparison

6.1 *Performance Metrics*

The parameter used to find the performance of our work is accuracy. The accuracy is the fraction of predictions made by the model that is correct. It may be described as the proportion of all correctly predicted values to all predicted values made by the model as given in Eq. (4).

$$Accuracy = \frac{\text{No of correct predictions}}{\text{Total number of predictions}} \tag{4}$$

6.2 *Training and Validation Results*

The training and validation accuracy and loss was found during every epoch to find the model's performance. The training results are got by using the dataset that have labels already predicted (Fig. 4).

The validation results are got by using the dataset which have labels but are not used in the training phase.

6.3 *Results Comparison*

As the performance metric used is accuracy, the final accuracy acquired from both the CNN without GRU model and CNN with GRU model is tabularized below (Table 2).

It is inferred from the above table that CNN along with GRU produces higher accuracy by 5.31% than CNN alone.

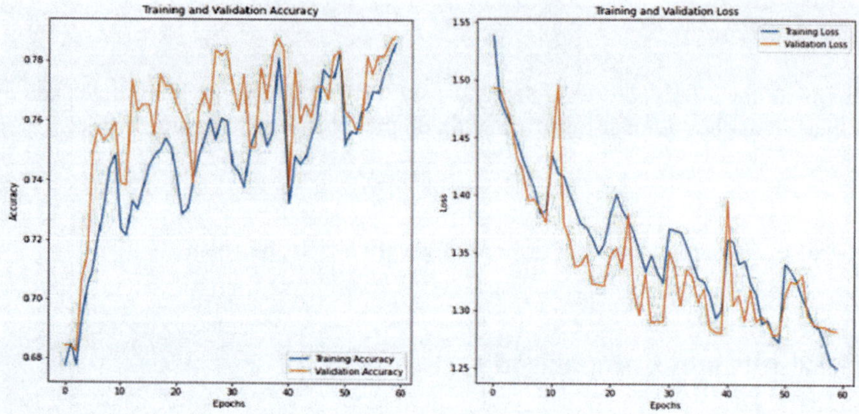

Fig. 4 Training vs validation accuracy & loss

Table 2 Accuracy vs architecture

S.No	Architecture	Accuracy
1	CNN	73.54
5	CNN with GRU	78.85

7 Conclusion and Future Work

Precision agriculture frequently needs accurate crop classification to estimate crop yield and crop area. In this study, a crop classification model was created using deep learning with CNN and GRU. The model accuracy is evaluated with dataset which has five different crop variations taken over one month. Furthermore, a comparison was also made between using a model with CNN and CNN with GRU. The comparison proved that using CNN along with GRU produces higher accuracy.

In future work, mix-up augmentation will be used to improve the model performance. In the mix-up augmented method, two images are mixed with weights and new virtual training samples can be created. These samples can give different scenarios of the dataset and expose many new variations to the model which can help create a better model. As a result of this the accuracy can be greatly improved.

References

1. Seydi ST, Amani M, Ghorbanian A (2022) A dual attention convolutional neural network for crop classification using time-series sentinel-2 imagery. Remote Sens 14:498 https://doi.org/10.3390/rs14030498

2. Wang L, Wang J, Liu Z, Zhu J, Qin F (2022) Evaluation of a deep-learning model for multi-spectral remote sensing of land use and crop classification. Crop J 2214–5141. https://doi.org/10.1016/j.cj.2022.01.009
3. Ji S, Zhang C, Xu A, Shi Y, Duan Y (2018) 3D convolutional neural networks for crop classification with multi-temporal remote sensing images. Remote Sens 10:75. https://doi.org/10.3390/rs10010075
4. Pott L, Amado T, Schwalbert R, Corassa G, Ciampitti I (2021) Satellite-based data fusion crop type classification and mapping in Rio Grande do Sul, Brazil. ISPRS J Photogrammetry Remote Sens 176:196–210 https://doi.org/10.1016/j.isprsjprs.2021.04.015
5. Wang Z, Zhao Z, Yin C (2022) Fine crop classification based on UAV hyperspectral images and random forest. ISPRS Int J Geo-Inf 11:252. https://doi.org/10.3390/ijgi11040252
6. Turkoglu M et al (2021) Crop mapping from image time series: deep learning with multi-scale label hierarchies. Remote Sens Environ 264. https://doi.org/10.1016/j.rse.2021.112603
7. Zhao H, Chen Z, Jiang H, Jing W, Sun L, Feng M (2019) Evaluation of three deep learning models for early crop classification using sentinel-1A imagery time series—a case study in Zhanjiang, China. Remote Sens 11:2673. https://doi.org/10.3390/rs11222673
8. Yang S, Gu L, Li X, Jiang T, Ren R (2020) Crop classification method based on optimal feature selection and hybrid CNN-RF networks for multi-temporal remote sensing imagery. Remote Sens 12:3119. https://doi.org/10.3390/rs12193119
9. Viskovic L, Kosovic IN, Mastelic T (2019) Crop classification using multi-spectral and multi-temporal satellite imagery with machine learning. In: International conference on software, telecommunications and computer networks, pp 1–5. https://doi.org/10.23919/SOFTCOM.2019.8903738
10. Murmu S, Biswas S (2015) Application of fuzzy logic and neural network in crop classification: a review. Aquatic Procedia 4:1203–1210. https://doi.org/10.1016/j.aqpro.2015.02.153
11. Orynbaikyzy A, Gessner U, Conrad C (2019) Crop type classification using a combination of optical and radar remote sensing data: a review. Int J Remote Sens 6553–6595. https://doi.org/10.1080/01431161.2019.1569791
12. Desai G, Gaikwad A (2021) Deep learning techniques for crop classification applied to SAR imagery: a survey. Asian Conf Innov Technol 1–6. https://doi.org/10.1109/ASIANCON51346.2021.9544707
13. Sun Z, Di L, Fang H, Burgess A (2020) Deep learning classification for crop types in North Dakota. IEEE J Sel Top Appl Earth Obser Remote Sens 13:2200–2213. https://doi.org/10.1109/JSTARS.2020.2990104
14. Zhang X, Sun Y, Shang K, Zhang L, Wang S (2016) Crop Classification based on feature band set construction and object-oriented approach using hyperspectral images. IEEE J Sel Top Appl Earth Obser Remote Sens 9:4117–4128. https://doi.org/10.1109/JSTARS.2016.2577339
15. Henning S, Francesco M, Giuseppe S (2011) Crop classification using short-revisit multitemporal SAR data. IEEE J Sel Top Appl Earth Obser Remote Sens 4:423–431. https://doi.org/10.1109/JSTARS.2011.2106198

An IoT-based Arduino System for Client Health Monitoring & Interpretation on Account of Basic Essential Markers

Shipra Varshney, Basant Kumar Verma, Prashant Vats, Ranjeeta Kaur, Tanvi Chawla, Siddhartha Sankar Biswas, and Ashok Kumar Saini

Abstract Throughout this research, the development of a patients monitoring system for core body temperature and breathing rate—two crucial pulse oximetry covered. The surveillance system was constructed on an IoT platform using the Mega 2560 Arduino board and ESP8266 Wi-Fi Components. Two sensing systems that each utilize temperatures measurements to identify each obtain data value. The program's goals are to develop a health - monitoring system that can identify, gather data readings, assess heart rhythm levels based on patient age, provide alarms for dangerous circumstances, and dis-play data remotely via Mobile applications. By making this effort, nursing staff would have less work to do while also having a far more practical way to monitor everyone's physiological parameters all throughout hospitalization. The conventional method, typically requires a clinician to examine everyone and

S. Varshney (✉)
Dr. Akhilesh Das Gupta Institute of Technology and Management, GGSIPU, New Delhi, India
e-mail: shipra_vin@yahoo.com

B. K. Verma
Panipat Institute of Engineering and Technology, Panipat, Haryana, India
e-mail: bkverma.3474@gmail.com

P. Vats · A. K. Saini
Department of CSE, School of Computer Science and Engineering, Faculty of Engineering, Manipal University Jaipur, Jaipur, Rajasthan, India
e-mail: prashantvats12345@gmail.com

A. K. Saini
e-mail: shksaini@gmail.com

R. Kaur
Kamal Institute of Higher Education and Advance Technology, GGSIPU, New Delhi, India
e-mail: ranjeetapopli@gmail.com

T. Chawla
Department of CSE, FEAT, SGT University, Gurugram, Haryana, India
e-mail: tanvi90.chawla@gmail.com

S. S. Biswas
Department of CSE, SEST, Jamia Hamdard University, New Delhi, India
e-mail: ssbiswas@jamiahamdard.ac.in

N. Hasteer et al. (eds.), *Decision Intelligence Solutions*, Lecture Notes in Electrical Engineering 1080,
https://doi.org/10.1007/978-981-99-5994-5_14

keep track of their physiological traits, is time-consuming. Caretakers utilizing this strategy might monitor patient's condition by installing Mobile application on any Android smartphone. Nurses or physicians may quickly assess the previous cardiac rhythm state by collecting the information gathered from the system in the form of spreadsheets. The results were very similar when the two key indicators of this approach were contrasted to advanced standards by visual examination or regular measuring tools.

Keywords Medical services · Insurance · Temperature sensors · Monitoring · Biomedical · CCU surveillance

1 Introduction

The construction of a patient surveillance system for breathing rate and core temperature—two essential pulse oximetry topics—was covered all through this study. The Mega 2560 Microcontroller board and ESP8266 Wi-Fi Modules were used to build the surveillance system on an IoT ecosystem. Sensation technologies that use temperature measurements to distinguish between each other and to acquire data values. The plan's objectives are to create a health-monitoring system that recognizes patients, collect data measurements, gauge irregular heartbeat levels based on the patient's age, issue warnings in case of precarious environments, and electronically visualize information via smartphone platforms. By making this attempt, registered nurses would have less work to perform and a much more useful means to monitor and manage each patient's physiological signals while they are in the facility. Make sure no one who has been transported to the institution suffers any harm by keeping an eye on them. The caregiver must sometimes keep an eye on too many individuals at once. To monitor patients and evaluate their conditions, the medical staff still needs to go to every hospital's emergency department separately today. Worryingly, the percentage is large [1]. A suggested regimen should be developed in order to improve the participant's work. Often, physiological signals are used to evaluate the person's entire physical makeup. Height, stance, pulse, body temperature, and breathing rate are the six key qualities [2]. Normal body temperatures and breathing rate are the main topics of this study. Normal internal average temperature is 37 °C. The jaw, the perineum, the armpits, and ultimately the auditory nerve are the four major areas of the human body where thermostats are commonly used to measure temperatures. Hyperthermia is described as a drop in body temperature under 35 °C, even though a fever is described as an increase in normal temperature that is slightly over normal. The rate of breathing or respiration rate is the total number of breaths an individual must take per minute when lying still. Table 1 displays the typical breathing and heart rate with each age group. An average person breathes approximately 12 or 20 times each minute. Consequently, it is assumed since it is uncommon to have a breathing function during slumber that goes beyond this range. Numerous

Table 1 The average age of the group, their regular age bracket, and their regular respiration rate

Group Age	Age	Regular Breathing Rate
0–1	*0–1.5 year*	*20 to 30*
2–3	*2.5–5.5 years*	*30 to 40*
4–5	*6.5–12.5 years*	*25 to 35*
6–7	*13.5–20.5 years*	*10 to 24*
8–9	*21.5–72.5 years*	*13 to 21*
10–11	*> 72.5 years*	*16 to 24*

illnesses, such as asthma, anxiousness, cardiovascular disease, respiratory disease, and numerous more, can affect the breathing function [3].

2 Related Work

Many researchers have looked at individual monitoring programs that include data on temperature ranges and oxygen consumption [4–7]. The ambulatory care monitoring software applied in this research is centered on monitoring the rate of breathing (RR) and electrodermal utilizing temperature measurements from MLX90614 using LM35 and the ESP8266 as the Wi-Fi component once again for Arduino Mega 2560 controller [8]. The physicians can enter the patient's member-ship number into a computer as well. Additional features include an LCD display that shows crucial information including the patient's condition, body temperature, breathing rate, and group emotional maturity.

For Technological Approaches for Remote Health monitoring Fig. 1 depicts the flowchart for the remote healthcare project's method of detecting core body temperature and respiratory rate. Connections consist of a touchpad and a temperature probe (LM35 and MLX90614). The individual's rate of respiration is being measured by two Digital millimeters, and core temperature is now being measured by a technology called an MXL90614 thermometer. The Arduino Mega MCU processes the data after which it is transmitted to a Led display and a mobile device. The information delivered to smartphones is supported by the ESP8266 Wi-Fi module. The technique for calculating each participant's RR is depicted in the schematic in Fig. 2. Following the caregiver's entry of the people's age range, the wristwatch will commence counting down for two min (loop counter for 20). The computer has now become capable of determining the temperature required for individual respiratory and weather conditions, in addition to operation of temperature sensor 1 (TS1) and 2 (TS2). These two degrees were again compared. If the divergence is much more than 2 °C, the RR count will only advance by one. Over there until the 20 h are up, the technique is repeated. Afterwards when, to obtain the RR in a single minute, the RR count is increased by 3 (to get the RR for 60 s or one min). After that, the procedure is performed once again.

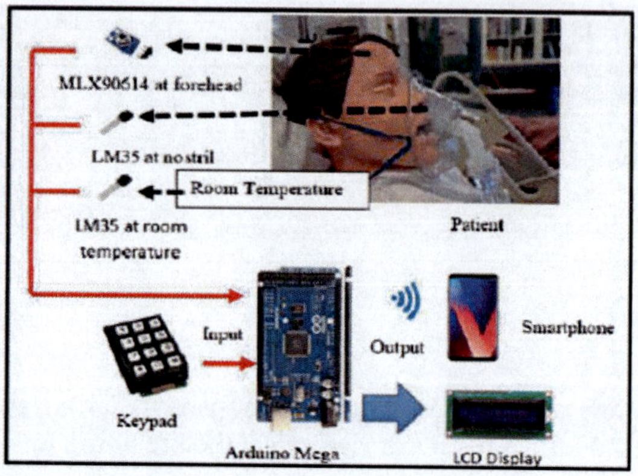

Fig. 1 Graphical Visualization of Patient Surveillance Equipment

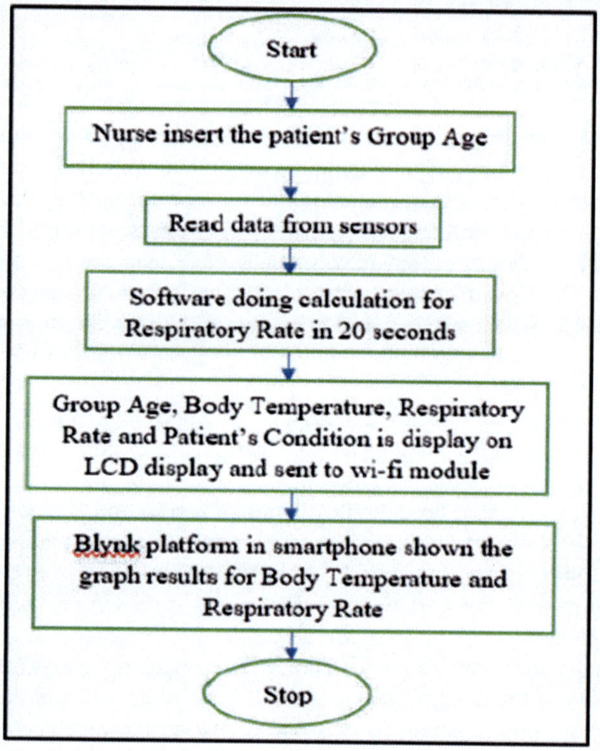

Fig. 2 Visualization of illustrations in system analysis

3 Proposed Work

There are different classifications for the supplies, processors, and results of outpatient wireless connectivity. This would include LM35, MLX90614, touch screens, Microcontroller 2560, Liquid crystal displays, and Wi-Fi ESP8266 components in addition to analog temperature measurements, LM35, quasi thermal monitors, and keypads. The Systems Architecture Information for the purposes depiction of the program is shown in Fig. 3. Input Sensors connections included a computer and heating elements. Two commonly used temperature type LM35 detectors are used to detect temperature sensor and individual respiratory characteristics. To calculate the differences between these two pieces of data, a calculation is run for 2 s. A different type of digital thermometer is an autonomous non-contact infrared cameras sensor, product code MLX90614 in this case. Through into the front of the head, the MLX90614 is employed to measure the participant's core body temperature. It is shown in Fig. 4 below. The gadget can accurately determine a patient's usual respiration rate within such a continuum with the help of a computer, and it can also better precisely indicate or alert about the patient's aberrant condition. Table 1 shows the company's age breakdown. Men between the ages of 11 and 18 have age ranges that seem to be equivalent to those of people over the age of 70, indicating that their oxygen concentration fall within a diverse configuration. An-other transmitter for the program is the LCD display. It is employed to show the individual's age range, body temperature, and respiratory rate. Additionally, depending on computations the software performed to use the input data, the Display represented the patient's status, either being "normal" or "anomalous." The LCD panel comprises four groups, so each evidence piece is shown within each row. An additional interface is the software program (GUI) on the participant's cellphone. The Bluetooth module app's GUI is supported by the ESP8266 Wi-Fi Shields. As indicated in Table 2, 13 participants of diverse ages took part in the study. For each measure, everyone is checked three times at a time. The four estimation techniques were then used to compute the coefficient of determination in order to deter-mine the precision and dependability of the outpatient surveillance system.

Using LED screen, the 4 columns on the Led display utilized in this study were as follows: the very first row displayed the respondents' central body temperatures; their second row displayed respective subgroups memberships; the third column displayed their computed oxygen consumption; and the final row displayed their circumstances. Figure 4 (a) shows a representation of a representative of the generational who is between the ages of 11 and 18 years old. The average breathing rate per second for the population is between 15 and 20, as seen in Table 1. Because the patient's respiration rate is within the expected range, the condition of the individual is good and the patient's breathing rate, meanwhile, is below average in Fig. 4(b).

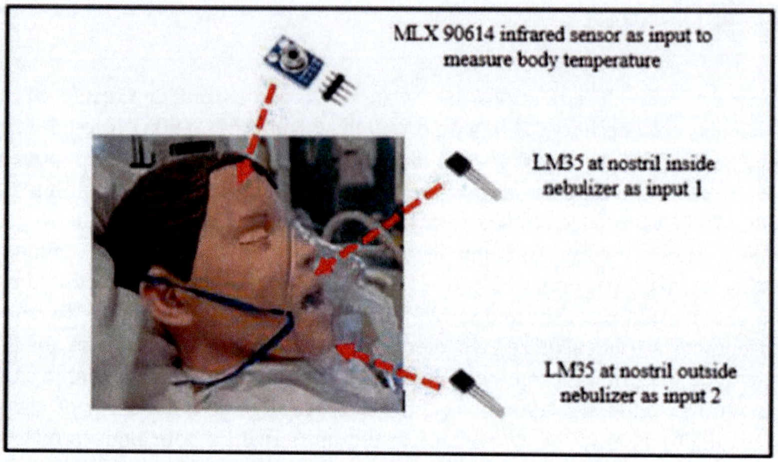

Fig. 3 Position of the Temperature Measurement Sensors

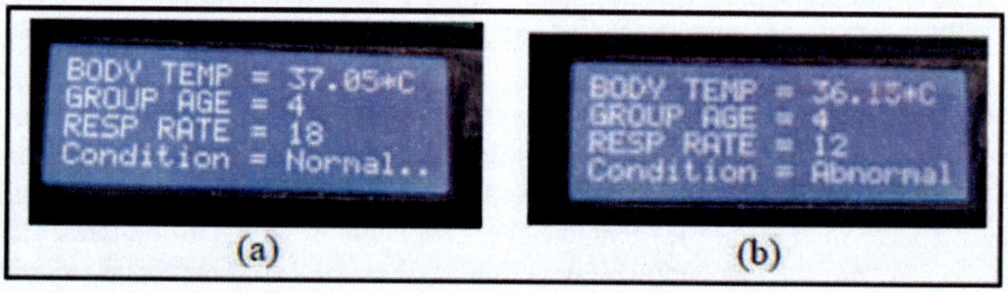

Fig 4. Depicts the outcome for an individual under (a) conditions to a tolerable extent and (b) problematic situations matches in GUI using Blynk Platform.

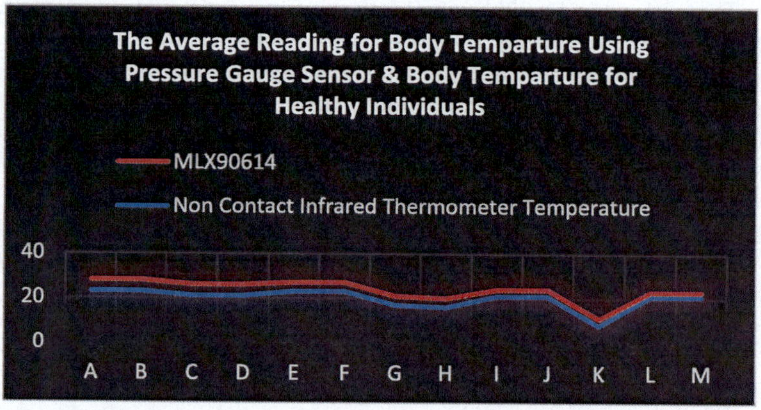

Fig 5. Contrasts the central current temperature obtained from the MLX90614 gadget with a body temperature gauge

Table 2 The Individual ages and group ages

Individual	Age	Group Age
A	*24*	*6*
B	*25*	*6*
C	*22*	*6*
D	*23*	*6*
E	*18*	*5*
F	*16*	*5*
G	*18*	*5*
H	*17*	*5*
I	*11*	*4*
J	*9*	*4*
K	*7*	*4*
L	*5*	*3*
M	*4*	*3*

4 Experimental Results

Utilizing the Display Screen for the Application Presentation. The anatomy was shown on the three columns of the LCD used for this application.

4.1 *Assessment of Quasi Temperature Mercury Thermometer and Sensor Components (MLX90614) for Bodily Temperature Readings (FI01)*

A non-contact MRI scanner (FI01). and measuring cylinder were utilized to take the central temperatures of 13 individuals five times every fifteen minutes (MLX90614). The link between both the MLX90614 sensing and the FI01 temperatures for the 13 participants is shown by comparing the average measurements on the graph.

4.2 *For Those with Respiratory Disease, a Comparison of the MLX90614 Detectors' Measurements of Internal Temperature with a Quasi-Infrared Camera (F101)*

The findings of measuring the usual core temperature of two asthma symptoms are shown in Table 2. Regardless of whether the two of the patients have bronchitis, neither of these persons has a temperature, hence the measures are unchanged by the blood temperature readings.

The Bluetooth module architecture was used to accept data from the ESP8266 module. A caretaker may monitor patients from such a distance using a mobile app called Bluetooth module operating systems. In Table 2, information from ESP8266 is gathered and shown on the Graphical screens and charting results of the Blynk product. The participant's indoor temperature, which is consistent at around 37.5 °C, shows that they are in good general health.

4.3 A Comparison of the MLX90614 Sensors' Measurements of Core Body Temperature with a Quasi-infrared Camera for Asthmatic Patients (F101). A Subsection Sample

Four of these patients suffer bronchitis, but the measures are influenced by the core temperature reading since none of the patients have a temperature like that. The MLX90614 and thermometers both offer excellent levels of precision, with all readings being within 1% of each other. Figure 5 additionally explains the relationship between the historic information for the System designed to allow and the surrounding temperature. While detecting body temperature with Bondholders may be claimed to provide solid data, heart and breathing measurement doesn't truly provide precise and reliable measurements. This is a result of the detectors' LM35, which are used to monitor respiration, having a protracted cooling trend therefore taking a significant amount of time to generate reliable results. This problem has a detrimental impact on how accurately the respiration rate is measured. The measurement's precision is increased by switching out the LM35 detection for a coverage changes digital thermometer.

5 Conclusions

Training and monitoring of an IOT-based health surveillance system. This innovation may be used to monitor core temperature and breathing techniques, and the measurement results may be remotely transmitted to Android applications via the Development platform. This innovation may alert the user and caretakers to any emergency circumstances using Mobile application. The outcomes of these procedures, as visible on the touchscreen display, are often trouble-free, but smartphone apps have often created surprising outcomes because unexpected circumstances are crucial to the acceptance of the Internet of things. For reliable results to be obtained while measuring oxygen intake, the type of digital thermostat being used to detect breathing temperature is essential. Employing fry's safeguards might lead to greater and more accurate performance soon.

References

1. Singh P et al (2023) Cloud-based patient health information exchange system using blockchain technology. In: Information and communication technology for competitive strategies (ICTCS 2021). Lecture Notes in Networks and Systems, vol. 401. Springer, Singapore. https://doi.org/10.1007/978-981-19-0098-3_55.
2. Sharma AK et al (2023) An IoT-based temperature measurement platform for a real-time environment using LM35. In: information and communication technology for competitive strategies (ICTCS 2021). Lecture Notes in Networks and Systems, vol 400. Springer, Singapore. https://doi.org/10.1007/978-981-19-0095-2_57
3. Phogat M et al (2023) Identification of MRI-based adenocarcinoma tumors with 3-D coevolutionary system. In: Information and communication technology for competitive strategies (ICTCS 2021). Lecture Notes in Networks and Systems, vol 401, Springer, Singapore. https://doi.org/10.1007/978-981-19-0098-3_57
4. Jain E et al (2023) A CNN-based neural network for tumor detection using cellular pathological imaging for lobular carcinoma. In: ICT with intelligent applications. smart innovation, systems and technologies, vol 311, Springer, Singapore. https://doi.org/10.1007/978-981-19-3571-8_51
5. Doja F et al (2023) A comprehensive framework for the IoT-based smart home automation using Blynk. In: Kaiser MS, Xie J, Rathore VS (eds) Information and communication technology for competitive strategies (ICTCS 2021). Lecture Notes in Networks and Systems, vol 401. Springer, Singapore. https://doi.org/10.1007/978-981-19-0098-3_6
6. Gupta A et al (2023) A sustainable green approach to the virtualized environment in cloud computing. In: Zhang YD, Senjyu T, So-In C, Joshi A (eds) Smart trends in computing and communications. Lecture Notes in Networks and Systems, vol 396. Springer, Singapore. https://doi.org/10.1007/978-981-16-9967-2_71
7. Vats P et al (2022) A novel approach for detection of intracranial tumor using image segmentation based on cellular automata. In: Nagar AK, Jat DS, Marín-Raventós G, Mishra DK (eds) Intelligent sustainable systems. Lecture Notes in Net-works and Systems, vol 334. Springer, Singapore. https://doi.org/10.1007/978-981-16-6369-7_54
8. Chauhan K et al (2022) A comparative study of various wireless network optimization techniques. In: Joshi A, Mahmud M, Ragel RG, Thakur NV (eds) Information and communication technology for competitive strategies (ICTCS 2020). Lecture Notes in Networks and Systems, vol 191. Springer, Singapore. https://doi.org/10.1007/978-981-16-0739-4_61

A Blockchain Based System for Product Tracing and Tracking in the Supply Chain Management

P. Nandal

Abstract Collection of information, resources, people, activities and organizations intended to transfer a service or a product from supplier to buyer is known as the supply chain. The purpose is to maintain the characteristics of the product or service. Therefore, fast and reliable tracking and tracing of the goods right from the source to the destination is needed. Most of the existing solutions depend on the centralized architecture and therefore have a single-point of collapse. Blockchain, being a distributed ledger as a decentralized system helps create applications that preserve data integrity and are secure. In this paper, the author proposes a blockchain-based management system to find any forged items that may be present in the supply chain.

Keywords Supply chain management · Blockchain · Product tracking and tracing

1 Introduction

Counterfeiting is defined as the deliberate alteration of a product's true nature or originality for deception purposes, in order to give a false impression of tests/facts/results. Many industries that manufacture and supply products along the chain face numerous issues in tracing their products, giving the intruders/infiltrators a chance to inject counterfeit products in the chain, damaging the business and perception of the industries. This issue has been observed globally. Counterfeit traffickers/infiltrators usually do not comply with labor, tax, or trade laws. A major hurdle in counterfeit protection is to ensure product tracing all through the supply network. The growth of global supply chains has made tracking goods from the manufacturing source, i.e. assembly line to the end destination, retail stores, more difficult. The reach of counterfeiting is what makes it so dangerous; counterfeit parts and goods affect every stage of the product life cycle. Supply chains currently involve a complex network

P. Nandal (✉)
Computer Science and Engineering Department, Maharaja Surajmal Institute of Technology, New Delhi, India
e-mail: priyankanandal@msit.in

N. Hasteer et al. (eds.), *Decision Intelligence Solutions*, Lecture Notes in Electrical Engineering 1080,
https://doi.org/10.1007/978-981-99-5994-5_15

of players, each with their own system. This makes coordination difficult and gives brand owners a limited view of their operations, damaging their reputation.

A significant amount of time and money is invested in tracing the products, validating their source, communicating with associates, completing extensive documentation to certify the authenticity of the goods, protect customers, and meet regulatory and compliance requirements. Existing solutions make use of electronic data interchange (EDI), QR codes, bar codes, data interfaces, business-to-business messaging and reports, global positioning system (GPS) devices, radio-frequency identification (RFID), and near-field communication (NFC). These technologies have limited abilities such as: restricted logistics activities, for e.g. identifying locations; data integrity and accuracy cannot be verified as in some instances the data captured is not in digital format; limited or no support to share secure, multiparty information. Most importantly, these tools fail to provide real-time, end-to-end transparency and visibility required for entire stakeholders of the supply chain. This is basically due to the centralized architectures that these techniques are based upon. This is where blockchain comes to the rescue [1–3].

A shared and a distributed data store as well as a computational framework is implemented by the blockchain technology. Blockchain, analogous to data structure, permits data insertion devoid of data updation and deletion which exists on the blockchain to avert revision and tampering. Blockchain facilitates decentralisation by introducing improved types of distributed software architecture. Without relying on a central integration point, the components can agree on historical shared states for decentralisation and transactional data which is shared over a wide chain of members [4]. Blockchain is a decentralized ledger system that registers all edits to a record in real time. It indicates the cumulative performance of a technical solution which, through decentralisation, manages a continuous record file like a trustworthy database. In blockchain, each node stores data and then these nodes exchange this data with one another over the entire network. Every single node contains the entire data present on the blockchain. Whenever a new transaction is initiated, each node verifies it and only then it is included in a new block. This system helps build accountability and makes the entire process secure.

In the current application, blockchain technology is used as the underlying approach to ascertain the genuineness of the products. The author propose a system which is designed with ethereum smart contracts to record product ownership to overcome all the limitations discussed above. This research paper is arranged as follows. Section 2 gives an insight of the literature reviewed. The proposed application is presented in Sect. 3 and the results are discussed in Sect. 4. The work done is then concluded in Sect. 5.

2 Literature Survey

There have been various approaches to stop forgery in the supply chain. Sunny et al. presents a comparative analysis of these existing traceability solutions [5]. They used a proof of concept in Blockchain to test the cold chain management. The study analyses how Blockchain technology affects the transparency in the distribution chain and how its impact can be elevated using technologies like smart contracts and Internet of Things (IoT). The study identifies that although blockchain cannot eliminate the introduction of forged products in the supply chain it can definitely help mitigate it. The authors of [6] list the important factors which are required for a successful supply chain application. Their main focus was to implement traceability systems in two distinct industries; first, the cobalt mining business and second, the pharmaceuticals business. According to the authors, the factors involved to implement a successful business traceability system are restraining illegal activities, improving coordination between various supply-chain components, improving sustainability performance, increasing operational efficiency, and sensing market trends. The research presented in [7] is focused upon deceptive counterfeits and the author advocates the role of government in increasing the utilization of blockchain technology to help both the manufacturer and customers to deal with the issue of deceptive products. The authors of [8] also propose a fully functional anti-forgery system. The details of every product are saved in the blockchain so that it is no more tampered and every block of the blockchain has a unique hash which will be the proof of its authenticity. The public key cryptography system is also presented in this paper to protect the customer, sellers, and manufacturer's details. In [9] the focus is on transparency and hence a single product id is created. Digital signatures are used to preserve the integrity and avoid repudiation. The smart contract also employs elliptic curve public-key encryption, which helps provide data confidentiality among the stakeholders. The framework is so created that it focuses on security and building confidence in the supply chain environment. In [10] the authors present this approach in the form of a software based solution. They conclude that blockchain helps provide various levels of information visibility and also checks the data integrity and authenticity. Toyoda et al. [11] propose a product ownership administration method that uses RFID tags and blockchain to meet its purpose. This solution makes cloning of tags redundant thereby reducing the chances of counterfeit products. At each point in the supply chain, the ownership of the product is relocated, which helps identify the product's authenticity. Using the protocol a proof of concept system has been constructed on the Ethereum network. The performance of the system is reasonably well for low transfers as it takes less than US $ 1 to manage the product ownership.

This generic concept of blockchain in the supply chains has drained down to specific supply chains as well. Some of the use cases are explored further. [12] proposes a traceability structure for the textile and clothing supply chain. They discuss how the textile and clothing industry is susceptible to prevalent issues of information tampering and low transparency in the supply chain. In [13], the author presents a distributed and safe architecture for the Food supply chain. They discuss how the modern centralized applications and inadequacy of traceability in the supply chain result in less secure applications. The authors [14] propose a syatem for the Agriculture and Food industry that ensures transparency and traceability. To ensure the data integrity and credibility, Interplanetary File Storage System (IPFS) is integrated with the system which stores all the data and returns a cryptographic value (hash) of the data to the application which is then stored on the blockchain to retrieve it back whenever needed. They have used the specific use case of the dairy industry to assess their architecture. [15] uses fuzzy logic in addition to IOT and blockchain to not only effectively track food but also ensure its quality along the chain. Feng Tian et al. [16] builds a new decentralized system that does real-time food tracing on the basis of HACCP (Hazard Analysis and Critical Control Points), alongside blockchain, and the Internet of things. Furthermore, he brings a new mutant concept of blockchain and distributed databases called BigchainDB. This helps to reduce the gap in the decentralized systems at scale. This system provides live information to all stakeholders on the safety condition of food products, mitigates the problem of centralized information systems, and brings more transparency, neutrality, openness, reliability, and security. In [17], the authors discuss at length the various ways in which blockchain can be applied to the healthcare sector and the advantages it brings. One, major benefit of the decentralized type of blockchain is the transparency and immutability it provides in the pharma supply chain. In [18] an access control policy model is designed for the medication industries to prevent any alteration in medication information. In [19], the author presents the Gcoin blockchain as the base architecture for the implementation of the drug supply chain management system for the transparent flow of drugs. Additionally, another system [20] uses chaincodes to control the collaboration among all the stakeholders. It is implemented using sequence diagrams. The transactions and records are stored in the immutable Medledger system.

There has been an attempt to combine hardware technologies with software to get an even secure and transparent solution. Alzahrani et al. [21] suggest a new decentralized supply chain which is a mix of blockchain and Near Field Communication (NFC) technologies, called Block-Supply chain. This novel idea uses a scalable consensus protocol which maintains both security and efficiency. The protocol is apt for larger networks but has a few performance issues with low hostility blockchains. In [22] the author proposes a framework for dNAS application which uses Blockchain technology and Near Field Communication technology to design and establish a decentralized anti-counterfeiting application for the wine industries supply chain. In [23] the author focuses on the current challenges faced by different industries due to increasing counterfeiting attacks and tries to propose a solution by combining blockchain and NFC-enabled systems, a modernized system that is decentralized, secure, trustworthy, provides transparency to its users and can prevent any counterfeiting attempt by the intruders.

3 Methodology

The implementation methodology of the system is mainly divided into two phases: first, carrying out research, evaluating requirements, and system designing; second, development, testing, and validation of the system.

3.1 System Design

In this phase, system requirements and functionalities are evaluated and how blockchain will be used to solve the problem is described.

Basic Overview. The fake product identification system comprises of four types of users: Admin, manufacturer, retailers and customer. Every user has different privileges and actions to carry out.

- **Admin:** This user has only one action to carry out which is to accept the account creation request of the manufacturers who want to register themselves in the system.
- **Manufacturer:** This user can add new products to its inventory and also can approve or reject the product buying requests coming from the retailers. A manufacturer can also access the list of registered products and the list of requests received from the retailers.
- **Retailers:** A retailer has the privilege to send product buying requests to the manufacturers. This user can access the list of his purchased products and this user can also view all the products of every manufacturer. Retailers can also view the various stages from which the product has passed before buying to verify its authenticity.
- **Customers:** A customer can view all the products of every retailer and can order any product from the list. This user can also access the list of his own orders. A customer can also view the various stages from which the product has passed before buying to verify its authenticity.

Operation Flow. The process starts with a manufacturer sending an account creation request to the admin. If the admin accepts the request then an account of the requested manufacturer is created. After getting approval from the admin a manufacturer can register his products on the blockchain. A retailer can view the products registered by the manufacturer. A retailer can send a product buying request to the manufacturer. If the manufacturer accepts the product buying request of the retailer the ownership of the product is transferred to the retailer and updated on the blockchain. A retailer can view his owned products. After that, a customer can access the list of products in the products inventory, and all these products are owned by retailers. Before buying a product a customer can trace the product to find all the checkout points from which the product has passed. A customer can buy the products and then the ownership is transferred to the customer and updated on the blockchain.

In this way, the product starting from the manufacturer's end reaches the customer's end and the problem of counterfeiting is resolved with the help of blockchain.

3.2 *System Implementation*

In this part, the system implementation is explained in detail including the functions and user interfaces.

The user interface that the user sees is represented by a web page. The http-server suite provided by node.js is used to build the server side of the web page, and web3.js serves as the interface between the smart contract and the user interface. Following server configuration, the Private Chain and Address information can be linked. The overview of the system is depicted in Fig. 1.

Smart Contract Structure: Firstly an ethereum smart contract is developed to implement the blockchain. Solidity language is employed to implement the smart contract. All the variables and functions present in the smart contract are provided in the Tables 1, 2 and 3.

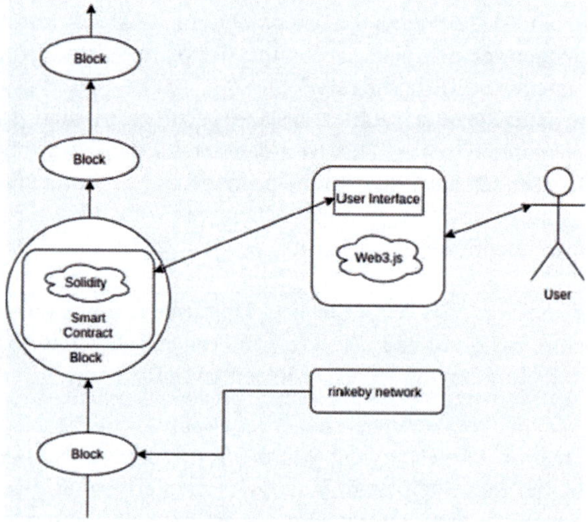

Fig. 1 Overview of the system

Table 1 Variables used in the smart contract

Admin	Address	Address of the manufacturer
States	Struct	Object for recording a particular product's state
Product	Struct	Object of a product
Products	Product[]	Mapping to store every product registered by a manufacturer
States	State[]	Mapping to store each state of a product

Table 2 State structure (struct)

Description	String	Log to show the details of the state
Person	Address	The person who added the state
Name	String	Name if the person who added the state
Timestamp	String	The timestamp when the new state has added
Description	String	Log to show the details of the state

Table 3 Functions created in the smart contract

Constructor	To set the admin address
Register Product	Called by admin to register a new product
Add state	Called by the person who is going to buy the product
Search product	Called by someone who wants to get the details of a product and its current state
Track Product	Called by someone who wants to watch the journey of a product starting from the point of manufacture
Constructor	To set the admin address

Algorithms of the all the functions of the smart contract are given in Tables 4, 5, 6, 7 and 8.

System Operations.

- **Login/Signup:** To sign up a user needs to enter their name, email id, password and select the type of account he/she wants to create. Only in the case of a user who wants to be registered as a manufacturer account request is sent to the admin for approval and after getting approval from the admin account of that user is activated. For logging in a user does not need to select a user type as the system identifies it automatically.
- **Add Product:** Manufacturers can add new products to their inventory and register them on to the blockchain. Manufacturers needed to fill a form to add a product and make them available for the retailers.
- **Buy Product for Reseller:** A retailer can trace the origin of a product by fetching the details from blockchain and hence can check the authenticity of the product. If the retailer is satisfied with the info he/she can send a product buying request to the manufacturer of the product. If the manufacturer of the product approves the request the products used to get credited in the retailer's inventory and becomes available for the customers.
- **Buy Product for Customer:** A customer can also view the various points from which the product has passed before reaching them. If the product tracing information satisfies the customer then the customer can opt for buying the product by entering the quantity that the customer required.

Table 4 Algorithm of the function register product

Algorithm 1 **Register Product**

Input: product Id, name, cost, description, manufacturer Name, quantity, timestamp

```
1.   Product memory newProduct = Product ({
                  name: _name,
                  cost: _cost,
                  description: _description,
         manufacturerName: _manufacturerName,
         manufacturerTimestamp:
         _manufacturerTimestamp,
         statesCount: 0
              });
2.   for(i=0;i<quantity;i++)
         Execute step 3 to 6
3.   newId = product id + i
4.   Products[newid] = newProduct
5.   message = "Product registered at manufacturer"
6.   Call addState(newId,message,manufacturerName,timestamp)
```

Table 5. Algorithm of the function add state

Algorithm 2 **Add State**

Input: productId, message, name, timestamp

```
1.   If Products[productId].cost !=0 then
2.   newState = State({
              person: sender address,
              description: info,
              name: name,
              date: timestamp
        });
3.   Products[productId].States = newState
4.   Products[productId].statesCount =
     Products[productId].statesCount + 1
```

The operations presented above are some of the major operations of the system. The operations related to the product e.g. product registration, changing the ownership and tracing of product, etc. are carried out on the blockchain. The operations like user login/signup, accepting and rejecting requests, etc. are not carried out on the blockchain to simplify the smart contract code, hence decreasing the transaction cost. An external no SQL database is used to store all the user and products related

Table 6 Algorithm of the function search product

Algorithm 3 **Search Product**
Input: productId 1. Count = Products[productId].statesCount-1; 2. Return Products[productId] and Products[productId].States[count]

Table 7 Algorithm of the function track product

Algorithm 4 **Track Product**
Input:productId 1. Length = Products[productId].statesCount 2. checkPoints = [] 3. for(i=0;i<Length;i++) Execute steps 4 to 5 4. Checkpoints[i] = Products[productId].States[i] 5. Return checkPoints

Table 8 Algorithm of the function add state to many

Algorithm 5 Add State To Many
Input: productId, countFrom, quantity, timestamp, name, message 1. for(i=countFrom;i<countFrom+quantity;i++) Execute steps 2 to 2. newId = product id + i 3. Call addState(newId,message,name,timestamp)

information. Only three operations in the proposed system require some ethers for execution and these operations are registration of products, accepting buying requests from retailers, and ordering the products. All other operations do not have any transaction cost and this way of implementation makes the system a fast and economically feasible solution to reduce product counterfeiting problems.

4 Results and Discussion

A decentralized architecture has been successfully developed to prevent products counterfeiting in the supply chain that uses blockchain technology which ensures decentralized, trustworthy, accountability, transparency, traceability and a secured supply chain. The author has also developed a platform (web application) which directly connects to EthereumRinkeby network which keeps a track of all transactions at all times and provides manufacturers, retailers and consumers to interact with each other. The application is also tested rigorously against many test cases and all issues have been resolved. The application is also deployed on the internet and is publicly available for use. Figures 2 and 3 represent the screenshots of the developed application.

The application successfully achieved all its objectives but still any human intervention such as replacing the labels of the fake products with the genuine ones can't be overruled.

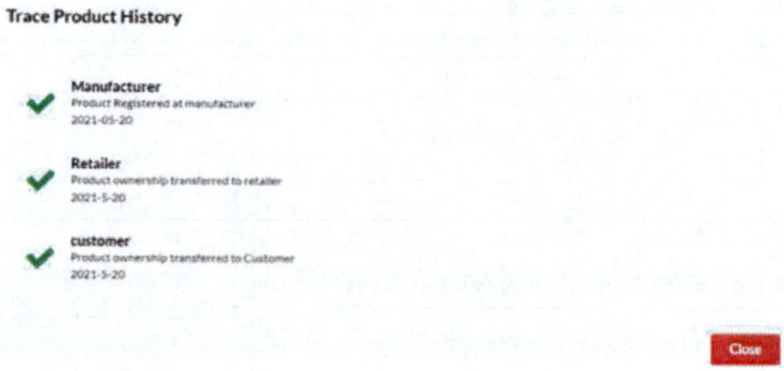

Fig. 2 Screenshot of the product tracking page of application

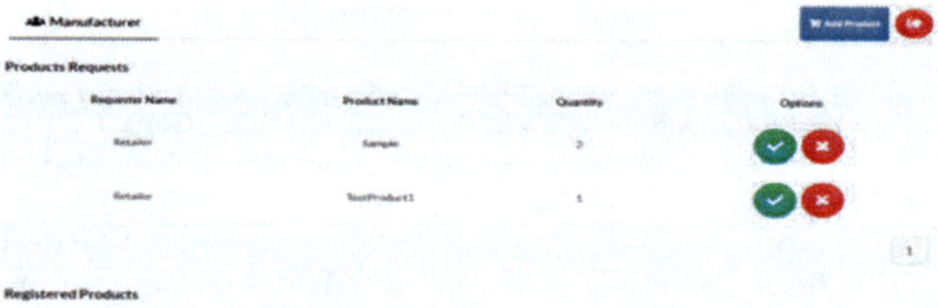

Fig. 3 Screenshot of the product request page

5 Conclusion and Future Work

This paper presents a fully functional fake product identification system as a solution to product counterfeiting. A decentralized web application is developed by using blockchain technology. Manufacturers can use the web application to store product-related information on the blockchain. All the products of every manufacturer are visible to all the retailers and they can purchase these products from manufacturers. A customer can view every retailer's products and order them using the application. The developed web application is deployed on Heroku (the cloud application platform). This solution can help manufacturers, retailers, and customers to prevent counterfeiting. This system can be used in preventing counterfeiting in products ranging from conventional consumer goods to high-end luxury items. The proposed system provided both an economical and easier approach to curb product counterfeiting with the assurance as each record on the blockchain cannot be forged.

The major point of concern in the application developed using blockchain is the cost of each transaction and the simpler the code in the smart contract the less the transaction will cost to the user. So, the future work of the system is to make the code as simple as possible and as a result, decrease the transaction cost so that a more economical and practically feasible solution can be provided.

References

1. Saberi S, Kouhizadeh M, Sarkis J, Shen L (2019) Blockchain technology and its relationships to sustainable supply chain management. Int J Prod Res 57(7):2117–2135
2. Chang SE, Chen Y (2020) When blockchain meets supply chain: a systematic literature review on current development and potential applications. IEEE Access 8:62478–62494
3. Lim MK, Li Y, Wang C, Tseng ML (2021)A literature review of blockchain technology applications in supply chains: a comprehensive analysis of themes, methodologies and industries. Comput Ind Eng 119:107133
4. Stranieri S, Riccardi F, Meuwissen MPM, Soregaroli C (2021) Exploring the impact of blockchain on the performance of agri-food supply chains. Food Control 119:107495
5. Sunny J, Undralla N, Pillai VM (2020) Supply chain transparency through blockchain-based traceability: an overview with demonstration. Comput Ind Eng 150:106895
6. Hastig GM, Sodhi MS (2020) Blockchain for supply chain traceability: business requirements and critical success factors. Prod Oper Manag 29(4):935–954
7. Pun H, Swaminathan JM, Hou P (2021) Blockchain adoption for combating deceptive counterfeits. Prod Oper Manag 30(4):864–882
8. Ma J, Lin S-Y, Chen X, Sun H-M, Chen Y-C, Wang H (2020) A blockchain-based application system for product anti-counterfeiting. IEEE Access 8:77642–77652
9. Sezer BB, Topal S, Nuriyev U (2021) An auditability, transparent, and privacy-preserving for supply chain traceability based on Blockchain. arXiv preprint arXiv:2103.10519
10. Longo F, Nicoletti L, Padovano A, d'Atri G, Forte M (2019) Blockchain-enabled supply chain: an experimental study. Comput Ind Eng 136:57–69
11. Kentaroh Toyoda P, Mathiopoulos T, Sasase I, Ohtsuki T (2017) A novel blockchain-based product ownership management system (POMS) for anti-counterfeits in the post supply chain. IEEE Access 5:17465–17477

12. Agrawal TK, Kumar V, Pal R, Wang L, Chen Y (2021) Blockchain-based framework for supply chain traceability: a case example of textile and clothing industry. Comput Ind Eng 154:107130
13. Casino F et al (2020) Blockchain-based food supply chain traceability: a case study in the dairy sector. Int J Prod Res 8:1–13
14. Shahid A (2020) Blockchain-based agri-food supply chain: a complete solution. IEEE Access 8:69230–69243
15. Tsang YP, Choy KL, Chun Ho W, Ho GTS, Lam HY (2019) Blockchain-driven IoT for food traceability with an integrated consensus mechanism. IEEE Access 7:129000–129017
16. Tian F (2017) A supply chain traceability system for food safety based on HACCP, blockchain & Internet of things. In: 2017 International conference on service systems and service management, pp 1–6. IEEE
17. Siyal AA, Junejo AZ, Zawish M, Ahmed K, Khalil A, Soursou G (2019) Applications of blockchain technology in medicine and healthcare: challenges and future perspectives. Cryptography 3(1):3
18. Zhu P, Jian H, Zhang Y, Li X (2020) A Blockchain based solution for medication anti-counterfeiting and traceability. IEEE Access 8:184256–184272
19. Tseng J-H, Liao Y-C, Chong B, Liao S-W (2018) Governance on the drug supply chain via gcoinblockchain. Int J Environ Res Public Health 15(6):1055
20. Uddin M (2021) BlockchainMedledger: hyperledger fabric enabled drug traceability system for counterfeit drugs in pharmaceutical industry. Int J Pharm 597:120235
21. Alzahrani N, Bulusu N (2018) Block-supply chain: a new anti-counterfeiting supply chain using NFC and blockchain. In: Proceedings of the 1st workshop on cryptocurrencies and blockchains for distributed systems, pp 30–35
22. Yiu NCK (2021) Decentralizing supply chain anti-counterfeiting and traceability systems using blockchain technology. Future Internet 13(4):84
23. Yiu NCK (2021) Toward blockchain-enabled supply chain anti-counterfeiting and traceability. Future Internet 13(4):86

Deadline Laxity and Load Imbalance Analysis for Energy Efficient Greedy, Semi-Greedy and Random Fog Scheduling

Savina Bansal, Rakesh K. Bansal, and Nikita Sehgal

Abstract Task scheduling plays a major role in effective utilization of fog platforms comprising of distributed, resource-constrained and energy-constrained computing devices. Most of the available heuristics in real-time task scheduling are being reassessed for their effectiveness in such a scenario. Researchers are formulating task scheduling problem for minimizing energy consumption of fog nodes and tasks' deadline violation time while meeting the quality of service (QoS) requirements of users. A recently published priority-aware semi-greedy (PSG) algorithm employs an alpha parameter for scheduling internet of things (IoT) tasks on heterogeneous fog nodes in an energy efficient manner using a semi-greedy approach. Since, energy efficiency and timeliness are two major parameters, in this work, we analyse the work further to assess amount of deadline laxity available in greedy, semi-greedy and random scheduling approach, by varying the alpha parameter, with the purpose of further exploiting the energy saving aspects. Further, balancing of workload among fog nodes has also been examined. Simulation based experimental results for the tested set-up, reflect huge amount of available deadline laxity with the highest and lowest being in the random and greedy approach respectively. Further, random approach results in a more balanced work load allocation as compared to the greedy and semi-greedy schemes.

Keywords Scheduling · Real-time tasks · Deadline laxity · Fog computing

1 Introduction

IoT is fast emerging as the next generation technology to bring pervasive computing to its ultimate realm. Immense range of applications are just awaiting the rapid development of miniaturised and energy efficient IoT hardware and software techniques.

S. Bansal (✉) · R. K. Bansal · N. Sehgal
Department of ECE, Giani Zail Singh Campus College of Engineering and Technology, MRSPTU, Bathinda, Punjab 151001, India
e-mail: savina.bansal@gmail.com

N. Hasteer et al. (eds.), *Decision Intelligence Solutions*, Lecture Notes in Electrical Engineering 1080,
https://doi.org/10.1007/978-981-99-5994-5_16

However, due to huge amount of data generation, cloud computing is unable to meet real-time applications requirements like low latency, location awareness, and mobility support. As a result, real-time applications desiring time bound responses and low latencies are leading to a wider interest of research community towards fog or edge computing platforms supplementing cloud computing for supporting real-time computation, data analytics and storage provisions near the edge device only. Cisco proposed fog computing platform in between the cloud computing and IoT devices [1, 2]. As the edge devices are resource-constrained, therefore job or task scheduling in fog or edge computing is a great challenge [10, 11, 14, 20]. Task scheduling plays a vital role in effective utilization of such platforms comprising of distributed, resource and energy constrained computing devices. An efficient scheduling scheme that aims at assigning a job to the most appropriate computing node can improve energy consumption of the network and response time of the application request significantly [8, 9, 12]

Most of the available heuristics in real-time task scheduling are being reassessed for their effectiveness in such a scenario [3, 5, 7, 15, 17, 19]. Being an upcoming hot research area, comprehensive survey of scheduling algorithms in fog computing is presented in [5, 12, 16], wherein it is concluded that about 57% of the scheduling algorithms in fog computing are related to solving task scheduling related issues that aims at judicious assignment of tasks to fog nodes to meet the user's QoS requirements, with energy consumption and makespan projected as the most important metrics for fog computing environment. For real-time tasks, however, minimizing deadline violation is more important than makes pan. As a result, finish time of a job/task can be delayed upto the maximum allowable response time or deadline of the task without compromising the user satisfaction. Accordingly, deadline laxities that refer to time by which a task's finish time can be delayed without violating the deadline condition become an important parameter for examining. This laxity can be exploited further for energy management or reduction by using DVFS schemes [6, 15]. Keeping it in mind, here in this work, we chose a recently proposed deadline and energy aware *priority-aware semi-greedy (PSG)* algorithm (PSG) [18] that has been shown to significantly outperform the FCFS [19], EDF [3], GfE [4] and Detour [13] scheduling algorithms for the purpose of deadline laxity analysis. In addition, we are presenting the load imbalancing analysis for the greedy, semi-greedy and random versions of the algorithm.

2 Related Work and Objective

A deadline-aware and energy-efficient priority based semi-greedy scheduling algorithm (PSG) for IoT plat form is recently proposed by Azizi et al. [18], wherein scheduling objective is to allocate and assign independent deadline-constrained real-time IoT tasks on to heterogeneous fog computing nodes so as to minimize overall energy consumption and deadline violations. Fog nodes in the network possess different computing capacities and power consumptions. The algorithm works as follows- real-time tasks are first sorted based on deadline in earliest deadline first (EDF) order. Next, possibility of scheduling highest priority task is explored on every fog node keeping a record of overall system energy consumption in each case subjected to fulfilment of task deadlines. In case, deadline is not satisfied on any of the nodes, then task gets scheduled onto a node resulting in minimum response time, and deadline violation time (called deadline penalty) is recorded. If task deadline is satisfied on more than one node, prospective nodes are first arranged in increasing order of their overall system energy consumption, and then a fixed percentage of minimum indexed nodes (based on an alpha parameter value (taken as 0.4)) is considered and the task is scheduled at random on any of these in a semi-greedy fashion. Based on an alpha parameter value, an algorithm can behave as greedy (alpha = 0), semi-greedy (0 < alpha < 1), or random (alpha = 1) scheduling algorithm.

Here in this work, we have analysed semi-greedy as well as random and greedy scheduling approaches- with the motive of assessing available deadline laxity under the three schemes for it's possible exploitation in further energy reduction without any deadline violations using dynamic power management or voltage and frequency scaling schemes [6, 7, 15]. In addition, since load balancing also plays a significant role to keep all nodes busy in a balanced manner so as to keep bottlenecks under check in a computing system, therefore, a load imbalancing factor is also proposed and assessed for the three scheduling approaches, which has not been considered in the base work. Source code of the base paper [18], is modified to incorporate the changes, wherever needed.

3 Simulation Set-Up

Simulation set-up for the fog platform is somewhat similar to the one used in base work and is comprising of heterogeneous fog nodes (FNs) interconnected with random mesh topology. The details are shown in Table 1. Number of FNs is varied from 10–100 in the intervals of 10, and number of independent IoT tasks are kept 10 times the number of fog nodes as 100 to 1000. Going by the parameters used in base work, some of the task deadlines were found to be not fitting well into the simulation scenario used, as these were even lesser than the minimum computation time of a task on a given fog node, hence unduly increasing the task rejection ratios. So, we modified the task deadline of a task to be randomly distributed from (2–10) times its average computation cost on the given fog network, so as to give a level playing field to all nodes in the network. Type of greedy, semi-greedy or random scheduling scenario is selected by keeping the alpha parameter $\alpha = 0$, 0.4, and 1 respectively. All simulations are coded in C/C++ programming language. The experiments were carried out on a desktop running on Intel(R) Core(TM) i7-6700 CPU @ 3.40 GHz 3.41 GHz, 8 GB of RAM, and Windows 10 operating system. Each experiment is repeated 5 times with different seed value for random number generation and average value is reported in the results reported.

Table 1 Simulation parameters used

Node Parameter	Value	Task Parameter	Value
Number of nodes	10, 20, 30,40, 50,60,70,80, 90,100	Number of Tasks	100,200,300,400,500,600,700,800,900,1000
Capacity (MIPs)	2000–4000	Size(MI)	100–10,000
Bandwidth (Mbps)	1000	Input file (kB)	100–10,000
PropagationDelay (ms)	1–3	Output file (kB)	1–1000
P^{max}(W)	80–200	Task Deadline	2–10 times average computing cost of the task on fog network
P^{min} (W)	(0.6 –0.7) * P^{max}		

4 Performance Metrics

The metrics used in the work are discussed here. It is pertinent to add that the Makespan parameter that refers to the maximum finish time of a task in a given job set is not taken up here, as used in the base work. Our rationale for the same is that the present job set is comprising of independent tasks with their own independent deadlines and hence overall makespan might not be of much concern to a particular task user. Though, in case of jobs comprising of precedence constrained tasks, makespan parameter become more relevant and also the main scheduling objective for minimization. Further, while calculating overall energy consumption of the fog network only active energy consumptions are being used after modification in the base work. Further, it is assumed that the free slots can be utilized for fitting in tasks from other users as fog networks could be serving a wide number of clients.

- **Energy Consumption** (EC)**:** Energy consumption of the system refers to total active energy consumed by all the m fog nodes in the system. Energy consumption of a fog node (f_j) is amount of energy consumed when it is actually computing a task (t_{ij}) scheduled on it and is given by the product of it's active power consumption and task execution time.

  Energy consumption of node

$$f_j = EC_j = \sum_{i=0}^{n} P_{max}^{j} * \mathrm{et}(t_{ij}) \tag{1}$$

  Total energy consumption of the network.

$$EC =_{(j=0)}^{m} EC_j \tag{2}$$

- **Deadline Laxity**: Deadline laxity of a task is the difference between task deadline and its response time. In real-time task scheduling, earliest completion of a task is not always desirable as the response can be delayed up to the task deadline without affecting user satisfaction. Total deadline laxity is calculated by summing up all deadline laxities for the tasks scheduled on different fog nodes.
- **Load imbalance factor (LIF):** Load balancing demands even distribution of work among the available computing nodes with the aim to optimize response time and to avoid creation of hotspots or bottlenecks in the system. For the purpose, variance of the load distribution on the given network is calculated under the current and worst scenario (when all tasks get scheduled on a single fog node).

$$LIF = \left(\frac{curr_{var}}{worst_{var}}\right) * 100;$$
$$curr_{var} = \sum_{j=1}^{m}\left(\frac{N}{m} - f_j^{cnt}\right)^2 \text{ and} \tag{3}$$
$$worst_{var} = \frac{N^2}{m^2} * (m-1)$$

where, N/m = Total number of tasks/Total number of fog nodes, and f_j^{cnt} = number of tasks scheduled on f_j

5 Results and Discussion

Simulation based experiments are done for the greedy, semi-greedy and random fog scheduling schemes by keeping the parameter $\alpha = 0$, 0.4, and 1 respectively using the simulation parameters as discussed above. Figure 1 presents the active energy consumption of the network with respect to different number of fog nodes. It is to be noted that number of tasks were 10 times the number of fog nodes, and hence, minimum and maximum number of tasks were 100 and 1000 respectively. For every network size the simulations were repeated 5 times with different fixed random seed values to maintain the confidence and uniformity in comparison. All tasks, except one, could meet their deadlines as gathered from the zerodeadline penalty parameter. As expected, greedy approach results in minimum active energy consumption as it chooses the fog node with lowest active power dissipation to schedule a task. Semi-greedy approach competes quite closely with 4.9% higher energy consumption. Random approach, which does not keep an eye on node's power dissipation while scheduling a task fair poorly with overall 26.8% higher average energy consumption (over 27,500 tasks).

From the deadline laxity metric, as shown in Fig. 2, it is collected that a huge amount of laxity is available for all the three tested scheduling schemes with Random_dl(R_dL) > Semi-greedy_dl(SG_dL) > Greedy_dl(G_dL). On an average, for all

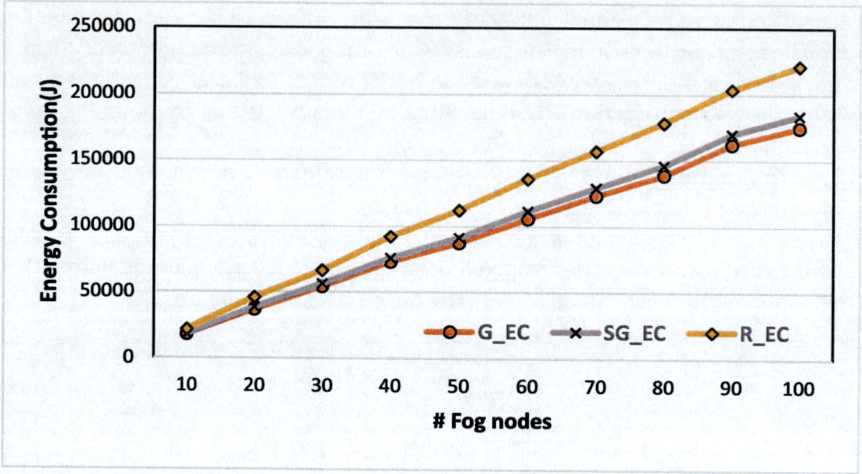

Fig. 1 Active energy consumption comparison for greedy, semi-greedy, and random fog scheduling

the tested scenarios, semi-greedy scheme have 151% higher available laxity while random scheme has 448% higher deadline laxity over the greedy approach. It reflects that greediness to execute a task at the earliest can be put on hold and still the task deadline can be easily satisfied and this period can be exploited for further energy reduction using various energy aware and management schemes as available in literature [6, 15].

Load imbalance factor as shown in Fig. 3 indicates random scheme with minimum LIF as the forerunner, resulting in a more balanced usage of resources while greedy scheme tend to confine the tasks on few selected nodes, giving highest LIF values. Semi-greedy approach falls in between these two extremes. On an average, semi-greedy approach gives 41% lower LIF while random approach gave 94% lesser load imbalance factor than the greedy approach.

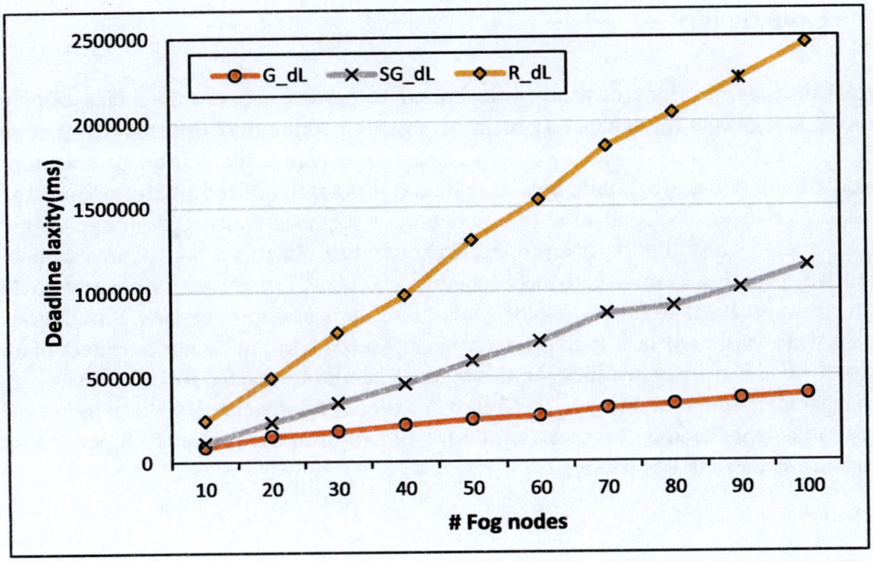

Fig. 2 Deadline laxity comparison for greedy, semi-greedy, and random fog scheduling schemes

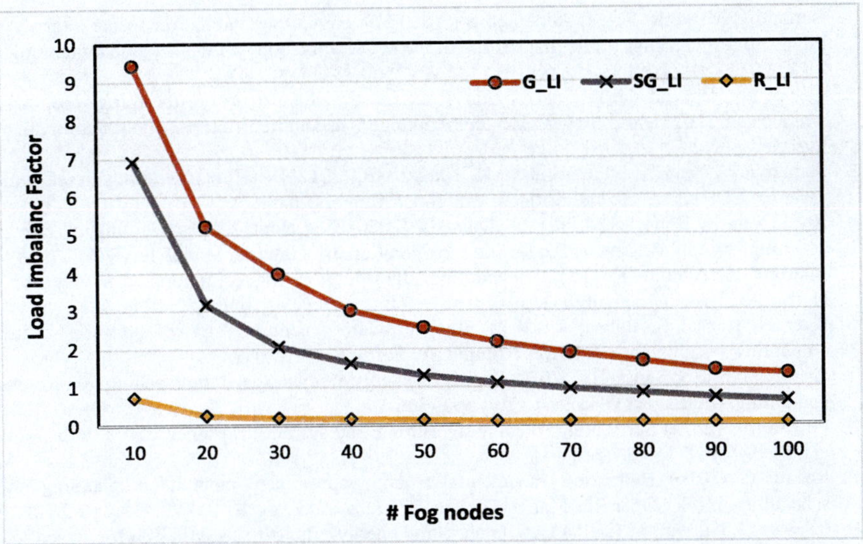

Fig. 3 Load imbalance comparison for greedy, semi-greedy, and random fog scheduling schemes

6 Conclusion

Real-time applications desiring time bound responses and low latencies offer a promising ground for enhanced interest of research community towards fog or edge computing platforms. Honoring task deadlines along with judicious energy consumption are the two major challenges to realize full potential of fog platforms for time critical applications. Real-time task scheduling schemes that keeps an eye on the total energy consumption are proposed in literature. In this work, comparison of recently proposed semi-greedy task scheduling scheme is taken up with random and greedy approaches to further explore the possibilities of improvements. Simulations on a wide variety of task load on heterogeneous fog computing nodes reflect there exist a huge margin of deadline laxities that can be further utilized for efficient energy management schemes. Further, it is found that greedy approaches lead to an increased load imbalance among the nodes of the fog network. In our future work, we intend to address these issues further.

References

1. Bonomi F, Milito R, Zhu J, Addepalli S (2012) Fog computing and its role in the internet of things. In: Proceedings of the first edition of the MCC workshop on mobile cloud computing – MCC 2012 (2012). https://doi.org/10.1145/2342509.2342513
2. Fog computing and the internet of things: extend the cloud to where the things are, cisco white paper. (2015). http://www.cisco.com/c/dam/en_us/solutions/trends/iot/docs/computing-overview.pdf
3. Stankovic JA, Spuri M, Ramamritham K, Buttazzo GC (2012) Deadline scheduling for real-time systems: EDF and related algorithms, vol. 460, Springer, Cham
4. Xu J, Sun X, Zhang R, Liang H, Duan Q (2020) Fog-cloud task scheduling of energy consumption optimisation with deadline consideration. Int J Internet Manuf Serv 7(4):375–392
5. Matrouk K, Alatoun K (2021) Scheduling algorithms in fog computing: a survey. Int J Networked Distrib Comput 9(1): 59–74. https://doi.org/10.2991/ijndc.k.210111.001
6. Kaur N, Bansal S, Bansal R (2013) Energy aware scheduling strategies for distributed computing systems. Int J Adv Res Comput Sci Softw Eng 3:280–283
7. Kaur N, Bansal S, Bansal RK (2021) Survey on energy efficient scheduling techniques on cloud computing. Multiagent Grid Syst 17(4):351–366
8. Pinedo ML (2012) Scheduling: theory, algorithms, and systems. Springer, Cham. https://doi.org/10.1007/978-1-4614-2361-4
9. Rabadi G (2016) Heuristics: metaheuristics and approximate methods in planning and scheduling. Int Ser Oper Res Manag Sci 236–271 https://doi.org/10.1007/978-3-319-26024-2
10. Rahbari D, Nickray M (2019) Low-latency and energy-efficient scheduling in fog-based IoT applications. Turkish J Electr Eng Comput Sci 27:1406–1427. https://doi.org/10.3906/elk-1810-47
11. Rao T, Khan MA. Maschendra M, Kumar MK (2015) A paradigm shift from cloud to fog computing. IJCSET 5(11): 385–389
12. Belmahdi R, Mechta D, Harous S (2021) A survey on various methods and algorithms of scheduling in fog computing. Int Inf Eng Technol Assoc (IIETA) 26(2): 211–224. https://doi.org/10.18280/isi.260208
13. Misra S, Saha N (2019) Detour: dynamic task offloading in software-defined fog for IOT applications. IEEE J Sel Areas Commun 37(5):1159–1166

14. Sarkar S, Misra S (2016) Theoretical modelling of fog computing: a green computing paradigm to support IoT applications. IET Networks 5(2):23–29. https://doi.org/10.1049/iet-net.2015.0034
15. Bansal S, Bansal RK, Arora K (2020) Energy-cognizant scheduling for preference-oriented fixed-priority real-time tasks. J Syst Archit 108: 101743. https://doi.org/10.1016/j.sysarc.2020.101743
16. Singh P, Dutta M, Aggarwal N (2017) A review of task scheduling based on meta-heuristics approach in cloud computing. Knowl Inf Syst 52(1):1–51. https://doi.org/10.1007/s10115-017-1044-2
17. Topcuoglu H, Hariri S, Wu MY (2002) Performance-effective and low-complexity task scheduling for heterogeneous computing. IEEE Trans Parallel Distrib Syst 13(3):260–274. https://doi.org/10.1109/71.993206
18. Azizi S, Shojafar M, Abawajy J, Buyya R: Deadline-aware and energy-efficient IoT task scheduling in fog computing systems: a semi-greedy approach. J Netw Comput Appl 201. https://doi.org/10.1016/j.jnca.2022.103333
19. Zhao W, Stankovic JA (1989) Performance analysis of FCFS and improved FCFS scheduling algorithms for dynamic real-time computer systems, real-time systems symposium. IEEE Comput Soc 156–157
20. Hao Z, Novak E, Yi S, Li Q (2017) Challenges and software architecture for fog computing. IEEE Internet Comput 21:44–53

Survey on the Effectiveness of Traffic Sign Detection and Recognition System

Pranav A. Patel, Vipul Vekariya, Jaimeel Shah, and Brijesh Vala

Abstract Advanced driving assistance systems - ADAS are essential for Recognition and Detection. The term "ADAS" refers to various cutting-edge in-vehicle technologies created to improve road traffic safety by assisting drivers in becoming more aware of the road, its potential hazards, and other vehicles around. Constructing reliable methods for recognizing traffic signs has been an important and exciting field of study. The research community has shown a more interest in investigating traffic sign detection and recognition (TSDR), primarily driven by detection and classification. In recent years various TSDR approaches have been introduced. This research provides a comprehensive look at methods for identifying and classifying traffic signs. It briefly reviews the most recent technology in this area and then discusses a future trend.

Keywords Deep learning · Traffic sign · Object detection and recognition · Machine-vision

1 Introduction

Road signs play an essential part in ensuring the safety of vehicles as well as communicating information about the road to them [1]. This is why the signs all look the same and include bright, primary colors: to grab people's attention [2]. It is vital for DAS and autonomous vehicles to include a sub-module that handles automatic detection and identification [3]. Commercial interests have prompted significant investment in traffic sign recognition (TSR) technologies in the past. Several methods, notably deep learning, provide satisfactory results on some traffic sign detection datasets.

The national regulations of each country guide the development of their respective nations' traffic signs [4]. Most nations use a consistent set of signs, which may be roughly categorized according to the form and color of their respective elements

P. A. Patel (✉) · V. Vekariya · J. Shah · B. Vala
Parul Institute of Engineering and Technology, Limda, Vadodara, Gujarat, India
e-mail: patelpranava6@gmail.com

N. Hasteer et al. (eds.), *Decision Intelligence Solutions*, Lecture Notes in Electrical Engineering 1080,
https://doi.org/10.1007/978-981-99-5994-5_17

[5]. Major distinguishable characteristics of the traffic sign, such as its form, color, text, or images, may be used for detection and recognition. In the context of the recognition system for traffic signs, the most common forms to appear are circular, triangular, rectangular, square, octagonal, and Pentagonal [6]. The particular purpose of the traffic sign may be recognized from its hue (i.e., red, blue, yellow, green, or white), and it ought to stand out from its chaotic surroundings.

Thorough research on the status of traffic object detection in the age of deep learning is provided here. Our study aims to delve deeply into such topics as small-sized traffic sign recognition, Real-time detection, under-demanding circumstance detection, Image segmentation, Integrated circuits, and so on in the context of traffic sign identification methods. We do a taxonomic analysis of existing sign identification methods and investigate datasets and object detection evaluation criteria. As we do so, we look closely at the efficacy of sign detection methods and bring attention to several promising new avenues of study.

Within a wide range of intelligent systems, the TSDR are essential components of enabling technology. This technology is essential to identifying the road sign's location and what it says. It may be used in a variety of contexts, some of which are outlined below [7].

1) Driver assistance system: The TSDR system is a driver aid system that helps in hazardous driving circumstances such as snow, ice, and abrupt turns by providing the driver with relevant data. So, it might assist the driver control the vehicle's speed and warn him of any possible hazards. Some researchers have suggested that the TSDR system may be utilized to help visually impaired pedestrians. Eventually, smart cars may be able to read the road ahead of them and adjust their speed, course, and other behaviors accordingly. So, TSDRs may soon be standard on all self-driving cars.
2) Intelligent unmanned vehicle: To help the autonomous car interpret the road signs, they are intended to seem like man-built landmarks. These marks will be used by mobile robots to help them find their way back to familiar spots. Despite the impossibility of doing so now, the TSDR system guides unmanned vehicles by controlling traffic signs.
3) Road sign maintenance: Keeping road signs in good condition There has been a long tradition of relying on manual operators to help with the upkeep of road signs. Because of extensive study in computer vision, traffic signs may now be recognized and identified automatically. As a result, computers may one day replace human operators since they can run computer vision algorithms more effectively and inexpensively than humans can.

We pick articles that are relevant or influential and have been published in major journals and conferences with great consideration. The main focus of this review is on the significant advancements made in TSDR during the past three to four years. An example of a flow diagram for TSDR is shown in Fig. 1. However, additional relevant works are also includedfor completeness and easier reading.

The following is the structure of this paper: The Traffic Sign and its dataset are discussed in Sect. 2, while the TSDR system and related work are summarized in

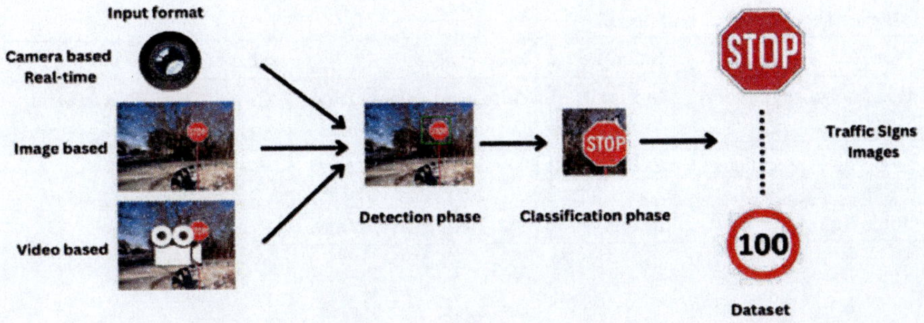

Fig. 1 Traffic Sign Detection and Recognition Flow Diagram

Sect. 3. Section 4 explains the evaluation system of the TSDR. In addition, we explore the difficulties in obtaining an accurate TSDR in Sect. 5. The last part of this work is the conclusion, which can be found in Sect. 6.

2 Traffic Sign and Dataset

The many different types of traffic signs are an essential and essential component of the traffic environment, and they provide a great deal of information that is essential to the present traffic environment [8]. They are designed to control the flow of traffic, alert drivers to hazards and obstructions in their immediate vicinity, aid drivers in navigating properly, and regulate the flow of automobiles, all of which contributes to an overall improvement in driving safety and convenience [9]. All traffic signs may be broken down into two basic categories: those based on ideograms and those based on text. The meaning of the signs in the first group is conveyed by using straightforward ideographs, whereas the meaning of the signs in the second group is communicated through words, arrows, and other symbols. The first category of traffic signs is the primary focus of most of the studies done so far.

The mandotory, cautionary, and Guide signs are the three kinds of traffic signs that are considered to be the most important. As can be seen in Table 1, the Indian government makes use of a variety of different traffic signs. Also, it's essential to keep in mind that various nations could have different signs. In most cases, the design of the traffic signs is satisfactory. They do this by using certain colors and forms, such as triangles, rectangles, circles, and octagons, to stand out against the natural environment's background and draw vehicles' attention [11]. This is done with the intention of improving the situation concerning the identification of traffic signs. Yellow, red, blue, and white are some of the colours they use. Black and white are also used.

It is possible that spotting a Traffic Sign will provide some of the same challenges as recognising objects in their natural environments outside:

Table 1 Types of signs and details

Type	Example	Color	Shape
Mandatory/Regulatory Signs	In Fig. 2	White, Red, Black, Blue	Circle, Triangle, Octagon
Cautionary/Warning Signs	In Fig. 3	White, Black, Red	Rectangle, Triangle
Guide Signs	In Fig. 4	Blue, Red, Black, White	Rectangle

Fig. 2 Mandotory/Regulatory Signs [10]

Fig. 3 Cautionary / Warning Signs [10]

Fig. 4 Guide Signs [10]

1) The presence of other items in the area, such as a moving automobile, bicycles, pedestrians, or even a store sign, might partially occlude the view of a sign or produce uncertainty about what the sign is trying to communicate.

2) Conditions of lightning are subject to change, It might vary according on the time of day and the weather at the moment, which can range from sunny to sandy to foggy to snowy.
3) The colour of traffic signs often progressively loses its vibrancy as a result of prolonged exposure to sunlight as well as the response of the paint to the surface with the surrounding air.
4) When a picture is captured from a moving vehicle, there is a good chance that it may suffer from motion blur as well as automobile vibration.
5) The Look of the sign may vary as a result of vandalism committed by humans or variations in size, visible angle, or placement within the image; however, these changes may also occur independently of one another.

Because of these factors, the accurate recognition of traffic signs in diverse settings might be considered rather difficult.

In Table 2 contains a synopsis of the most important aspects of the existing traffic sign datasets. These aspects include the number of video sequences, the number of annotated images, the number of annotated signs, the annotation information, the resolution, the number of sign types, the size of the annotated sign, the location of the acquisition. When a given classification does not apply to a given dataset, the "N/A" notation indicates this. As can be seen from the table, only three datasets use video. In addition to the standard sign types and bounding box coordinates used in annotations, some datasets include additional information on 3D location, pixel map, visibility status, occlusion condition, and road status. Additionally, these datasets may also include information on the condition of the road. Most of the datasets were collected in Europe; however, participants from China, India, and the United States have been included in more recent research. The LISA dataset is one of a kind since it contains black-and-white photographs taken by a camera mounted in vehicles. The other data sets' scope is restricted only to color photographs.

3 Sign Detection and Recognition

We will discuss about TSDR in this section. Table 2 depicted many sorts of datasets. For research purposes, the majority of the authors used BelgiumTSC [12], STS [13, 14], GTSRB [12, 15–22], LISA [20, 23–25], GTSDB [1, 2, 14, 17, 23, 26–30], TT100K [3, 14, 26, 28, 31–37], CTSD [27, 38], CURE-TSD-Real, and CCTSDB [2, 23].

A. DETECTION

Traditionally, the detection strategies utilized during this period have been classified into two categories which are mentioned in Fig. 5.

This model [2] utilizes the SE-ResNet for detecting high-visibility traffic signs. First, CCFE is used to minimize information loss of coarse features in order to detect traffic signs. After that, carry out a CHFR using a SE-ResNet module starting at a

Table 2 Summary of datasets

Names of Dataset	Number of Videos	Number of annotated image	Number of annotated signs	Annotation information	Resolution	Number of sign types	Annotated sign size	Acquisition Location
BelgiumTS	4	9,006	13,444	sign type bounding box 3D location	1,628 × 1,236	62	9 × 10 to 206 × 277	Belgium
BelgiumTSC	N/A	7,125	7,125	sign type bounding box	22 × 21 to 674 × 527	62	11 × 10 to 562 × 438	Belgium
Sterepolis	N/A	273	273	sign type bounding box	1,920 × 1,080	10	25 × 25 to 204 × 159	France
STS	N/A	3,488	3,488	sign type bounding box visibility status road status	1,280 × 960	7	3 × 5 to 263 × 248	Sweden
GTSRB	N/A	51,840	51,840	sign type	15 × 15 to 250 × 250	43	15 × 15 to 250 × 250	Germany
LISA	17 tracks	6,610	7,885	sign type bounding box occlusion status road status	640 × 480 to 1,024 × 522	47	6 × 6 to 167 × 168	USA
GTSDB	N/A	900	1,206	sign type bounding box	1,360X800	43	16–128 longer edge	Germany
TT-100 K	N/A	1,00,000	30,000	sign type bounding box pixel map	2,048 × 2,048	45	2 × 7 to 397 × 394	China
CTSD	N/A	1,100	1,574	sign type bounding box	1,024 × 768 & 1,270 × 800	48	26 × 26 to 573 × 557	China
CURE-TSDReal	2,989	8,96,700	6,48,186	bounding box challenge type challenge level	1,628 × 1,236	14	10 × 11 to 206 × 277	Belgium
CCTSDB	N/A	10,000	13,361	sign category bounding box	1,024 × 768 & 1,270 × 800 & 1,000 × 350	3 classes	10 × 11 to 380 × 378	China

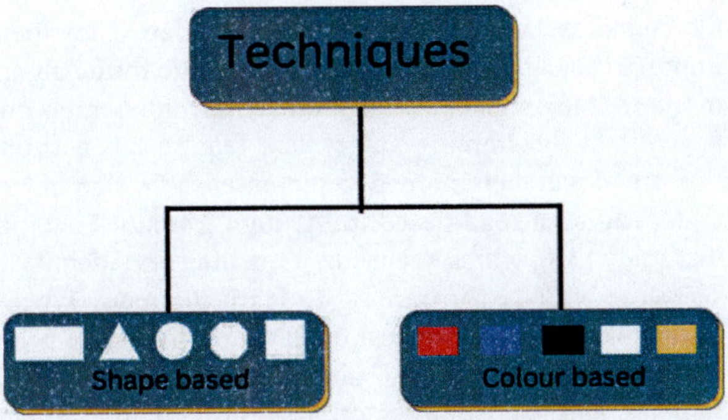

Fig. 5 Detection techniques

high level and working your way down to further investigate the possible specifics. For, Challenging conditions, CHFR plays an essential role. Finally, HFMF is used to build the final saliency map, complete with in-depth details and cues.

The author [3] suggested a Multipath Feature Aggregation Network (FAMN). FAMN is a comprehensive system for detecting objects at a high level of detail. To extract finegrained, scale-invariant features from input images is the goal of FAMN. To do this, they suggest two different structures. First, they build a pyramid of features and suggest a Feature Aggregation (FA) structure to get features independent of size. Features from various tiers of the feature pyramid are aggregated using the element-wise max operation in the FA structure. This allows objects to choose a feature that works best for them, which is very helpful for enhancing the accuracy with which tiny objects can be detected. Second, they suggest the MultipathNetwork (MN) structure to extract a detailed representation of traffic signs. Specifically, they employ a multipath network topology to extract features at the component, instance, and context levels for each area proposal they create. By fusing the three types of features, they obtain a finegrained representation of traffic signs, improving the finegrained classification effectively.

In [26], to detect signs, they created a Deconvolutional SSD for multi-scale analysis. An attention method is utilized to concentrate on local context information to enhance traffic sign detection. The author suggests recurrent attention, which uses late fusion to integrate types of information from adjacent receptive fields.

In [13], the system's primary process consists of two components. The multi-resolution feature learning module is a technique for combining features from various semantic levels with added densely connected deconvolution layers. The VSSA module, another option, clearly encodes the vertical spatial context for correct traffic sign classification. During training, an input image is multi-scale converted to learn the scale-invariant in-depth features. MRFeature network is then used to construct multi-resolution feature maps. LSTM networks represent the spatial sequence along each column of multi-resolution feature maps. Finally, the results of classification and detection are produced concurrently by using an attention LSTM decoder layer.

This study proposed a multiscale cascaded R-CNN [23] and a multiscale feature in pyramids. Positive samples are used to train the cascaded network, which modifies the bounding boxes with the highest overlap. They then use multiscale attention to improve the capacity to recognize authentic traffic signals rather than false ones. To minimize false sign detection increase the number of samples that are unambiguously negative at the training stage. The feature taken from the ROI for each scale is fine-tuned and then merged to increase traffic sign detection accuracy.

Managing a road-based traffic-sign inventory is a problem that was explored in this study [39], which focused on learning and identifying several traffic-sign types. They use a CNN for training to learn the many types of road signs. The authors of this work improved Mask R-CNN by including prior knowledge about types of traffic signs. Online Hard-Example Mining, Sample weighting, Distributing training samples, and Adjusting region pass-through upon detection are used to improve the network.

A coarse-to-fine traffic sign detection technique is proposed in this study [32]. The suggested technique pre-processes the input picture before using the target grid prediction network to determine if the grid includes a target. Grid prediction extracts possible target regions to cover real targets. The first four tiers of the VGG-16 network are used for grid prediction. The SSD is then used to identify traffic signs in the target areas. Predicting the target grid relies on a concise CNN that can rapidly and reliably detect the target's approximate location and eliminate the background from the searching area. To improve the detection speed of small objects, they proposed an improved SSD algorithm by jointly exploring MF-SSD [18].

To improve both detection accuracy and processing speed, they developed a combination of the ROI-based technique with the YOLO architecture [40]. They preprocessed the images for salience detection and utilized prior knowledge of the locations of traffic sign instances to focus in on the regions of the images where the signs were most likely to be present. MSA YOLOv3 [33] is another real-time method for detecting small traffic signs. During model training, the model's generalization is enhanced using the picture mixup approach. Next, a multiscale spatial pyramid pooling module is added to Darknet53 to increase detection accuracy by using features from many smaller-scale local regions. In Darknet53, the enhanced path is developed to enhance the localization capability of a small traffic sign by making use of the fine-grained features of the lower convolutional layer. The YOLOv5 [22] algorithm is utilized for Mobile GPU platefarms-based real-time traffic sign detection.

SADANet [28] is a network that combines a MSPN with a DAN. By examining the robust, domain-invariant features, DAN helps to reduce domain conflict. At the same time, MSPN extracts multiscale features from input images.

In [35], proposed a improved sparse R-CNN. In order to enhance the performance of the sparse R-CNN backbone ReNest, the authors build a hierarchical residual-like link from inside each radix block. Moreover, they deployed a branch network for adaptively recalibrating the channel feature response using the GAP operation and a completely linked layer.

In [36], the light-weight improved YOLOv4-Tiny algorithm is proposed. The K-mean clustering algorithm is used to improve the anchor box generation. The K-man clustering algorithm's distance calculation method was modified to use generalized IoU instead of IoU. In order to improve smallsign recognition, they employ an optimization method to map the large scale of the feature map, transforming the YOLOv4-Tiny two-scale output feature map from its original 19 × 19 and 38 × 38 to a 76 × 76 format. To lower the rate of missing detections, they replace non-maximum suppression with soft NMS.

It is shown in this research [14] that the size and class imbalance problem of proposed IO traffic signs can be addressed with the use of integrated mapping ROI. FA structure aggregates features to get features with high representation capacity. Then, they suggest Light-FA by switching out FA's ensemble-like structure for a residual shape. At last, an IFA-FPN network was proposed for detection.

This paper proposes a two-stage CNN traffic sign detection algorithm based on improved Faster R-CNN [37]. They used HRNet to improve the feature extractor of Faster R-CNN and the attention mechanism of Faster R-CNN. PISA is a simple and effective sampling strategy using a simple weighting scheme to impact the training result and optimize their model significantly.

The author presents a modular and resilient architecture for TSDR using deep CNNs in various challenging conditions in this study [38]. Using the VGG16 architecture, a challenge classifier correctly identifies and classifies the challenge, sending the picture to the appropriate Enhance-Net, which recovers valuable features for recognizing traffic sign regions. They propose a new loss function and training workflow for Enhance-Nets that uses MAE for traffic signs as a constraint both in the pixel and feature domains.

In this research [30], they introduced an algorithm for selecting anchor boxes based on sorting and scaling, and it extracts optimal anchor boxes by using the dimension density of bounding boxes. By utilizing this, they reduce the number of false positives and the log-average miss rate. Last, the YOLOv3 network layers have been proposed for pruning, patch-wise training, and testing method for minor traffic indications.

B. RECOGNITION

Currently, most algorithms use CNN for feature extraction and classification. As data volumes increase over time, CNN has the disadvantage that training takes longer. M. A. Mahmoud, et al. [12], introduced a novel method called DCGAN and MLP with PILAE. DCGAN is utilized as a features extractor after CNN or SVM with backpropagation has been employed for training. DCGAN is used to extract features, and then PILAE is trained to use those features as a model for feature extraction.

The author of [24] presented a visually-based TSR system. In order to maximize the usefulness of visual inspections, the author created a method for extracting potential areas derived from the first picture input using content analysis and crucial information identification. The authors solved the problem of projection distortion by developing a HOG technique for real-world pictures. A Convolutional CapsNet-based traffic learning system is created to effectively capture the poses and direction of the traffic sign.

C. HYBRID

For fast detection on the FPGA, the author of [15] uses an OpenVINO and incorporates many lightweight models, including SqueezeNet-v1.1SSD, MobileNet-v2SSDLite, ResNet-18-SSD, and MobileNet-v1-SSD. They employ Neon and MXNet as their classification systems on the GPU in order to get the greatest possible training speed and inference accuracy. They test for detection and classification on different platforms, such as FP11, FP16, and GPU, for comparison.

Author [4] proposed a model for TSDR on degraded traffic signs. This paper uses a novel linear mapping technique for detecting degraded traffic signs. By using a linear transformation, we may minimize the dimensions of the original RGB data while simultaneously increasing the distance between the rim and background pixels in Euclidean space. After transformation, to estimate the parameters of the mixture expectation maximization algorithm used. Segmentation of traffic sign GMM-based technique served. The multiscale CNN (pre-trained Alexnet) is augmented with dimensionality reduction layers in order to use only highly relevant and least redundant features during multi-class classification.

To improve detection accuracy, they propose the fusion of feature representations in the TSDR phases of a MR-CNN [31]. As part of the detection process, the multi-scale deconvolution operator up-samples the results from the more advanced convolution layer. In L2 normalization, the features from the shallow convolution and deconvolution layers are scaled to the same level and then concatenated along the channel axis. This method compresses the concatenated features with pointwise convolution to create the fused feature map. They hypothesize that small traffic sign recognition may benefit significantly from contextual information during the classification phase. As a result, the combined feature gives a clearer picture of small traffic signs. It does this by making use of the information in and around an item proposal area via the usage of contextual regions. RoI-pooling, concatenation, and compression are employed to build the final fused feature.

In this [31] study, initially introduced, ENet a network for classification, and EmdNet a network for detection. They also apply the ENet's principles to the design of EmdNet's core network. They employed depthwise separable convolution to accelerate the process (DSC).

In order to improve the performance of the detection process, they integrate and optimize faster versions of R-CNN and MobileNets [17]. In an effort to improve quality of detection and the precision of conventional traffic sign classification, a straightforward method based on color and shape information is presented. The CNN's asymmetric kernels make it an effective and reliable classifier.

In [27], SegU-Net is a Fully CNN architecture motivated by SegNet and U-Net and used to create the localizer. The downsampling stage of the network includes an encoder, whereas the upsampling stage includes a decoder. The CNN used to classify the regional traffic signs is based on the famous VGG-16 architecture, designed to recognize images.

This study [34] presents a two-stage CNN-based technique for efficiently detecting small traffic signs in images with a high pixel count. Their approach adds two distinct convolutional networks, or"locator" and"classifier," for both rapid localization and precise classification. This locator achieves excellent recall on traffic signs of varying sizes by combining the concepts of FPN, the fire module of CornerNet-Squeeze, and the center-point estimation of Centernet. The classifier can extract fine-grained local features, which significantly enhances classification accuracy.

In [29], an improved Mask R-CNN is proposed for detecting text-based traffic signs. Mask R-CNN shows significant results replacing RoIPool with RoIAlign. Then, localization refinement is performed to obtain the Mask RCNN score class and mask. This filtering technique reduces the number of false objects passing to the classifier. They employed two convolutional neural networks (CNNs) for classification: one to learn traffic signs (the Cat-CNN) and another to identify traffic sign class (the Class-CNN).

This study [41] provides a unique and adaptable traffic sign identification technique by integrating a lightweight superclass detector with a refinement classifier. Their method can identify and recognize traffic signs from various subclasses in complicated scenarios. The benefits of their method are highlighted in three ways. The first is to enhance computing performance by employing previous geographical statistics as traffic sign probability distribution. The second step is to include the Inception and Channel Attention mechanisms into their detector, resulting in a high-performance light model. The lightweight variant is easier to install on mobile devices. The third approach proposes the classifier, which enhances the effectiveness of distinguishing hard-related subclasses.

In this section we study about traffic sign detection, recognition and Hybrid techniques briefly. Here is a chart illustrating the key aspects of their research, as shown in Fig. 6. In Table 3 we summarize research papers algorithms and drawbacks. And We also describe key points of models used in papers In Fig. 6 Chart.

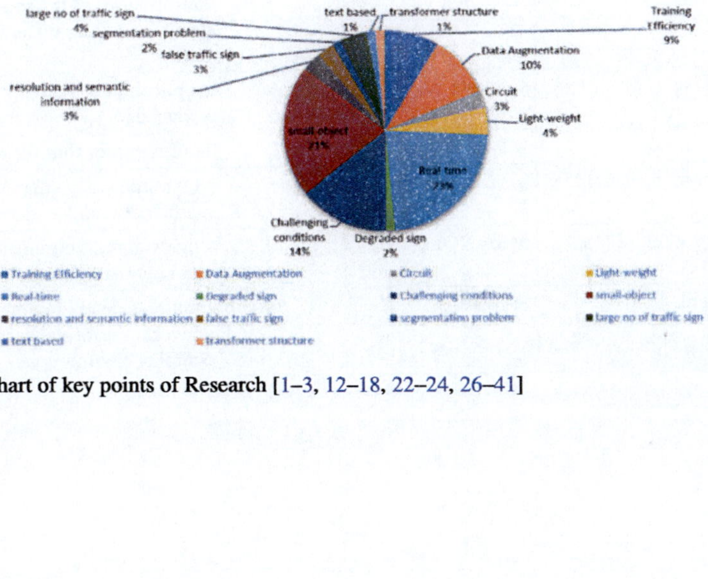

Fig. 6 Chart of key points of Research [1–3, 12–18, 22–24, 26–41]

Table 3 Summary of research paper algorithms and drawbacks

Research paper	Algorithms	Drawbacks
M. A. Mahmoud, et al. [3]	DCGAN and MLP with PILAE	Dataset is imbalance
Z. Lin, et al. [6]	MobileNet-v1-SSD, VGG-SSD, ResNet-50-SSD	Need to train and compare different plafarm like ONNX format
A. Mannan, et al. [17]	Flexible Gaussian mixture model based technique with automatic split and merge, Multiscale CNN	Non availability of Degraded Traffic sign Dataset
C. Li, et al. [18]	CCFE, CHFR, HFMF	
Z. Ou, et al. [24]	FAMN	Need improvement on extracting more fine-grained features
Z. Liu, et al. [25]	MR-CNN	training algorithm to ifferentiate hard negative samples from positive samples
X. Bangquan, et al. [7]	ENet, EmdNet	- performance,- input video based, - multitask learning
J. Li, et al. [8]	Faster R-CNN, MobileNet	- performance,- localization refinement of bounding box
Y. Tian, et al. [19]	Multi-scale and recurrent attention	- Uneven lighting, - perspective irregular or the blurry image
Y. Yuan, et al. [4]	Multi-resolution feature fusion network, Vertical Spatial Sequence Attention network	false positive detection
J. Zhang, et al. [14]	Cascaded R-CNN, multiscale attention	Dataset imbalance problem
U. Kamal, et al. [20]	SegU-Net, Modified Tversky Loss Function with L1-Constraint	Robustly perform task
D. Tabernik, et al. [34]	Mask R-CNN	Performance of missed detection of classification network
L. Liu, et al. [26]	Corse-to-fine	Application of spatial prior knowledge
Y. Jin, et al. [9]	MF-SSD	improve algorithm for real time
A. Avramović, et al. [35]	YOLO	- no traffic sign, - fine-tunning neural network, - video-based
H. Zhang, et al. [27]	MSA YOLOv3	- lightweight, - eliminate useless and redundant feature
Z. Liu, et al. [21]	SADANet	- multi-scale prediction network, - erformance in more complex environment

(continued)

Table 3 (continued)

Research paper	Algorithms	Drawbacks
L. Wei, et al. [28]	Improved fire-modules	- speed up model, - more discriminative and more featurer
C. G. Serna, et al. [22]	Mask R-CNN, CNN	- GAN, - data augmentation
J. Cao, et al. [29]	Improved Sparce R-CNN	- reduce complexity, - weight sharing feature pyramid network, - attention
L. Wang, et al. [30]	YOLOv4-Tiny, improved K-means clustering	Evalution indicators
S. He, et al. [15]	ROI extraction, HOG feature design, stacked CapsNet	performance
Q. Tang, et al. [5]	IFA-FPN	- Lightweight, - real-time, - performance
J. Cao, et al. [31]	Improved Faster R-CNN with HRNet, Attention: PISA, ISR, CARL	Speed up model
E. Guney, et al. [13]	YOLOv5	- performance, - challenging condition
S. Ahmed, et al. [32]	CNN, EnhanceNet encoder-decoder, VGG16	Optimum network
Y. Rehman, et al. [23]	YOLOv3	validation
Z. Wang,, et al. [36]	Combination of superclass detector with the refinement classifier	optimize algorithm

4 Performance Evalution of the TSDR

A wide range of metrics is available to choose from when evaluating machine learning models for diverse purposes. Let's have a look at the many metrics that can be used to evaluate the performance of a machine-learning model. This is a very important phase in any data science project since it seeks to predict the generalization accuracy of a model based on data that will be collected in the future.

1) **Precision**

When there is a disparity between the classes of people in our workforce, accuracy as a performance indicator may become unreliable. As a result, we also need to consider the performance indicators that are unique to each class. One of these measures is called precision, and it refers to the ability to accurately forecast positive results.

$$\text{Precision} = \left(\frac{TruePositive}{TruePositive + FalsePositive}\right) \tag{1}$$

2) **Recall**

The percentage of true positive instances that are identified correctly is referred to as the recall percentage, and it is another one of the essential measures.

$$\text{Recall} = \left(\frac{TruePositive}{TruePositive + FalseNegative}\right) \tag{2}$$

3) **F1-Score**

The F1 score relies on two important error measurements, Precision and Recall. In light of binary data categorization, it may be seen as the Harmonic mean of Precision and Recall error metrics for an unbalanced dataset.

$$\text{F1} - \text{score} = \left(\frac{2 * Precision * Recall}{Precision + Recall}\right) \tag{3}$$

4) **Accuracy**

The accuracy of the model is one metric that may be used to evaluate its overall performance across all classes. It is advantageous when all classes are on an equal footing in terms of importance. Calculating it requires taking the number of correct forecasts and dividing it by the total number of predictions.

$$\text{Accuracy} = \left(\frac{TruePositive + TrueNegative}{TruePositive + TrueNegative + FalsePositive + FalseNegative}\right) \tag{4}$$

5) **mean Average Precision**

mAP is a well-liked assessment metric for object detection in computer vision. An instance's localization locates its precise position, whereas its categorization identifies its nature

$$\text{mAP} = \frac{1}{N}\sum_{i=1}^{N} APi \tag{5}$$

5 Challenges in Accurate Tsdr

Despite enormous investments in TSDR technology made by both the academic community and the private sector, several components of these systems continue to struggle due to the multiple obstacles that are mentioned in the following paragraphs:

A. Dataset imbalance

The classification problem known as imbalanced classes occurs when there is an imbalance in the class distribution in a training dataset. The existence of an issue with unbalanced classes is one of the typical issues observed in the datasets used for classification. In [12], Development is still needed to fix the issue with the GTSRB dataset's uneven distribution without using any new procedures.

B. Integrated circuit

FPGA is utilized in [15] for the purpose of TSDR so that improved accuracy and detection speed could be achieved. However, research on FPGA is still need to be conducted. In this study, FPGA was implemented using an SSD model, however it was not successful when implemented using a non-SSD model. Examine the impact on FPGA and GPU performance of employing non-SSD models and approaches such as deep compression and deploying models trained by frameworks that are compatible with ONNX but are not directly supported by OpenVINO.

C. Features extraction

One of the most notable uses is image processing, in which algorithms are used to find and extract desired parts or shapes (features) from a digitized picture or video stream. This application is one of the most prominent applications. There are many different characteristics, each of which has the potential to be retrieved using one of the many other machine-learning techniques. The results might be enhanced by employing pre-configured collections of app-specific features, which are often configured by specialists. With deep learning, we do not need to extract features manually from the photos we are given. In [3, 34], required development in TSDR, The development of a robust model was made possible by the extraction of more fine-grained features for the object detection framework.

D. Performance

Problems such as the ease with which conventional traffic sign detection can be affected by the environment and the poor real-time performance of the Deep Learning-based TSR Methodology. In [16, 18, 39] despite the fact that the strategy in question has good performance, there is still potential for advancement. There is a requirement for the SADANet [28] to have its strength increased so that it can keep its performance constant in contexts that are more complicated. In [35, 37] is necessary to accelerate the model while preserving a high level of accuracy and to undertake in-depth analysis on the feature optimization machine of the object identification neural network. The performance can be improved by developing an effective training algorithm [31] that can discern between difficult negative samples and straightforward positive data. In [32], increasing performance may be accomplished by researching the application of spatial prior knowledge in the quicker filtering of the foreground region and by designing processing algorithms that are particular to those densely grouped miniature traffic signs.consider using data augmentation for the Mask R-CNN model because the training set for the GTSDB is relatively small [29]. Next, use GANs to produce features for extremely tiny traffic signs that are equivalent to those for clear and big signs in order to increase our TSDR performance. These features should be able to differentiate between different types of signs.

E. Localization and bounding box

In [17], the retention of information regarding traffic signs needs more work. According to the findings of the evaluation, we have reason to think that the performance will be significantly improved in the event that additional annotated data is utilized. It has been determined that the method used for the localization refining of the bounding box is insufficiently robust and will be the subject of additional research as part of our ongoing efforts.

F. Challenging condition

The identification of objects can be hampered in environments that are ordinary or autonomous if adverse weather conditions are present. Driving in inclement weather can be hazardous for a variety of reasons, such as when roads become icy or when there is an optical illusion of dense fog. In [27, 42–44], the work has to be expanded further so that the model can reliably carry out the responsibilities of TSDR even when faced with difficult situations.

G. Dataset

Objects may be under any hard circumstance (blur, colour fading, bad light, occlusion, rain, snow, etc.), but the system will still be able to recognize them. It is important for us to have a good dataset so that our model can recognize the item regardless of the context. In order for the model to be able to recognize a greater variety of traffic signs, the class quantity in needs to be increased [45]. In addition to this, there is a requirement to incorporate tough condition pictures into datasets so that the system can perform effectively under that circumstance [22, 46]. In [47], there is a requirement for comparison regarding the combining of synthetic pictures of various sizes for the purpose of training.

H. Not contain sign

Taking into consideration the fact that the regions frequently lack any instances of traffic signs is such way in which our approach might be improved further [40]. It would be useful to determine the likelihood that an area has no traffic sign instances; this would allow for the elimination of empty regions, which would result in the savings of both time and hardware resources. In addition, there is still opportunity for development in the process of fine-tuning the neural network in order to attain higher levels of detection accuracy. It could also be interesting to modify the algorithm so that it can be used to video sequences of traffic imagery, in which a single instance of a traffic sign might occur in a number of frames that follow one another. We could increase the accuracy of our detection by taking use of this temporal coherence.

I. Light-weight

The vast majority of the time, when we discuss methods based on deep learning, we refer to GPU-based processing units.GPU-based hardware is often rather pricey, which is why we need to make our solution lightweight so that it can be adapted to a wide variety of devices. This system has a number of drawbacks, many of which have an impact on the effectiveness and accuracy of the system. We want to integrate the suggested technique into the ADAS or ADS of a real car after completing the work necessary to lighten the load on the network [14, 41] In order to achieve greater levels of performance.

Exploring a wide range of structural and compressive techniques strategies to create a nearly optimal network for each module in order to ensure that all of The difficult circumstances that are included in the CURE-TSD dataset may be tackled most effectively by focusing on both performance and the amount of time needed for inference [38].

6 Conclusion

The primary purpose of the paper was to analyse the primary path that the research was taking in the field of TSDR. The paper also aimed to classify the primary methods into specific sections in order to make the topics simple to comprehend and to visualise the overall research in order to determine where it could go in the future. The paper discusses topics such as the many forms of traffic signs, the different sources of data, the detection and recognition challenges, and so on. In addition, this article has made an effort to examine as many research as it can in order to offer a comprehensive evaluation of the many different technologies that are suited and available for the TSDR.

According to the report, there has been an increase in research on TSDR. Despite this, it is widely accepted that picture quality in real-world traffic scenes is often poor due to low resolution, fluctuating lighting, occlusion, motion blur, scale and rotation, and other variables. Furthermore, traffic signs often have a diverse appearance, with a high degree of inter-class similarities and complicated backgrounds. Despite recent improvements in the overall performance of TSDR systems, further research is needed to build a TSDR system that is accurate, robust, and reliable. It has been proposed that integrating the detection and classification operations into a single phase, rather than doing them separately, might increase the overall performance of the TSDR system. It's feasible that doing so will boost detection and decrease categorization. Another idea for more TSDR improvements is to use standard, sufficient, and substantial datasets for learning, testing, and analyzing the recommended algorithms. Researchers who want to learn about the current state of research in the field of TSDR and identify the research issues that need to be addressed will find this publication a great resource.

References

1. Mannan A, Javed K, Rehman AU, Babri HA, Noon SK (2019) Classification of degraded traffic signs using flexible mixture model and transfer learning. IEEE Access 7:148 800–148 813
2. Li C, Chen Z, Wu QJ, Liu C (2018) Deep saliency with channelwise hierarchical feature responses for traffic sign detection. IEEE Trans Intell Transp Syst 20(7):2497–2509
3. Ou Z, Xiao F, Xiong B, Shi S, Song M (2019) FAMN: feature aggregation multipath network for small traffic sign detection. IEEE Access7:178 798–178 810
4. Fagnant DJ, Kockelman K (2015) Preparing a nation for autonomous vehicles: opportunities, barriers and policy recommendations. Transp Res Part A Policy Pract 77:167–181
5. Mogelmose A, Trivedi MM, Moeslund TB (2012) Vision-based traffic sign detection and analysis for intelligent driver assistance systems: perspectives and survey. IEEE Trans Intell Transp Syst 13(4):1484–1497
6. Khan JF, Bhuiyan SM, Adhami RR (2010) Image segmentation and shape analysis for road-sign detection. IEEE Trans Intell Transp Syst 12(1):83–96
7. Gudigar A, Chokkadi S et al (2016) A review on automatic detection and recognition of traffic sign. Multimedia Tools Appl 75(1):333–364
8. Balado J, González E, Arias P, Castro D (2020) Novel approach to automatic traffic sign inventory based on mobile mapping system data and deep learning. Remote Sens 12(3):442
9. Hancock P, Parasuraman R (1992) Human factors and safety in the design of intelligent vehicle-highway systems (IVHS). J Safety Res 23(4):181–198
10. "Road signs in India." https://en.wikipedia.org/wiki/RoadsignsinIndia
11. Arnheim R (1956) Art and visual perception: a psychology of the creative eye. University of California Press
12. Mahmoud MA, Guo P (2019) A novel method for traffic sign recognition based on dcgan and mlp with pilae algorithm. IEEE Access 7:74 602–74 611
13. Yuan Y, Xiong Z, Wang Q (2019) VSSA-NET: vertical spatial sequence attention network for traffic sign detection. IEEE Trans Image Process 28(7):3423–3434
14. Tang Q, Cao G, Jo K-H (2021) Integrated feature pyramid network with feature aggregation for traffic sign detection. IEEE Access 9:117 784–117 794
15. Lin Z, Yih M, Ota JM, Owens JD, Muyan-Özçelik P (2019) Benchmarking deep learning frameworks and investigating fpga deployment for traffic sign classification and detection. IEEE Trans Intell Veh 4(3):385–395
16. Bangquan X, Xiong WX (2019) Real-time embedded traffic sign recognition using efficient convolutional neural network. IEEE Access 7:53 330–53 346
17. Li J, Wang Z (2018) Real-time traffic sign recognition based on efficient cnns in the wild. IEEE Trans Intell Transp Syst 20(3):975–984
18. Jin Y, Fu Y, Wang W, Guo J, Ren C, Xiang X (2020) Multi-feature fusion and enhancement single shot detector for traffic sign recognition. IEEE Access 8:38 931–38 940
19. Li Y, Xu X, Xiao J, Li S, Shen HT (2020) Adaptive square attack: fooling autonomous cars with adversarial traffic signs. IEEE Internet Things J 8(8):6337–6347
20. Yang X, Liu W, Zhang S, Liu W, Tao D (2020) Targeted attention attack on deep learning models in road sign recognition. IEEE Internet Things J 8(6):4980–4990
21. Kim J, Kang J-K, Kim Y (2022) A low-cost fully integer-based cnn accelerator on fpga for real-time traffic sign recognition. IEEE Access 10:84 626–84 634
22. Güney E, Bayilmiş C, Çakan B (2022) An implementation of real-time traffic signs and road objects detection based on mobile gpu platforms. IEEE Access 10:86 191–86 203
23. Zhang J, Xie Z, Sun J, Zou X, Wang J (2020) A cascaded R-CNN with multiscale attention and imbalanced samples for traffic sign detection. IEEE Access 8:29 742–29 754
24. He S et al (2021) Automatic recognition of traffic signs based on visual inspection. IEEE Access 9:43 253–43 261
25. Lopez-Montiel M, Orozco-Rosas U, Sánchez-Adame M, Picos K, Ross OHM (2021) Evaluation method of deep learning-based embedded systems for traffic sign detection. IEEE Access 9:101 217–101 238

26. Tian Y, Gelernter J, Wang X, Li J, Yu Y (2019) Traffic sign detection using a multi-scale recurrent attention network. IEEE Trans Intell Transp Syst 20(12):4466–4475
27. Kamal U, Tonmoy TI, Das S, Hasan MK (2019) Automatic traffic sign detection and recognition using segu-net and a modified Tversky loss function with L1-constraint. IEEE Trans Intell Transp Syst 21(4):1467–1479
28. Liu Z, Shen C, Qi M, Fan X (2020) SADANet: integrating scale-aware and domain adaptive for traffic sign detection. IEEE Access 8:77 920–77 933
29. Serna CG, Ruichek Y (2019) Traffic signs detection and classification for european urban environments. IEEE Trans Intell Transp Syst 21(10):4388–4399
30. Rehman Y, Amanullah H, Shirazi MA, Kim MY (2022) Small traffic sign detection in big images: searching needle in a hay. IEEE Access 10:18 667–18 680
31. Liu Z, Du J, Tian F, Wen J (2019) MR-CNN: a multi-scale region-based convolutional neural network for small traffic sign recognition. IEEE Access 7:57 120–57 128
32. Liu L, Wang Y, Li K, Li J (2020) Focus first: coarse-to-fine traffic sign detection with stepwise learning. IEEE Access 8:171 170–171 183
33. Zhang H et al (2020) RealTime detection method for small traffic signs based on YOLOv3. IEEE Access 8:64 145–64 156
34. Wei L, Xu C, Li S, Tu X (2020) Traffic sign detection and recognition using novel center-point estimation and local features. IEEE Access 8:83 611–83 621
35. Cao J, Zhang J, Jin X (2021) A traffic-sign detection algorithm based on improved sparse R-CNN. IEEE Access 9:122 774–122 788
36. Wang L, Zhou K, Chu A, Wang G, Wang L (2021) An improved lightweight traffic sign recognition algorithm based on YOLOv4-Tiny. IEEE Access 9:124 963–124 971
37. Cao J, Zhang J, Huang W (2020) Traffic sign detection and recognition using multi-scale fusion and prime sample attention. IEEE Access 9:3579–3591
38. Ahmed S, Kamal U, Hasan MK (2021) DFR-TSD: a deep learning based framework for robust traffic sign detection under challenging weather conditions. IEEE Trans Intell Transp Syst
39. Tabernik D, Skočaj D (2019) Deep learning for large-scale traffic-sign detection and recognition. IEEE Trans Intell Transp Syst 21(4):1427–1440
40. Avramović A, Sluga D, Tabernik D, Skočaj D, Stojnić V, Ilc N (2020) Neural-network-based traffic sign detection and recognition in high-definition images using region focusing and parallelization. IEEE Access 8:189 855–189 868
41. Wang Z, Wang J, Li Y, Wang S (2020) Traffic sign recognition with lightweight two-stage model in complex scenes. IEEE Trans Intell Transp Syst
42. Chen R, Hei L, Lai Y (2020) Image recognition and safety risk assessment of traffic sign based on deep convolution neural network. IEEE Access 8:201 799–201 805
43. Guo J, You R, Huang L (2020) Mixed vertical-and-horizontal-text traffic sign detection and recognition for street-level scene. IEEE Access 8:69 413–69 425
44. Temel D, Chen M-H, AlRegib G (2019) Traffic sign detection under challenging conditions: a deeper look into performance variations and spectral characteristics. IEEE Trans Intell Transp Syst 21(9):3663–3673
45. Liu L, Tang X, Xie J, Gao X, Zhao W, Mo F, Zhang G (2020) Deep-learning and depth-map based approach for detection and 3-D localization of small traffic signs. IEEE J Sel Top Appl Earth Obs Remote Sens 13:2096–2111
46. Satti SK, Devi KS, Sekar K, Dhar P, Srinivasan P (2022) ICTS: Indian cautionary traffic sign classification using deep learning. In: 2022 IEEE international conference on distributed computing and electrical circuits and electronics (ICDCECE). IEEE, pp 1–7
47. Dewi C, Chen R-C, Liu Y-T, Jiang X, Hartomo KD (2021) YOLOv4 for advanced traffic sign recognition with synthetic training data generated by various GAN. IEEE Access 9:97 228–97 242

Detection of Ductal Carcinoma Using Restricted Boltzmann Machine and Autoencoder (RBM-AE) in PET Scan

J. Lece Elizabeth Rani, M. P. Ramkumar, and G. S. R. Emil Selvan

Abstract The most common type of cancer among women is breast cancer. Recently, breast cancer has maximum mortality rate next to lung cancer. Breast cancer detection at an early stage can lower the risk of mortality. As per oncologists, detecting cancer at the beginning stage will avoid the spread of cancer tissue which can increase survival time. Here, the main issues are identifying the malignant tumors and distinguishing the malignant from benign, finally finding whether the cancer stage is mild or advanced. Therefore, the researchers of cancer study accomplished the challenges through the deep learning application. The restricted Boltzmann machine and autoencoder (RBM-AE) are proposed for detecting breast cancer in the PET imaging dataset in this research work. The performance evaluation of the RBM-AE is good when compared with existing methods. The proposed RBM-AE model detects the malignant tumor of the breast with high specificity and accuracy.

Keywords Restricted Boltzmann Machine (RBM) · Autoencoder (AE) · Malignant · Benign · Image extraction · Image classification

1 Introduction

Cancer is a toxic genetic disease in humans where abnormal growth or damaged human cells form a malignant tumor, which is an inflammation of tissues. These tumors are invading and can bypass more areas in the human body to form a new

J. L. E. Rani (✉) · M. P. Ramkumar · G. S. R. E. Selvan
Department of Computer Science and Engineering, Thiagarajar College of Engineering, Madurai, India
e-mail: lecethomson@gmail.com

M. P. Ramkumar
e-mail: ramkumar@tce.edu

G. S. R. E. Selvan
e-mail: emil@tce.edu

N. Hasteer et al. (eds.), *Decision Intelligence Solutions*, Lecture Notes in Electrical Engineering 1080,
https://doi.org/10.1007/978-981-99-5994-5_18

tumor which is called metastasis [1]. Cancer is produced by variations in genes that have the power to manage cell functioning mainly in cell growth and partition. Depending on genes cancer will occur. Normally, cancer cells are different from normal cells in terms of cell growth, cell movement, depending on the immune system, increased chromosomes, and also fast growth. The human body has 100 more types of cancers such as lung, brain, skin, breast, prostate, basal cell, melanoma, colon, lymphoma, etc. [2]. Breast cancer is the type of cancer that affects 99% of women and 1% of males. Recent statistics on breast cancer reveal that 2.3 million women were detected globally and death rates up to 70,000 as stated by the world health organization (WHO). Mostly, women above 40 ages were affected in common [3]. By identifying breast cancer early and receiving appropriate care, it can be decreased. For the identification and prevention of breast cancer, enormous research is now being conducted [4, 5]. Furthermost, the researchers use medical image datasets for detection purposes. So, in this work, breast cancer detection is carried through deep learning techniques such as restricted boltzmann machine and autoencoder (RBM-AE) in medical imaging dataset - PET scan is used.

Popular research, diagnosis, and prevention of all types of cancer are analyzed through the collection of medical imaging datasets. Mammograms, breast thermography (BTD), magnetic response imaging (MRI), positron emission tomography (PET), computed tomography (CT), ultrasound (US), and histopathology (HP) are some of the techniques for detecting breast cancer in the early stage. Among these modalities, the more rarely used modality is the PET scan, because this type of scan is not highly recommended by oncologists due to its view in 2D or 3D [6]. Therefore, in this research, the authors concentrated on this collection of PET imaging datasets of cancer people. So, the authors generated a synthetic dataset for breast cancer detection. Immense methods have existed for breast cancer detection such as computer-aided detection (CAD), determining statistical methods, and traditional image processing methods [7, 8]. However, these methods consume more time and require complex computations when it comes to high-resolution images [9]. Therefore, the advanced techniques in deep learning help the researchers to cross the traditional methodology hurdles in breast cancer detection and also it plays an important role in image processing applications. Deep learning is a branch of machine learning that operates with a big dataset with ease. It is less expensive and time-consuming compared to the traditional method, easy to extract important information from high-resolution images. It includes more stage abstractions and is capable to deliver the important features of the image dataset without the need for manpower [10]. There are different types of deep learning techniques such as autoencoder (AE), Artificial neural network, probabilistic neural network (PNN), deep neural network (DNN), restricted boltzmann machine (RBM), long short-term memory (LSTM), convolutional neural network (CNN), etc. The proposed (RBM-AE) model performance is compared with existing methods which can accomplish high accuracy and sensitivity rate.

2 Literature Survey

The review of related studies is covered in this section, with a particular emphasis on the proposed framework (RBM-AE) for the detection of breast cancer using a Restricted Boltzmann machine (RBM) and extracting features from the dataset using autoencoder (AE). The major role of this proposed model is feature extraction and malignant tumor detection in the breast. To detect a malignant tumor, huge research has been studied. Moreau, Noémie, et al. have proposed a dual deep learning strategy in a U-Net architecture (i.e.) CNN in PET/CT images. It is first carried out for bone and bone lesions then it was drawn for breast cancer detection with data from 24 patients which depends on EPICURE (seinmeta). The performance metrics for evaluating the proposed model are precision and recall and special segmentation accuracy for bones lesion is achieved through a dice score. This method supports both monitorization and decision-making for the PET dataset [1]. However, this method cannot be fixed for precision because of network formation for bone and bone lesions. The proposed deep learning U-net architecture achieves low fixing for bone lesions segmentation is not advisable [11]. Papandrianos, Nikolaos, et al. have suggested a CNN framework for the diagnosis of breast cancer that is caused due to bone metastasis. The researchers worked on images of the whole human body to detect whether the patients are affected by bone metastasis or not. The proposed CNN model provides high flexibility mainly in medical image classification and the performance evaluation achieve 93% of accuracy in breast cancer detection. The authors compared the performance evaluation of the proposed CNN with some other benchmark medical imaging datasets (ResNet50, VGG16), etc. However, the proposed model doesn't reveal any evidence that it can apply to all types of medical imaging datasets. Therefore, the system doesn't prove whether it is acceptable for being a clinical decision support system [12].

Moreau, Noémie, et al. have proposed a deep learning-based automatic medical image segmentation concept for bone metastasis-based breast cancer detection. The collection of 60 patients' whole-body PET/CT scan images was derived through EPICURE (seinmeta) and is trained and tested through dual deep learning concepts. The first deep learning evaluates and divides the data into 4 sections such as SULpeak, TLG, PBI, and PLI which can achieve 0.66 as a dice score for baseline images. The second deep learning network for 4 sections of the dataset achieves a 0.58 dice score for follow-up images. All these working of dual deep learning architecture are achieved through the built-in tools. Nevertheless, the proposed work should experiment with whether it can achieve the prognostic value of medical images during prediction [13]. This overall study of existing work of breast cancer detection using PET scans using deep learning techniques had majorly concentrated on convolutional neural network (CNN). The performance of CNN for this application achieves at a greater level. However, it fails in maintaining time complexity and model stability, and flexibility in training and validating datasets. This is a huge issue in using CNN for breast cancer detection in PET scan images. Some studied existing works of feature extraction and breast cancer detection are summarized in Table 1.

Table 1 Literature Survey

Authors &Year	Deep Learning Technique	Dataset & its Modality	Description	Performance Evaluation
Shoshan et al. (2021)	Bone Lesion detection of breast cancer using CNN	3D DBTex Challenge Digital Breast Tomosynthesis (DBT)	It accepts more models and detects the presence of malignant formation particularly of the patient's breast within minimal time with minimum false positives [14]	91% Sensitivity
Zhang et al. (2021)	bimodal distribution removal (BDR)-(CNN)-graph convolutional network (GCN)	Mini-MIAS Mammograms (2018) (MM)	The proposed model is composed of 8 layered CNN architecture that was raised with two techniques: batch normalization and dropout. It is composed of rank-based stochastic pooling and is an amalgam with GCN [15]	96% Accuracy 96% Sensitivity 96% Specificity
Hossain et al (2022)	With data augmentation and CNN was proposed for breast cancer detection through the lesion	2D DBTex Challenge DBT CAD-based DBT dataset	The ensemble detection model was combined with the lesion detection algorithm for the DBT dataset to increase the performance detection algorithm [16]	82% Sensitivity
Li et al. (2021)	CNN and RNN for classification, ResNet for feature extraction	DDSM Mammography	CNN is used for network metric reduction and had high network speed. To view the spatial aspects of mammograms, an RNN's gate recurrent unit is employed. [17]	95% Accuracy 94% recall 97% AUC
Bai et al. (2022)	GCN is used for both feature extraction and breast cancer detection	3D-BCS-DBT synthetic dataset based on DBT modality	The self-practiced graph is drawn for extracting the reduction of unwanted mammogram images and GCN is used for detecting a malignant tumor in the breast [18]	84% Accuracy 87% AUC

3 Materials and Methods

3.1 *Restricted Boltzmann Machine (RBM) for Image Extraction*

RBM was founded by Paul Smolensky in the year 1986 by naming it harmonium and then it was called "Contrastive Divergence" and was coined by Geoffrey Hinton as a method to train the dataset taken. RBM is widely used for feature extraction and classification. The visible layer and hidden layer are the two layers of RBM. The visible layer is also called the input layer. When input as an image, pixels of the image dataset are processed through the visible layer. It supports both unsupervised and supervised learning methods [19]. The structure of the images is read by RBM and is processed for extracting the important features from the input image by reconstructing the input images. To reconstruct the input data, the RBM needs both a forward pass and a backward pass. The weight of the neural network is altered based on the input image features. It helps to find and ignore the same features. Once the training is completed, the image is reconstructed. The visible layer shows the difference between the obtained results and the original image [20, 21]. The training process consists of three steps forward pass, backward pass, and determining the difference between the obtained results and the original image.

$$\text{Forward pass: p(h\_((input)) | v\_output)} = \sigma\left(\text{V\^{}t*wgt} + \text{b}\right)\text{n} \tag{1}$$

$$\text{Backward pass: p (v\_((input)) } \left| \text{h\_output)= } \sigma \text{ (} [\![\text{wgt}]\!] \text{ \^{}t*h\_((output))+b)n} \right. \tag{2}$$

where h is the hidden layer input and v is the output of the visible layer. σ defined the stochastic probability, wgt is the weight of the neural network and b stands for network biases. The forward pass and backward pass are vice versa in the working of a neural network. The neural network's operating system will continue until the network obtains network convergence. Both the visible and hidden layers had energy functions,

$$\text{E(V, H)} = -\text{a\^{}n Vis - b\^{}n hid -V\^{}n*wgt} * \text{h (output)} \tag{3}$$

where a and b are the subset of the visible and hidden layers with weight wgt. Based on Contrastive Divergence

$$\text{Mean(F(Visible(Original))} = \text{Mean (F(Visible (reconstructed))} \tag{4}$$

Thus, the reconstructed image of RBM can have more information than the original data.

3.2 *Autoencoder (AE) for Detection*

AutoEncoder (AE) is a neural network that is trained to detect the latent image feature of input that is said to be encoded and then rebuild the latent image feature of output which is similar to the input image. AE aims to reduce the errors in the classified image of extracted data which leads to training only the important image data. AE is composed of dual networks as an encoder and a decoder. The encoder converts higher dimensional data into low dimensional data. This converted low-dimensional data is given as input to the decoder which forms a rebuilt input image. While training, unsupervised learning is followed for AE. And the cost function helps to choose the best parameter [22, 23].

Encoder: hidden=∅ (〖wgt〗 _((m))*input+bias) (5)

Decoder: 〖rebuild〗 _((input))=∅ (〖RE(wgt〗 _((m)))*hidden+RE(bias)) (6)

Here, "hidden" is the hidden layer of the encoder, 〖wgt〗 _((m)) is the weight matrix and bias is the bias vector of the neural network. 〖rebuild〗 _((input)) is the reconstructed data of decoder, 〖RE(wgt〗 _((m))) is the weight matrix of rebuild data and RE(bias) is bias vector of rebuild image. In the case of image detection, the SSIM loss function is used to find the similarity between the input and rebuild the image to detect the abnormality. This SSIM should be larger for abnormality data.

4 Materials and Methods

The proposed RBM-AE framework (Fig. 1) is aimed to detect a malignant tumor in the breast by achieving high classification accuracy and specificity. It includes three steps (1) Synthetically generated PET scan and its data normalization (2) Image extraction using RBM and (3) Detection of the malignant and benign tumors using AE.

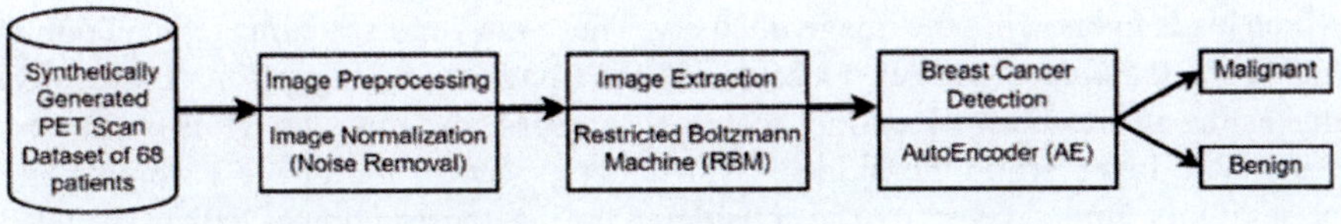

Fig. 1 RBM-AE Proposed Framework

Table 2 Stages of PET scan in Breast cancer detection

Stages	Description
Stage 1	Diagnose of tumor whether it is normal or benign or malignant
Stage 2	Begin treatment
Stage 3	Frist therapy treatment
Stage 4	Second therapy treatment
Stage 5	Final treatment (Before surgery)

4.1 Synthetically Generated PET Scan and Data Normalization

Data Collection
The image dataset of breast cancer using PET scan was generated synthetically by collecting images of 60 patients through the 3 stages. The modalities included are MRI, PET, and CT scans. These PET images are drawn from MRI where a predefined device is used to view the flat position of patients. In the beginning, the MRI images were collected quantitatively, and then longitudinally collected images were viewed in PET. To diagnose breast cancer, longitudinal-based quantitative medical imaging data was gathered. These data are a form of FDG-PET/CT is predefined [24].

Data Preprocessing
The collected PET scan breast images are preprocessed to remove their distortion and compressed for increasing the training speed whereas the PET images are similar in their feature. The DICOM of PET is verified to deal with bone lesions of breast image. The dynamic frames are created by fixing time based on image size. The breast images are normalized over spatial orientation by removing the noisy data and unwanted information through the scanner-specific filter function which gives a uniform-resolution image [25] (Table 2).

4.2 Image Extraction Using RBM

The PET scan of breast cancer with spatial resolution size varies from 4 to 5 mm. The image dimensions are determined by the image pixel with its height and width which leads to knowing the image intensity. These raw gray scale image pixel points define the features of images. Considering the images in a range of matrix formats makes the neural network extract features. In RBM, the input image is carried by the visible layer, while the hidden layer is comprised of the image's weights and biases. Only limited layers can be considered for the construction of RBM. The need for hidden layers cannot be added while constructing the model, so it is called a restricted boltzmann machine. The presence of RBM avoids image overfitting and makes the generated classification model present with better accuracy [26]. Making small vector images and avoiding overfitting makes the performance of detection and image visualization processed in faster way. It provides exact inter-dependent features compared to original images. It makes the classification model robust and feasible to train.

4.3 Detection of Malignant and Benign Tumors Using Autoencoder

Autoencoder receives the input from RBM which consist of extracted image data of the patient's PET image. It is an unsupervised neural network, where the encoder collects the input image and compresses it. The high-resolution and dimensional data are compressed to low-dimensional data in the encoder. Then, it gets decompressed in a decoder to obtain the image detection of the tumor then it will classify whether it is malignant or benign (Fig. 2).

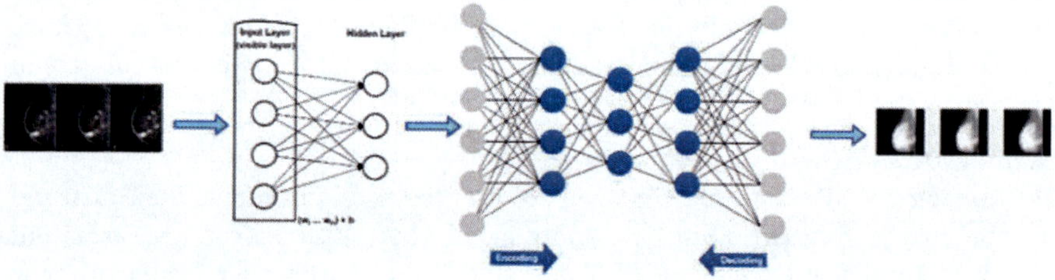

Fig. 2 Working of RBM-AE model

```
Algorithm of RBM-AE
Input: 60 patients PET scan with 10.324GB
Output: detecting tumor and classifying tumor as benign
or malignant
Start RBM image extraction
  Constructing one visible layer V and hidden layer H
  Two neurons for V & three neurons for H
  Weights W and Bias B for H
  Convert image into pixel-defined frames
  Frames have midpoint pixel for grayscale image
  Accumulate energy function
   Compressed PET image
End RBM
Start
  Input: the extracted image of PET
  Autoencoder with built-in encoder & decoder
  Input layer input, Hidden Layer, output layer O
  Encoder:  hidden=Ø ( 〖wgt〗_((m))*input+bias)
  Decoder: 〖rebuild〗_((input) )=Ø ( 〖RE(wgt〗_((m)
))*hidden+RE(bias))
  Separating benign & malignant
End
```

5 Experimentation and Results

5.1 Experimentation Setup

The proposed RBM-AE model was implemented in Python 3 with the Tensorflow framework and also evaluated using the PyTorch library. The experiments are conducted in PET scans of breast images and are plotted using the matplotlib library.

Python 3 – It's a programming language of version 3. It is discordant compared to version 2. Migration is simple in python3. Free in supporting AL, ML, and data science and data analytics.

Tensorflow framework – TensorFlow is a Python-friendly open-source toolkit for numerical calculation that speeds up and makes machine learning and the development of neural networks simple. With TensorFlow, programmers may construct dataflow graphs or structures that depict how data flows across a graph or a set of processing nodes. Each node in the graph represents a mathematical procedure, and each edge or link between nodes represents a tensor or multidimensional data array.

PyTorch library - To develop and train deep learning models based on neural networks, you may utilize the free machine learning package PyTorch. It was developed primarily by Facebook's AI research division. Both Python and C++ are compatible with PyTorch. The Python interface is more polished. In research labs, Pytorch, which is supported by giants like Facebook, Microsoft, SalesForce, and Uber, is extremely popular. Pytorch is not yet picking up quickly on many production servers, which are controlled by frameworks like TensorFlow (backed by Google).

5.2 *Results and Discussion*

The proposed RBM-AE model has taken RBM for image extraction and AE for breast cancer detection. This RBM-AE model is evaluated using a PET scan of 60 cancer-affected patients. For image extraction, the RBM was constructed with one input layer with three neurons and one hidden layer with weights and biases. The value 0 and 1 are taken as input variables and the variables are defined in probability-based parameters. The middle point pixel of the image depends on the nearby pixel value. The vision of the input image of RBM is occur in the middle column of the image while expanding. The expanding column in the image determines the black pixel and white pixel by RBM. Based on binary values, the RBM is induced for feature extraction in images. The extracted images of the PET dataset were forwarded to the autoencoder network for breast cancer by classifying the present tumor in the breast as benign or malignant. Complex pictures, high-resolution images, and the ranking of image alteration may all be managed using AE. This model achieves high precision, accuracy, and detection rate in performance evaluation to show a perceptual loss. This autoencoder has both an encoder and a decoder in its network. The performance evaluation was carried out to successfully determine the values of precision, recall, accuracy, and f-score.

$$\text{Accuracy} = ((\text{TP} + \text{TN}))/((\text{TP} + \text{FP} + \text{FN} + \text{TN})) \tag{7}$$

$$\text{Precision} = ((\text{TP}))/((\text{TP} + \text{FP})) \tag{8}$$

$$\text{Recall} = \text{TP}/((\text{TP} + \text{FN})) \tag{9}$$

$$\text{Fscore} = \text{TP}/(\text{TP} + 1/2(\text{FP} + \text{FN})) \tag{10}$$

Based on the performance evaluation (Table 3), the hyperparameters of RBM-AE are compared with the existing work of CNN, RBM, and AE [29,301]. Usually, CNN performs best for image applications. However, it failed to address the issues in some of the advanced medical images. So, the performance of CNN in PET images is not satisfied. Similarly, RBM and AE fail to achieve good accuracy when executed

Table 3 Performance Evaluation of RBM-AE compared with existing methods

Deep learning Techniques	Accuracy	Precision	Recall	F-score
CNN	0.88	0.92	0.87	0.91
RBM	0.90	0.94	0.91	0.93
AE	0.92	0.95	0.94	0.95
RBM-AE	0.98	0.99	0.98	0.98

separately. So, this paper presented the concept of combining RBM and AE as neural network structures to form RBM-AE for medical imaging data. Especially, PET of the breast provides good results.

6 Conclusion

Currently, enormous research is practicing in medical imaging datasets, especially in cancer. However, only countable and defined research is carried out in breast cancer imaging datasets such as MM, MRI, DBT, etc. and there is a lack of datasets mainly in PET images. So, this paper concentrated on the synthetic generation of the PET scan dataset and implemented a Restricted Boltzmann Machine (RBM) for image extraction and Autoencoder (AE) for the detection and classification of the tumor as benign or malignant (RBM-AE). The performance evaluation of RBM-AE achieves high accuracy, precision, recall, and f-score compared with some existing methods of breast cancer detection. This study is beneficial to deep learning, medical imaging, and cancer researchers.

References

1. Hanahan D (2022) Hallmarks of cancer: new dimensions. Cancer Discov 12(1)
2. Chen Y, Lin H, Miao L, He J (2022) Role of N7-methylguanosine (m7G) in cancer. Trends Cell Biol
3. Guida JL, Green P, Clague-DeHart J, Eshraghi L, Love S (2022) The Health of Women (HOW) study®: a web-based survey of breast cancer risk factors, diagnosis, and treatment. J Cancer Surviv
4. Devericks EN, Carson MS, McCullough LE, Coleman MF, Hursting SD (2022) The obesity-breast cancer link: a multidisciplinary perspective. Cancer Metastasis Rev, 1–19
5. Yu X, Zhou Q, Wang S, Zhang Y-D (2022) A systematic survey of deep learning in breast cancer. Int J Intell Syst 37(1):152–216
6. Azam MA, Khan KB, Gandomi AH (2022) A review on multimodal medical image fusion: compendious analysis of medical modalities, multimodal databases, fusion techniques and quality metrics. Comput Biol Med 144:105253
7. Burt JR, Torosdagli N, Khosravan NU (2018) Deep learning beyond cats and dogs: recent advances in diagnosing breast cancer with deep neural networks. Br J Radiol 91:20170545

8. Sharma S, Mehra R (2020) Conventional machine learning and deep learning approach for multi-classification of breast cancer histopathology images—a comparative insight. J Digit Imaging 33:632–654
9. Li W, Li J, Arnold C (2022) High resolution histopathology image generation and segmentation through adversarial training. Med Image Anal 75:102251
10. Serte S, Serener A, Al-Turjman F (2022) Deep learning in medical imaging: a brief review. Trans Emerging Telecommun Technol 33(10):e4080
11. Ramkumar MP, Mano Paul PD, Maram B, Ananth JP (2022) Deep maxout network for lung cancer detection using optimization algorithm in smart Internet of Things. Concurrency Comput Pract Experience 34(25):e7264
12. Moreau N, Rousseau C, Fourcade C (2020) Deep learning approaches for bone and bone lesion segmentation on 18FDG PET/CT imaging in the context of metastatic breast cancer. In: 42nd annual international conference of the IEEE engineering in medicine & biology society (EMBC). IEEE, pp 1532–1535
13. Papandrianos N, Anagnostis A, Feleki A (2020) A deep-learning approach for diagnosis of metastatic breast cancer in bones from whole-body scans. Appl Sci 10(3):997
14. Moreau N, Rousseau C, Fourcade C (2021) Automatic segmentation of metastatic breast cancer lesions on 18F-FDG PET/CT longitudinal acquisitions for treatment response assessment. Cancers 14(1):101
15. Shoshan Y, Zlotnick A, Ratner V (2021) Beyond non-maximum suppression—detecting lesions in digital breast Tomosynthesis volumes. In: Medical image computing and computer assisted intervention—MICCAI 2021. Springer, Cham, Switzerland, pp 772–781
16. Zhang Y-D, Satapathy SC, Guttery DS (2020) Improved breast cancer classification through combining graph convolutional network and convolutional neural network. Inf Process Manag 58:102439
17. Hossain B, Nishikawa RM, Lee J (2022) Developing breast lesion detection algorithms for digital breast tomosynthesis: leveraging false positive findings. Med Phys
18. Li H, Niu J, Li D, Zhang C (2020) Classification of breast mass in two-view mammograms via deep learning. IET Image Process 15:454–467
19. Bai J, Jin A, Jin A, Wang T, Yang C, Nabavi S (2022) Applying graph convolution neural network in digital breast tomosynthesis for cancer classification. In: Proceedings of the 13th ACM international conference on bioinformatics, computational biology and health informatics, Northbrook, IL, USA, 7–10 August 2022, pp 1–10
20. Lü X, Long L, Deng R, Meng R (2022) Image feature extraction based on fuzzy restricted Boltzmann machine. Measurement 204:112063
21. Zhang J, Ding S, Sun T, Guo L (2022) A Gaussian RBM with binary auxiliary units. Int J Mach Learn Cybern, 1–9
22. Fachechi A, Barra A, Agliari E, Alemanno F (2022) Outperforming RBM feature-extraction capabilities by "Dreaming" mechanism. IEEE Trans Neural Netw Learn Syst
23. Chen X, Ding M, Wang X (2022) Context autoencoder for self-supervised representation learning. arXiv preprint arXiv:2202.03026
24. Mujkic E, Philipsen MP, Ravn O (2022) Anomaly detection for agricultural vehicles using autoencoders. Sensors 22(10):3608
25. Covington MF, Koppula BR, Morton KA (2022) PET-CT in clinical adult oncology: II. primary thoracic and breast malignancies. Cancers 14(11):2689
26. Das KP, Chandra J (2022) A review on preprocessing techniques for noise reduction in PET-CT images for lung cancer. In: Congress on intelligent systems. Springer, Singapore, pp 455–475
27. Cui Z, Yang C, Wang S (2022) Insulator semantic segmentation in aerial images based on multiscale feature fusion. Complexity 2022
28. Prakaash AS, Sivakumar K, Surendiran B, Jagatheswari S, Kalaiarasi K (2022) Design and development of modified ensemble learning with weighted RBM features for enhanced multi-disease prediction model. New Gener Comput, 1–39
29. Ren T, Lin S, Huang P, Duong TQ (2022) Convolutional neural network of multiparametric MRI accurately detects axillary lymph node metastasis in breast cancer patients with pre neoadjuvant chemotherapy. Clin Breast Cancer 22(2):170–177

30. Gaona Y, Jiménez MJ, Rodríguez-Álvarez D, Malla C, Lakshminarayanan V (2022) Radiomics, deep learning, and breast cancer detection. Artif Intell Cancer Diagn Prognosis 2. Breast and bladder cancer
31. Mirbagheri S, Momeni M (2022) A hybrid deep learning methodology for breast cancer diagnosis using magnetic resonance images

Novel Approach for Network Anomaly Detection Using Autoencoder on CICIDS Dataset

Richa Singh, Nidhi Srivastava, and Ashwani Kumar

Abstract Through widespread internet protocols and network standards, networking services are accessible. Insecure communication and assaults on traffic networks must be taken into consideration moreover the special benefits of networking services. There are several methods for defending against network assaults. In order to combat network abnormalities, network anomaly detection systems are often employed. Systems for detecting network anomalies have drawn a lot of interest in terms of applying machine learning techniques to monitor network traffic intelligently. In this study, an efficient autoencoder-based model for anomaly detection in network traffic is presented. The autoencoder picks up a fundamental depiction of data and how to rebuild it with the least amount of error employed as an anomaly metric. A novel framework is proposed for malicious data detection in network congestion. Workflow for this proposed framework is mentioned in this paper. Also gives a standard formula for setting the threshold and compare these values with anomaly score i.e., error. Anomaly and normal sample are categorized on the basis of threshold value. As a well-used dataset in the literature, utilize the CIDDS-2017 dataset.

Keywords Anomaly detection · Autoencoder · CICIDS dataset

R. Singh (✉) · N. Srivastava
AIIT, Amity University, Lucknow Campus, Lucknow, U.P., India
e-mail: richa.singh@kiet.edu

N. Srivastava
e-mail: nsrivastava2@lko.amity.edu

A. Kumar
CSE, SRM Institute of Science and Technology, Ghaziabad, U.P., India
e-mail: ashwanik@srmist.edu.in

N. Hasteer et al. (eds.), *Decision Intelligence Solutions*, Lecture Notes in Electrical Engineering 1080,
https://doi.org/10.1007/978-981-99-5994-5_19

1 Introduction

The most significant computer concepts in recent years have been the idea of anomaly in network traffic. Access to processing and computing resources has been transformed by cloud computing [1]. In network, people and businesses may instantly uses a controlled, scalable network of IT resources including servers, storage, and apps [2]. Users can get platforms and applications as a service. It has made significant strides in recent years and now serves as a platform for a variety of users' technologies. Data and apps may be stored in the network and accessible from anywhere at any time using any internet-enabled device [3]. The security implications in network are among the biggest difficulties [4]. Numerous cyberattacks are conducted on the network for a variety of reasons as a result of the growing anomaly [5]. Attackers frequently target open and dispersed architecture while conducting cyber-attacks. Communication and information security continue to be major concerns since services are delivered through the internet. However, security breaches and catastrophes can have a significant impact on network users [6]. Users that depend on network services for day-to-day operations may have serious issues when they go offline or when software or websites fail. For security monitoring in networks, several strategies and techniques have been presented. Machine learning methods are frequently employed in anomaly detection to find data that differs significantly from regular data and may be suspicious [7, 8]. In the literature, autoencoders for anomaly detection have already been discussed [9, 10]. An encoder and a decoder in a neural network called an autoencoder that are used to study reconstructed output that are similar to the input sample. The reconstruction error of every element or feature of the input vector is totalled in the current approaches. In fact, the reconstruction error vector is produced rather than employing the summation of the reconstruction error in a single number in Fig. 1.

The performance of anomaly detection and classification is enhanced by this method. For this CICIDS-2017 dataset are used to assess the suggested technique [11, 12].

The following is a list of this paper's contributions, in brief:

- A workflow of a novel approach is proposed that provide a different way by utilising autoencoder for network anomaly detection on dataset.
- A standard formula is used for setting the threshold and generate a vector reconstruction error for each feature, unlike the previous techniques. This results in a threshold vector in anomaly detection that boosts performance in comparison to current techniques.
- The CIDDS-2017 dataset which is available on public domain is used and compare the accuracy of the proposed approaches with existing random cut forest that is also one of the unsupervised technique.

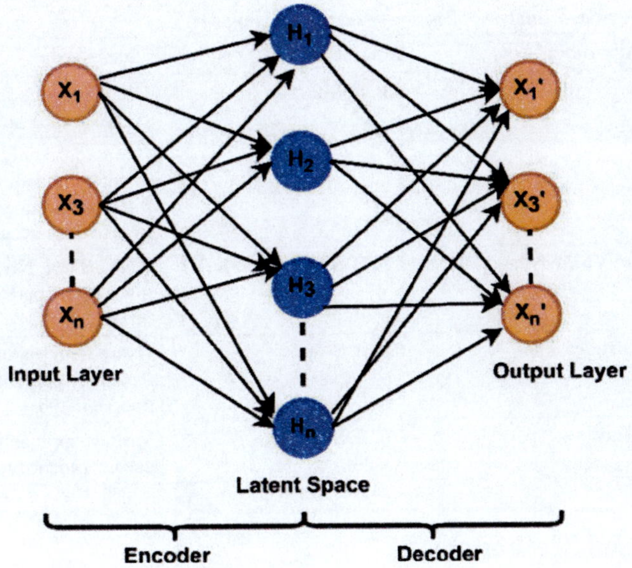

Fig. 1 Basic Diagram of Autoencoder [13]

2 Literature Survey

To categories the many assaults that may be made against cloud and internet of things (IoT), utilize Deep network. Utilizing cross-validation and in order to assess their suggested strategy, they performed several cross-validations using the CIDDS-001 dataset. This paper [14] employ various search approach to optimizing parameters in order to identify the best DNN parameter.

NIDS introduces anomaly in [15] also uses long short-term memory [LSTM] for various-class intrusion detection. This paper demonstrate that the LSTM is more accurate than previously described current approaches. Attained accuracy of LSTM is 85.5%, although the sequence lengths and several values of parameter are not described well.

This paper [16] suggest utilizing several machines and deep learning classifiers to compare the two popular datasets i.e., CIDDS-001 and NSL-KDD. The evaluation findings demonstrate that on the NSLKDD dataset, KNN, SVM, NN, and DNN algorithms perform better, whereas DAE and Nave Bayes methods work well on the dataset.

This paper proposed [17], autoencoder as an anomaly-based intrusion detection system. They assess the effectiveness of their suggested strategy using the CIDDS-001 dataset (Table 1).

Table 1 Comparative analysis of anomaly detection using ML

Ref	ML Algorithm	Dataset	Challenges
[14]	DNN	CICIDS-001	Use different optimized parameter for better result
[15]	LSTM	NSL-KDD	Accuracy is only 85.5 which is not good enough. Use different parameters, can improve the accuracy
[16]	KNN, SVM, NN, and DNN	CICIDS-001, NSL-KDD	KNN, SVM, NN, and DNN gives better performance only on NSL-KDD
[17]	Autoencoder	CICIDS-001	Using Grid search method, Accuracy percentage not good enough
[18]	Nave Bayes	CICIDS-001	Various approaches can be used for identify the attacks

3 Proposed Methodology

To understand the proposed workflow, need to understand the basic computer model of autoencoder, a brief description about the dataset and after that a detail summary of the proposed workflow.

3.1 Computer Model

The input and output of an autoencoder (AE) are the same, making it a unique kind of feed-forward neural network. As seen in Fig. 2, the output layer has the same dimensions as the input layer. These networks train input vectors to be rebuilt as output vectors using an unsupervised methodology. Encoders and decoders are the basic components of AEs. Figure 2 shows the architecture of autoencoder. The hidden representation H shown in Eq. 1 is created by the encoder by transforming the input vector A.

$$\mathrm{H} = \sigma\,(\mathrm{W} * \mathrm{A} + \mathrm{b}) \tag{1}$$

where, W: weight matrix, b: bias vector, σ: activation function, X: input and H: hidden layer. A decoder is used to apply the transformation operation on the hidden representation H in order to recover the initial input space shown in Eq. 2.

$$\mathrm{A}' = \sigma\,(\mathrm{W} * \mathrm{h} + \mathrm{bh}) \tag{2}$$

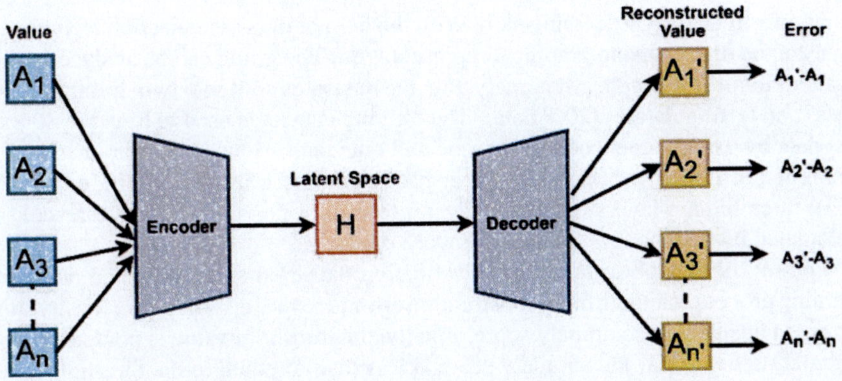

Fig. 2 Block diagram of autoencoder with reconstructed value and error [18]

The difference represents the original and reconstructed output is known as reconstruction error in Eq. 3.

$$\mathrm{RE} = |\mathrm{A}' - \mathrm{A}| \tag{3}$$

3.2 *CICIDS-2017 Dataset*

The CICIDS2017 (Canadian Institute for Cybersecurity Intrusion Detection System) shows the actual data of real world. It is used to compare all the data that which one is malicious and which is non-malicious. It gives the information about network data which lead the congestion or traffic in the network. The out of network traffic data is measured by CICFlowMeter, having information like timestamp of the data send and received, source IP address and destination IP addresses, its ports, different protocols, and various types of attacks. The abstract behaviour of human interactions using our proposed B-Profile system, which also creates naturalistic benign background traffic. This dataset has 72 columns and last column contains the 11 values. On the basis of the target column, anomaly is identified by using various machine learning techniques.

3.3 *Proposed Workflow*

In the proposed framework, five-layer autoencoder is used in which one layer is used as input layer, second layer is used as encoder, third layer is hidden layer, fourth layer is for decoder and last layer is used as output layer. The workflow for the framework to detect anomaly is shown in Fig. 3. CICIDS 2017 dataset is used and preform the pre-processing step. With the help of data visualization techniques, analyze the dataset using graphical representation. The graph can be analyzed using scatter plot or line graph. After analyzing, the dataset is split into two: training (80% data) and testing dataset (20% data). The training dataset is used to train the 80% of the data by using autoencoder technique and tune the various hyper-parameters that are required during process. The hyper-parameters such as epoch (30), batch size (64), three hidden layers (encoding/2) and learning rate is 1e−7. The threshold is calculated by threshold = mean + standard deviation.

During the model evaluation, the validation dataset that are obtained using the training process calculate the reconstruction error for every epoch. The reconstruction error is considered as anomaly score. Finally, the threshold value is compared with anomaly score and if the anomaly score is less than or equal to the threshold value then it is considered as normal and if anomaly score is greater than threshold value then it is considered as anomaly. If the anomaly is successfully detected, then the model is deployed and predict the outcome.

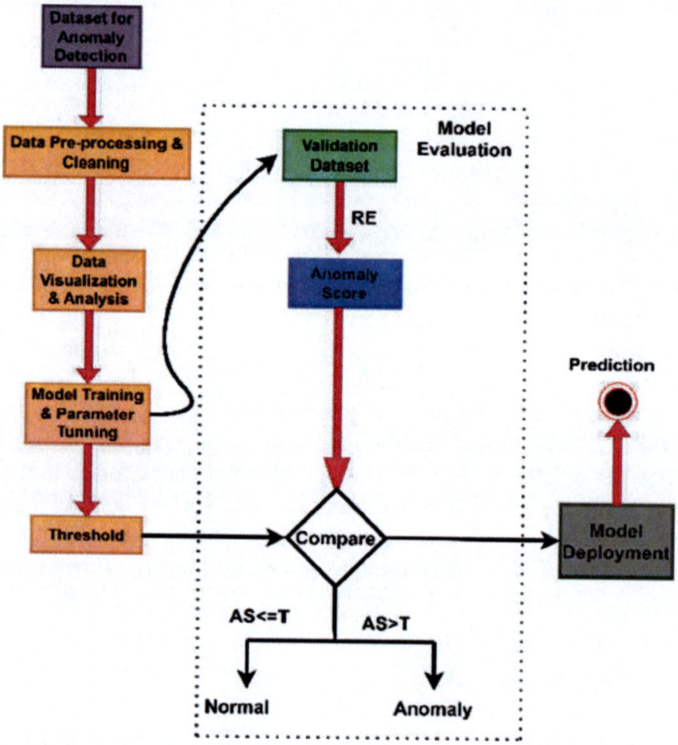

Fig. 3 Proposed framework for anomaly detection using five-layer autoencoder

4 Performance Evaluation

During the compilation of autoencoder, there are various metrics available to calculate the accuracy. The loss is calculated using mean squared error and ADAM optimizer is used. ADAM optimizer is an algorithm used for weight adjustment or change the learning rate of every network independently during training process. If there is a lot of data or factors for calculation then it makes the dataset more complex then in that case, the strategy is incredibly effective and uses little memory. Mathematically, the ADAM optimizer is represented with the below formula represented with Eq. 4 and 5.

$$m_t = \beta_1 m_{t-1} + (1 - \beta_1)[\delta L/\delta w_t] \tag{4}$$

$$v_t = \beta_2 v_{t-1} + (1 - \beta_2)\,[\delta L/\delta w_t]^2 \tag{5}$$

In the above equation epsilon is the small positive constant used to prevent divide by zero condition, β_1 and β_2 is the rate of average gradients and α is the learning rate parameter. By calculating "bias-corrected" mt and vt, this optimizer resolves the issue. After training the dataset using autoencoder, the validation accuracy is about 88.9%.

Now the plot for training and test loss is represent in the Fig. 4 shown above. This graph is a plot between the model loss and total epoch before early stopping. The blue line shows the training data and the orange line represents the test data. Now detect the anomalies on test data where the reconstruction error is high. This paper shows how the suggested model performs when applied to the CIDDS-2017 dataset. The performance of the suggested model is measured using the following factors: accuracy, detection rate and precision.

5 Result and Discussion

The data has 72 columns and unsupervised technique is applied on every column to find the anomaly value. The last column has approximately 11 values or categories. So there are one category and other are considered as the subcategory and then try to find out the anomaly in that particular dataset. To assess how well the anomaly detection works, the accuracy of this model is 86.3%. The confusion metrics gives the accuracy percentage of 86.39 along with the value of recall and precision which is shown in Fig. 5. The dataset which is used in this research paper is quite unbalanced, thus while accuracy is excellent but recall and precision are low. Adding additional pertinent features, changing the autoencoder's architecture, altering the hyperparameters, or using a different method are all ways to further enhance precision and recall. The result is compared with random cut forest technique which is also the unsupervised

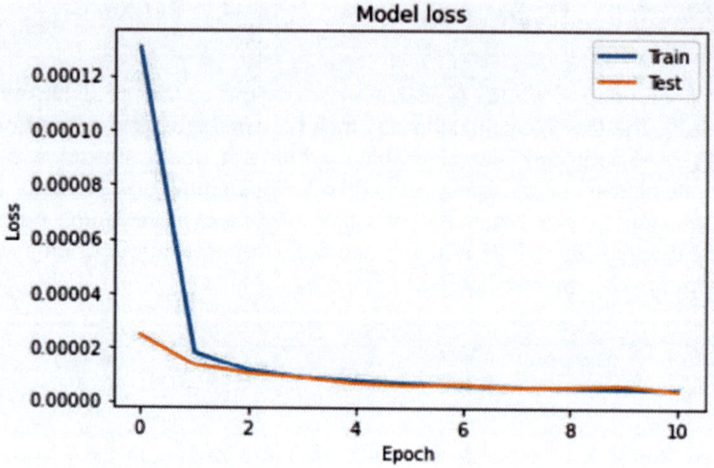

Fig. 4 Plot of model loss vs epoch

technique the accuracy is 81% but by using autoencoder the accuracy percentage of the proposed model is 86%.

Fig. 5 Confusion metrics using autoencoder technique

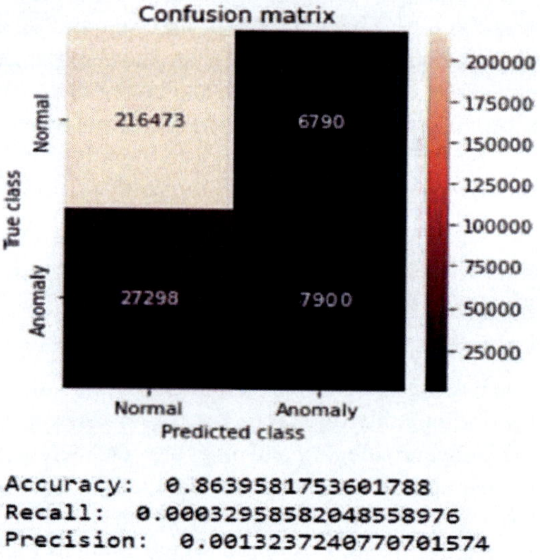

6 Conclusion and Future Scope

This research examines the use of autoencoders for network anomaly detection. This method differs from earlier autoencoder-based method for anomaly detection strategies in terms of the mechanism used to calculate reconstruction error, the method used to determine the threshold the hyper-parameters, and the RE as compared to the threshold method. It was demonstrated that the suggested anomaly detection performed very well than the current ones on the CICIDS-2017 dataset by assessing the evaluation findings and differ from some of the more recent proposed methods/techniques. To improve the performance various ensemble techniques are used in future. The accuracy percentage can be increased by using PCA and other unsupervised machine learning techniques.

References

1. Singh R, Singh A (2018) Challenges of various load balancing algorithms in distributed environment. IJITEE, 9–13
2. Singh R, Singh A, Bhattacharya P (2018) Challenges of load balancing techniques in grid environment. IJITEE, 1–5
3. Singh R, Singh A, Bhattacharya P (2022) A machine learning approach for anomaly detection to secure smart grid systems. In: Research anthology on smart grid and microgrid development. IGI Global, pp 911–923
4. Alam T (2020) Cloud computing and its role in the information technology. IAIC Trans Sustain Digital Innov (ITSDI) 1(2):108–115
5. Shamim N, Asim M, Baker T, Ismail Awad A (2023) Efficient approach for anomaly detection in IoT using system calls. Sensors 23(2), 652 (2023)
6. Kilincer IF, Ertam F, Sengur A, Tan R-S, Rajendra Acharya U (2023) Automated detection of cybersecurity attacks in healthcare systems with recursive feature elimination and multilayer perceptron optimization 1:30–41
7. Lei X, Xia Ye, Wang Ao, Jian X, Zhong H, Sun L (2023) Mutual information based anomaly detection of monitoring data with attention mechanism and residual learning. Mech Syst Signal Process 182:109607
8. Chaoyue D, Shiliang S, Jing Z (2023) MST-GAT: a multimodal spatial–temporal graph attention network for time series anomaly detection. Inf Fusion 89:527–536
9. Singh R (2022) Performance optimization of autoencoder neural network based model for anomaly detection in network traffic. In: 2nd international conference on advance computing and innovative technologies in engineering (ICACITE). IEEE, pp 598–602
10. Singh R, Srivastava N, Kumar A (2021) Machine learning techniques for anomaly detection in network traffic. In: 2021 sixth international conference on image information processing (ICIIP), vol 6. IEEE, pp 261–266
11. Singh R, Srivastav G (2021) Novel framework for anomaly detection using machine learning technique on CIC-IDS2017 dataset. In: 2021 international conference on technological advancements and innovations (ICTAI). IEEE, pp 632–636
12. Torabi H, Mirtaheri SL, Greco S (2023) Practical autoencoder based anomaly detection by using vector reconstruction error, no 1, pp 1–13
13. Tama BA, Rhee K-H (2017) Attack classification analysis of iot network via deep learning approach. Res Briefs Inf Commun Technol Evol (ReBICTE) 3:1–9
14. Althubiti SA, Jones EM, Roy K (2018) LSTM for anomaly-based network intrusion detection. In: 28th international telecommunication networks and applications conference

15. Rashid A, Siddique MJ, Ahmed SM (2020) Machine and deep learning based comparative analysis using hybrid approaches for intrusion detection system. In: 2020 3rd international conference on advancements in computational sciences (ICACS). IEEE, pp 1–9
16. Fenanir S, Semchedine F, Harous S, Baadache A (2020) A semi-supervised deep auto-encoder based intrusion detection for IoT. Ingénierie des Systèmes d'Information 25(5)
17. Javaheri D, Gorgin S, Lee J-A, Masdari M (2023) Fuzzy logic-based DDoS attacks and network traffic anomaly detection methods: classification, overview, and future perspectives. Inf Sci

Predictive Analysis of Road Accidents Using Data Mining and Machine Learning

Abhinav Kumar, Harshit Arora, and Prasuk Jain

Abstract Today, one of the top concerns for governments is road safety. Given the significance of the subject, reducing the harm caused by traffic accidents has shifted its focus to finding the root causes of incidents on the road. On the other hand, it is impractical to analyze the components in this regard due to the exponential growth of data volume. Despite everything that has been done so far to increase road safety in India, many areas continue to be vulnerable to these flaws. Due to the diversity of these vulnerability-causing factors, a thorough analysis is required in order to significantly reduce the worrying numbers. The aim of this paper is to use data mining and machine learning approaches to help develop a model using LR and RF that not only smooths out the heterogeneity of the data by grouping like objects together to identify the regions of the nation that are most likely to experience accidents in relation to various accident-factors, but also assists in establishing the relationship between these parameters, factors, and casualties. The outcomes of data mining will assist businesses like transportation in exploring the police information system's accident data, finding trends, forecasting future behavior, and making wise judgments to reduce accidents.

Keywords Data Mining · Machine Learning · Road Accidents · Predictive analysis

1 Introduction

With the growing population of the world, the need for transportation also increases whether it be public or private means of transport, and to meet the growing demand for it, more vehicles are added to the traffic. But with the increase in the number of vehicles on the roads, the number of accident cases also increases. In India, accidents claim the lives of more than 150,000 people annually [4]. That works out to over 400

A. Kumar (✉) · H. Arora · P. Jain
Department of Information Technology, ASET Amity University Uttar Pradesh, Noida, India
e-mail: abhinav9091@gmail.com

N. Hasteer et al. (eds.), *Decision Intelligence Solutions*, Lecture Notes in Electrical Engineering 1080,
https://doi.org/10.1007/978-981-99-5994-5_20

fatalities daily. On average, more than 1 million additional vehicles are added to the traffic each year. According to research by the WHO, the three most vulnerable road user-motorcyclists, cyclists, and pedestrians-are responsible for half of all accident-related deaths worldwide, with two-wheeler drivers responsible for roughly 31% of fatalities [4]. It continues by stating that 59% of all fatalities worldwide are adults.

WHO collaborates with governmental and nongovernmental partners worldwide to increase awareness of how easily injuries from traffic accidents may be avoided and to promote healthy driving habits such as wearing a seatbelt, and a helmet, and being visible in traffic [4]. However, the driver is not always held accountable. Since several other elements that may have contributed to the accident include weather conditions, road conditions, vehicle defects, etc. To prevent accidents and improve road conditions, transport system managers must balance the demands for increased road safety with their limited resource availability. In fact, the fundamental objective is to minimize accidents, which can be accomplished by traffic planning, driver education, and enforcement. Huge amounts of data are saved to aid management in making strategic decisions. Without the aid of sophisticated data analysis tools, it is increasingly difficult to evaluate and extract usable information from databases as their size in terms of both time and space increases.

In this study, we used various supervised machine learning (ML) classification techniques such as Logistic Regression (LR) and Random Forests (RF) on variables like weather, accident type, causes, road features, and road conditions that are significant contributors to accidents. also provides a list of essential elements from the best model, offers recommendations for safe driving to reduce accidents, and looks at how the models are presented. True positives, false positives, accuracy, precision, recall, rate of error, and the number of accurately categorized models are the exhibition measurements.

Finding the major causes of a road or traffic accidents is one of the main goals of our accident data analysis. But the heterogeneous structure of the data on road accidents, makes the analytical work challenging.

Due to the importance of road safety in the transportation system as well as in our daily lives, it is considered as everyone's major concern. The main aim of this research is to study and implement predictive analysis techniques or statistical analysis techniques, as well as some approaches related to data mining, including linear regression and random forest for road accident analysis.

With the use of data mining techniques, organizations like transportation sector can examine accident data logged by the police information system, which will help in finding patterns, predicting future behavior in similar scenario, and help them to take smart decisions to reduce road accidents.

2 Related Work

Tirumuru Ketha et al. [8] research effort creates models based on the factors that cause accidents, such as weather conditions, cause of accidents, road characteristics and conditions, and accident type. Likewise, they selected a number of significant elements from the best model in order to build a model for describing the cause of accidents. They used different supervised machine learning techniques, such as Logistic Regression, KNN, NB, DT, and Random Forest, to analyze accident data in order to understand how each factor affects the variables involved in accidents happen.

Ayushi Jain et al. [2] used data mining technique to analyse the road accidents in India. The main goal of their study is to use data mining technique to help develop a model that not only helps balance out the diversity of the data by categorising like objects together to identify the regions of the nation that are most likely to experience accidents in relation to various accident parameters and factors, but also assists in establishing the relationship among these factors and casualties. They used K-MEANS Clustering technique and examined how cluster analysis enables us to identify the Indian states and regions that are most prone to accidents.

Using Ada boost, LR, NB, and RF, Rabia Emhamed et al. [3] developed a system for categorising the severity of an accident using machine learning techniques. This study demonstrates that Random Forest performs better, with 75% accuracy. This project's shortcoming is that it only takes into account a small number of features, even if other parameters may also have considerable impact on the seriousness of the accident.

With the help of Logistic and Linear Regression, Decision Tree, NB, KNN, SVM, Random Forest, gradient boost and Naive Bayes, Ms. E. Suganya et al. [5] studied the road accidents and determined the year when there is an increase in the accidents in different regions of India. Based on experimental findings, KNN is superior to other algorithms with accuracy rate of 93.7%.

Using machine learning approaches that take into account both external and internal parameters, such as weather conditions, Saumya Suman et al. [9]. determined the severity of the accident. Their research aims to assist authorities in making decisions on weather and traffic patterns in accident zones.

Liling Li et al. [6], utilised algorithms including Naive Bayes, Apriori, and K-Means on Fatality Analysis Reporting System (FARS) data in order to provide safe driving recommendations. Investigations on the effects of the weather, surface, light, driving under influence of alcohol, and fatality rate revealed a correlation. They have recommended the areas where driving is risky using K Means algorithm.

Through a review of the literature, it was made obvious that road accident analysis is essential to minimize traffic accidents, which encouraged us to take on this project. Our project's goal is to overcome all the drawbacks of the prior road accident analysis system.

3 Methodology

3.1 Defining Problem Statement

3.2 Data Collection

We gathered the datasets of road accidents from the official government sites. Which were https://morth.nic.in/ and https://data.gov.in/ all the data that is published here is verified and authenticated and is published by a government official.

3.3 Data Cleaning and Preprocessing

Splitting of the datasets: Splitting of data plays an important role because these are the categories on which our analysis will be performed. For our classification, we classified the data on various factors like:

- Number of accidents happening per state from 2006 to 2019.
- Accidents happen according to the time of the day.
- Offenders/victims who died according to gender.
- Whether the driver had a license or not.
- Type of vehicles involved in accidents, etc.

3.4 Data Analysis (on Various Identified Parameters)

Analysis of road accident data is very important as it gives us an insight into the factors that are involved and to what extent they contribute to the total number of cases that happen across the country. It helps authorities such as traffic police and medical professionals to stay prepared to counter the predicted number of accidents by deploying more traffic police force in the accident-prone area and medical staff can be put on alert during that time.

We had our government verified dataset of road accidents in India, which told us about the various factors that contribute to the road accidents such as Over Speeding, Driving under Influence of Drugs/Alcohol, Defect in the Mechanical Condition of Motor Vehicle, Lack of Road Infrastructure, Physical Fatigue of Drivers, Dangerous or Careless Driving/ Over-taking, time at which the accidents took place, etc. doing analysis on all these factors at the same time was not possible due to heterogeneity of data so we split the dataset with some common factors and then used it for analysis and prediction [10].

With the help of the collected dataset, the machine learning (ML) technique can be used to predict the severity of the accident and analyse parameters such as accidents occurring per year, accidents occurring in each state, accidents occurring according to different timeframe of the day and in months, the age and gender of the individuals involved, types of vehicles involved, accidents according to gender, and the riskiest period to drive. This will work well to enhance safety precautions and minimize road accidents.

We also used MatLab functions, basically, it helps us in analyzing and visualizing the results in the form of Pie chart, histograms, line graphs, bar graphs, etc. pie charts tell us the percentage of contribution of each component in a whole.

3.5 *Building Prediction Model*

3.6 *Model Validation, Accuracy Test and Monitoring*

4 Building Prediction Model

Predictive modelling uses mathematical and statistical processes to make predictions about what will happen in the future by examining the patterns in a given set of inputs. Predictive modelling is a type of data mining/machine learning approach that analyses recent as well as historic data with the purpose of detecting patterns or trends and then using those findings to anticipate future events. It tends to find or predict future behavior of the dataset (Fig. 1).

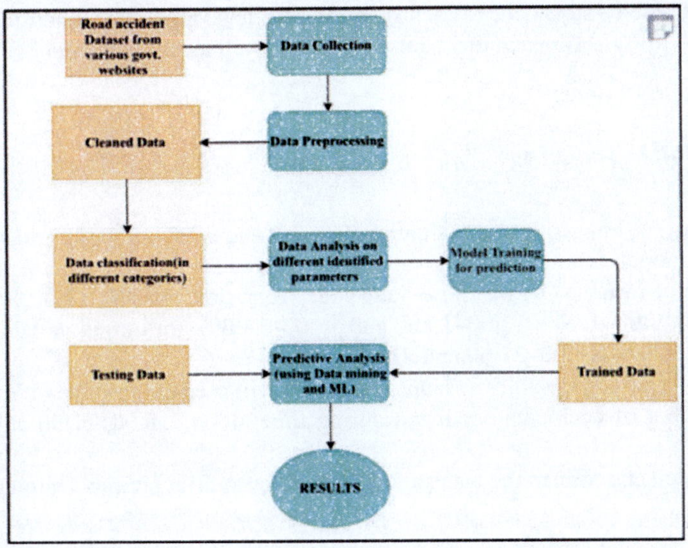

Fig. 1 Working Framework

Fig. 2 Distribution according to time slots

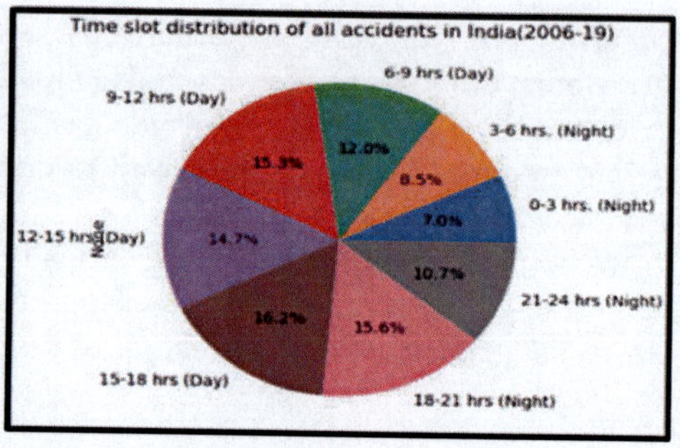

Predictive analysis makes some use of data from current datasets modifications to discover new trends and patterns. Then, utilising these patterns and trends, future actions, results, and trends are projected. For our prediction model we have used two Machine Learning techniques:

i. Linear Regression
ii. Random Forest

4.1 Parameter Identification

In this we first identified the factor on which prediction will be done (according to time slots of the day). 24 h format was divided into 8 slots od 3 h each as shown below.

We assigned a state code to each state. And we have government verified data for year 2006 to 2019 for each time slot and state as well as union territory (Fig. 2).

4.2 Model Training

In this part we have to give input parameters (X) and what output (Y) do we expect from the model. Input and output variables are given for model training.

For input variable (X) we chose State code, Year, time slots (00–3:00, 3:00–6:00, 6:00–9:00, 9:00–12:00, 12:00–15:00 and 15:00–18:00), for output variable (Y) we chose time slots (18:00–21:00 and 21:00–24:00).

So, if we pass X variables as input, our model will predict the value of Y variables (i.e., number of accidents occurring during time slots - 18:00–21:00 and 21:00–24:00).

We tested and trained the dataset for X and Y by splitting it into Training Set and Test Set.

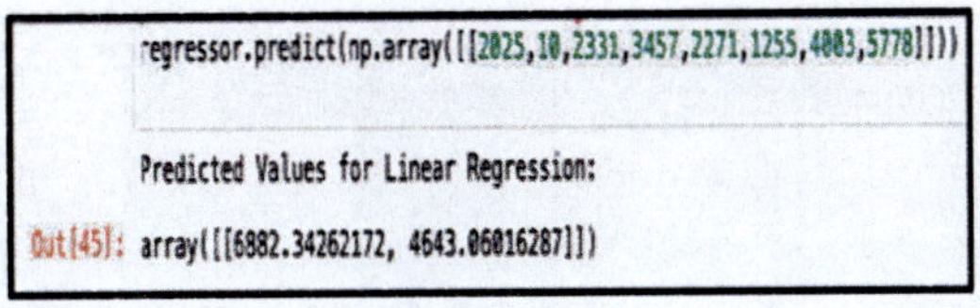

```
regressor.predict(np.array([[2025,10,2331,3457,2271,1255,4003,5778]]))

Predicted Values for Linear Regression:

Out[45]: array([[6882.34262172, 4643.06016287]])
```

Fig. 3 Prediction using Linear Regression

4.3 Prediction Using Linear Regression

In Linear Regression model we have passed values for X i.e., state code, year, time slots (00–3:00, 3:00–6:00, 6:00–9:00, 9:00–12:00, 12:00–15:00 and 15:00–18:00) to predict the values of Y (Fig. 3).

4.4 Prediction Using Random Forest

We imported 'RandomForestClassifier' and fitted it to the training set.

Then we have passed values for X to predict the values of Y (Figs. 4, 5, 6, 7, 8, 9, 10 and 11).

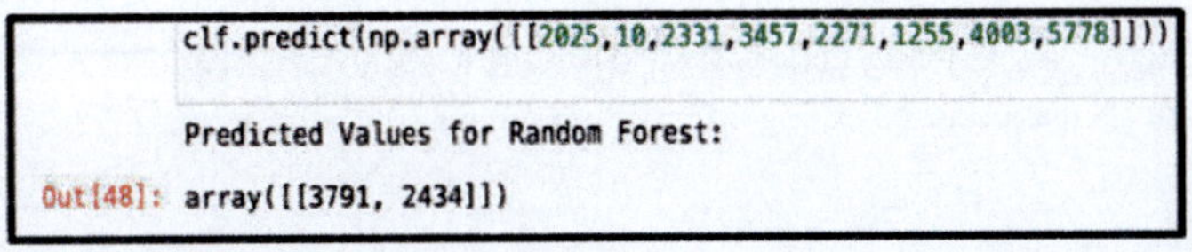

```
clf.predict(np.array([[2025,10,2331,3457,2271,1255,4003,5778]]))

Predicted Values for Random Forest:

Out[48]: array([[3791, 2434]])
```

Fig. 4 Prediction using Random Forest

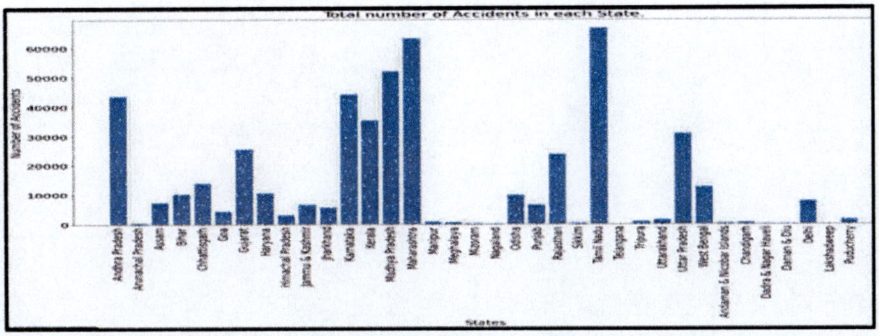

Fig. 5 Number of accidents occurred in each state

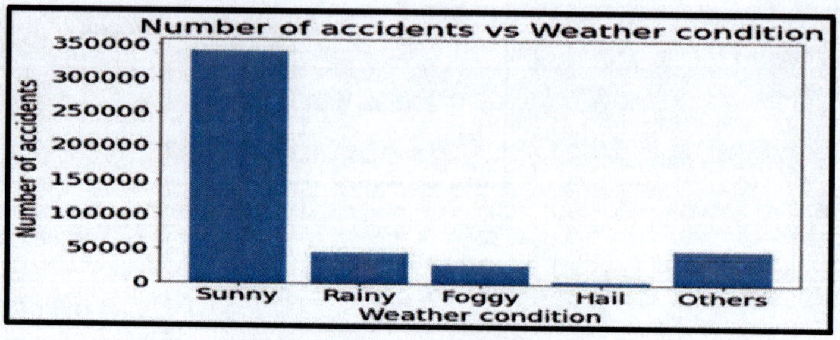

Fig. 6 Weather conditions under which maximum accidents occur

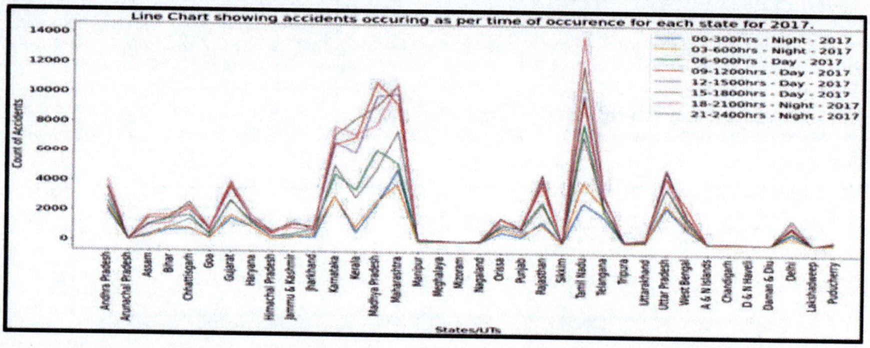

Fig. 7 Accidents that occurred during different time frames during the year 2017

Fig. 8 Total People who died

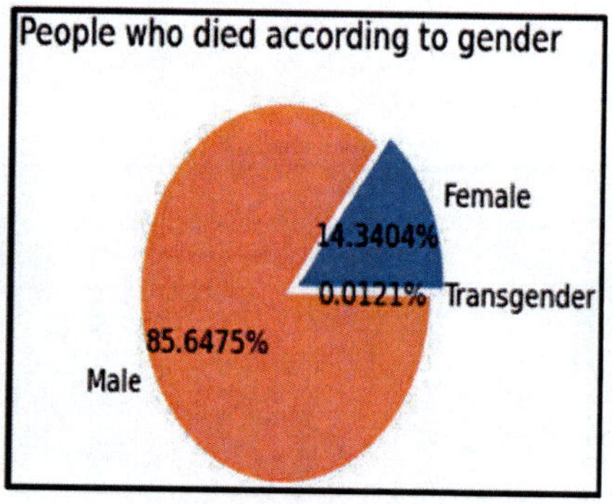

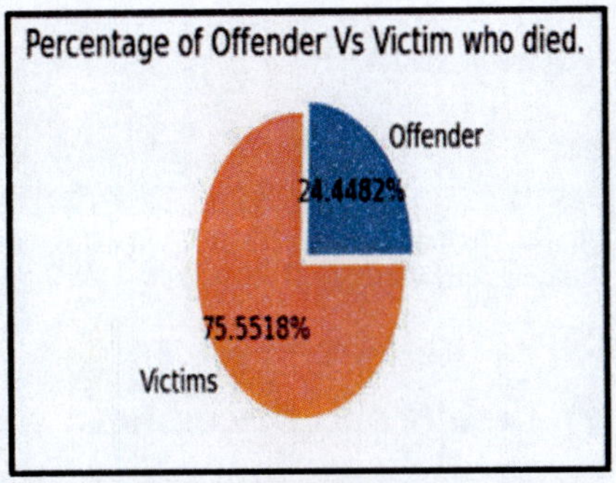

Fig. 9 Victims v/s offenders who died

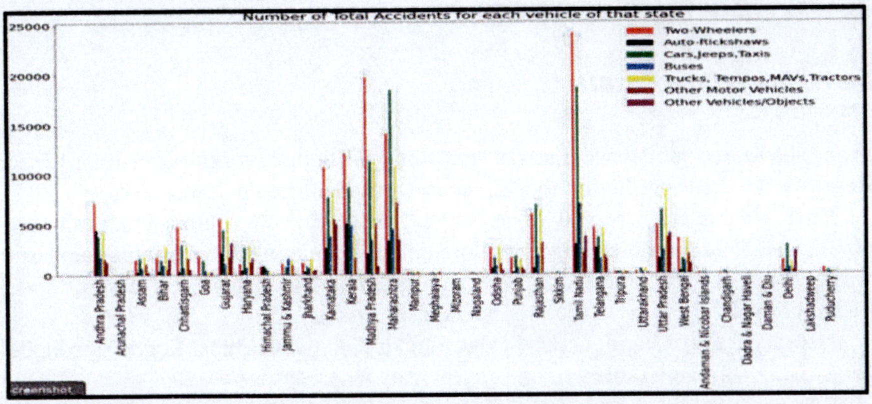

Fig.10 Histogram showing state wise comparison categorized on the type of vehicle responsible for the accident

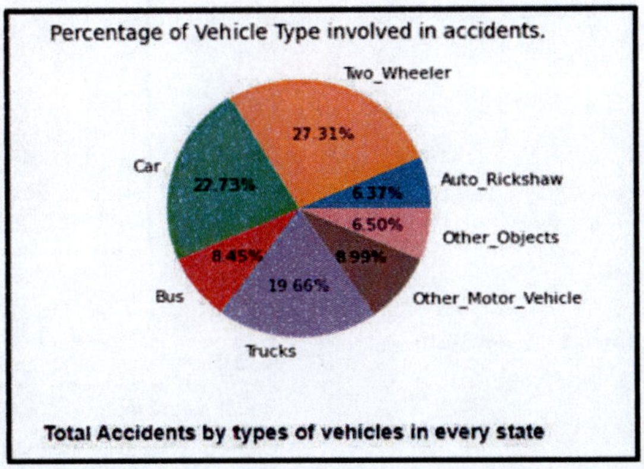

Fig. 11 Pie chart showing a comparison between vehicle types involved in accidents

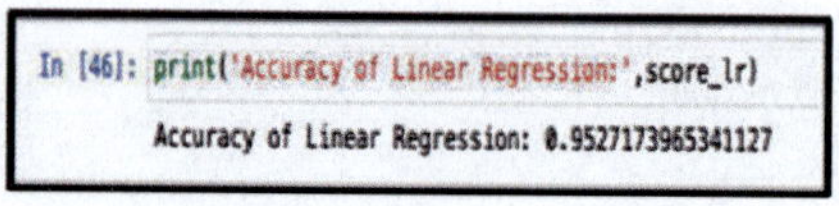

Fig. 12 Accuracy of Linear Regression Model

Fig. 13 Accuracy of Random Forest model

5 Results and Discussions

5.1 Analytical Observations

5.2 Prediction Result

From above two model we can say that Random Forest works better than Linear Regression for our prediction model, dataset and identified parameter.

When we passed historical values of year 2007/2011 of Andhra Pradesh or any other state, Random Forest algorithm predicted the output values almost same as it was given in original dataset, but the values predicted by Linear Regression Model was slightly different from the original dataset values.

Also, accuracy of Random Forest algorithm is better than Linear Regression model (Figs. 12, 13, 14 and Table 1).

Fig. 14 Accuracy of Random Forest Model and Linear Regression Model

Table 1 Accuracy Table

Model	Accuracy
Linear Regression	95.27%
Random Forest	96.68%

6 Conclusion

For our paper, we used data mining techniques and ml approaches which proved to be very efficient and helpful for us to analyze the data that we had with us, with the analysis that we conducted we were able to conduct a detailed study of all the factors that are involved in the road accidents that happen throughout India in each state and union territory. We were able to find out what factors contributed the most and what factors contributed the least and, we got to know about the states that have the highest no of road accidents throughout the year. We also implemented prediction models such as linear regression and random forest on our data set to predict the number of accidents that are going to take place between the time duration of 6:00 PM to 12:00 AM.

The critical insight of the analysis that we did and the values that we predicted was that it could be used by the government to make policies and improve road infrastructure and also by authorities such as traffic police and medical professionals. More police force can be deployed in the areas which are prone to accidents, barricading can be done, medical professionals can be kept on high alert so that they can reach the accident spot faster and more beds in the hospital can also be arranged for the purpose of saving lives.

References

1. Sridevi N, Keerthana MV (2020) Road accident analysis using machine learning. Int J Res Eng Sci Manag 3(5): ISSN 2581–5792
2. Jain A, Ahuja G, Mehrotra D (2016) Data mining approach to analyse the road accidents in India In: 2016 5th international conference on reliability, Infocom technologies and optimization (Trends and Future Directions) (ICRITO)
3. AlMamlook RE, Kwayu KM, Alkasisbeh MR, Frefer AA (2019). Comparison of machine learning algorithms for predicting traffic accident severity. In: 2019 IEEE jordan international joint conference on electrical engineering and information technology (JEEIT), IEEE
4. Ketha T (2019). Analysis of road accidents to indentify major causes and influencing factors of accidents – a machine learning approach. Int J Adv Trends Comput Sci Eng 8:3492–3497. https://doi.org/10.30534/ijatcse/2019/127862019
5. Suganya E, Vijaya Rani S (2017) Analysis of road accidents in India using data mining classification algorithms. In: 2017 IEEE international conference on inventive computing and informatics (ICICI)
6. Li L, Shrestha S, Hu G (2017) Analysis of road traffic fatal accidents using data mining techniques. In: 2017 IEEE 15th international conference on software engineering research management and applications (SERA)
7. Bačkalić S, Matović B, Jovanović D (2014) Identification of hotspots road locations of traffic accidents with pedestrian in urban areas. International co-operation on theories and concept in traffic safety
8. Kumar S, Toshniwal D (2016) A data mining approach to characterize road accident locations. J ModTransp 24(1):62–72
9. Sonal S, Suman S (2018) A framework for analysis of road accidents. In: Proceedings of 2018 international conference on emerging trends and innovations in engineering and technological research (ICETIETR), pp 1–5. IEEE. https://doi.org/10.1109/ICETIETR.2018.8529088

10. Open government data (OGD) platform India [Online]. Available 2016: https://data.gov.in/catalogs/sector/Transport-9383
11. Sharma B, Katiyar VK, Kumar K (2016) Traffic accident prediction model using support vector machines with Gaussian kernel. In: Proceedings of fifth international conference on soft computing for problem solving, Springer, Berlin, pp 1–10
12. Shruti S et al. (2021) Machine learning approach towards road accident analysis in India. In: Proceedings of second international conference on smart energy and communication

Digital Shopping Cart with Automatic Billing System

Puja Cholke, Phalesh D. Kolpe, Sanket D. Patil, Abhishek M. Pote, and Siddhesh S. Patil

Abstract Now a day's most of the people face the issue of overcrowding and long queues in supermarkets. Due to these long queues people have to spend a lot of time waiting in queues which is frustrating. In this post COVID era everyone has become health conscious and overcrowding may lead to spreads of disease and viral infection. This research builds up the solution to deal with the problem of overcrowding and long queues into the supermarkets. In this research work we proposed a model to solve this problem by creating a Digital shopping cart which works on the concept of image reorganization and used various IoT protocols. The customer just has to put the products which they want to buy in the cart and the product would be added to their bill. We have used various technologies like Image Processing by using OpenCV and TensorFlow library in Python and integrating them with microprocessor, cameras, and load cell to create the digital shopping cart.

Keywords Image Processing · IoT · Python · ESP32 camera module · Load cell · HX 711 · Machine learning

P. Cholke · P. D. Kolpe (✉) · S. D. Patil · A. M. Pote · S. S. Patil
Department of Multidisciplinary Engineering, Vishwakarma Institute of Technology, Pune, Maharashtra 411037, India
e-mail: phalesh.kolpe21@vit.edu

P. Cholke
e-mail: puja.cholke@vit.edu

S. D. Patil
e-mail: sanket.patil21@vit.edu

A. M. Pote
e-mail: abhishek.pote211@vit.edu

S. S. Patil
e-mail: siddhesh.patil21@vit.edu

N. Hasteer et al. (eds.), *Decision Intelligence Solutions*, Lecture Notes in Electrical Engineering 1080,
https://doi.org/10.1007/978-981-99-5994-5_21

1 Introduction

People are consistently innovating to ease our living from the very start of humankind. The main fundamental of innovation is problems we face in real life. Most of the supermarkets still use the traditional method of billing and checkout in which customers have to wait in long queues. This way of billing and checking out has become very inefficient and time consuming with the increase in the amount of people going to supermarkets. The fear of standing in long queues also stops people coming into purchase small or single items as it is very hectic process and if we can bring in those customers, we can increase the profit margin of the supermarket owners.

To tackle with this problem we came up with an idea of creating a "Digital shopping cart with automatic billing system" which helps the customer in easy bill creation and hassle-free checkout. The cart will save time of the customers and they won't have to stand in long queues. This would also attract more people to the supermarket. The proposed system uses ESP 32 cam module, load cell, NodeMcU for recognizing the object and add the product to the bill. Implementation of such system will lead to improve consumers experience and we can avoid the hectic process of billing using barcode scanning. Also we can save the valuable time of consumer which is wasted in the queue. The system can also help supermarket operators to focus on the growth of business rather than scanning every product and wasting time on billing counter.

2 Literature Review

In 2020, [1] Muhammad Aitf Sarwar et al. proposed a self-checkout system using a Linux based cloud server with the help of YOLOv2 which is a deep learning network. Recently in 2022 [2] Rakshit Shetty et al. built-up a digital shopping cart, in which a model was set using machine learning and Raspberry pi which is highly beneficial to the buyer to avoid inconvenience. Similar such system was built by Meghana T. K. et al. [3] and Anitha R. et al. [7] with the concept of RFID and the microcontroller board used in it was Arudino UNO. For high speed camera imaging CNN based recognition a method was used to detect accurate objects by performing data cleansing and data ensemble operations by Shigeaki Namiki et al. [4]. Takaaki Oka and Masakazu Morimoto [5] developed a system for recognition of overlapped object by using an RGB-D sensor. This technique helps to divide objects into segments and merge them by consulting partial recognition. According to a research done by Jerome Paul N. et al. [6] Artificial Neural Network classify the object accurately which was utilized by three steps namely object recognition, image processing, ANN processing and interpretation. Taking into consideration the research done by Ragesh N et al. [9] we can not place RFID tags or bar code on fruits and vegetables so they implemented image processing using YOLOv2 for identifying the fruits and vegetables and Load Cell for detecting the weight of the fruits and identifying the price of it. Everything is then computed by using Arduino Microcontroller. The research done by Sakorn Mekruksavanich [10] suggest that we can use the Barcode scanning for shopping by using android application it uses load cell and ESP8266 NodeMcU for detecting the weight and for computing. The products is scanned using the application and the weight of the product is fetched from database for that particular barcode number then it is cross checked using the load cell it makes sure no product goes in without scanning. Similar such research was done by Viswanadha V et al. [16] in which they make use of RaspberryPi to implement a smart shopping cart. The research work consists of use of barcode – barcode scanner system in it. Also if the customer wants to remove any product he/she have to make use of push button while rescanning the product. The research work done by T.R. Lekhaa et al. [17] also depicts the same but by using the Bolt ESP8266 Board. According to a research done by V. N. Prithvish et al. [11] the product can be scanned and verify using barcode - barcode scanner system which is used in their research work along with it the research work also make use of image processing to reassure about the product put into the cart. Also the research work make use of load cell to detect the weight of the product inserted into the cart. This research work makes use of Wi-Fi adapter for connection of the overall system and the dataset is stored in the cloud (in tabular format).The mentioned research work used Beagle Boneblack microcontroller for their processing. Use of multiple object recognition systems is the biggest advantage of this research work. Another paper describing a full fledged and functional a system for a smart cart is "Smart Shopping Cart" [12] in which authors described a smart cart using a camera for capturing the video images and Intel Neural Compute Stick 2 as a computational module as the computing power of Raspberry Pi's processor which is Broadcom processor is quite low. Basically the Raspberry Pi is used for integrating all the components. Load Cell is used to validate the products scanned by the visual detection system. In 2020, Rajeev Ratna Vallabhuni et al. [18] purposed a research work based on RFID system for the same. The uniqueness in this work was, the researchers used RBAC algorithm for verifying the user. Also the encryption – decryption phenomenon was used to access the data. The research work makes use of LED's of different colour to show weather the product is expired or not. Similar such research was done by Thomas Arciuolo and Abdel-shakour Abuzneid [19]. Khalid Yusuf et al. [20] also purposed a model, in which they also used decision tree ID3 algorithm in addition to RFID system. Subhanvali Shaik et al. [21] also did work on RFID in which Zigbee is serving the purpose of transreciver.

3 Methodology

3.1 Existing System

The current system for shopping is very old and has many scope for improvements. In the current system people have to shop for the product they want and they have to stand in queues for the billing of that product. The queues increase even more at the time of the holidays and at the days of the sale. The employee scans the barcode on the products and creates the bill. But this system leads to unnecessary wastage of time for the customer. To tackle the problem of the long queues, the on cart barcode scanning system was introduced. The system of barcode scanning shopping cart consists of a barcode scanner and load cell. The consumer or customer needs to scan the barcode of each product for every time and then put it into the cart. After adding the product into the cart then the system is going to check whether the added product's weight is matching with the net weight of that product, which is stored into the cloud database, so for each product, system is going to fetch the product's weight from the cloud database and then comparing it with the added product's weight, to ensure the product's weight load cell is mounted in the cart. Then again to avoid malpractice by the customer or consumer image processing is used in the system to ensure the added product [10, 11]. Still this lacks in the feasibility of using the system, the customer has to search for the barcode on every product and scan the barcode which might get hectic for the customer and sometimes the barcode is not clearly visible which might lead to faulty output of the system. The other existing system is RFID tag scanner shopping cart system for which RFID tag reader is mounted on every shopping cart. For every product customer needs to scan the RFID tag with the help of an RFID reader. Every Product in the supermarket is attached with a unique RFID tag which consists of a chip and an antenna. While adding any product to the cart, the user has to scan the RFID tag for further billing purposes. Initially the system is powered up and displays the initial data on an LCD screen. Furthermore the system scans the RFID membership card which fetches the details stored in that membership card and displays it on the LCD screen. Now the system is ready to scan the product using RFID. Whenever the product code is detected it displays the details of the product on the LCD screen and this process repeats for each product. Similarly if the user wants to remove the product from the cart he/she must have to rescan the RFID tag on the product which will remove the product from the microcontroller's memory and from ongoing bill. Lastly after the overall shopping is done the user has to scan the RFID membership card again which will finally display the total billing amount on LCD screen [3, 7]. Again, in both of these systems there is one big disadvantage is high cost for developing and maintaining each shopping cart, to develop the automated billing shopping cart, there must need one microcontroller board. So, mounting high specification boards like raspberry pi, beagle bone black on every shopping cart is very costly so this is not a feasible solution. So, to tackle this problem (Figs. 1 and 2).

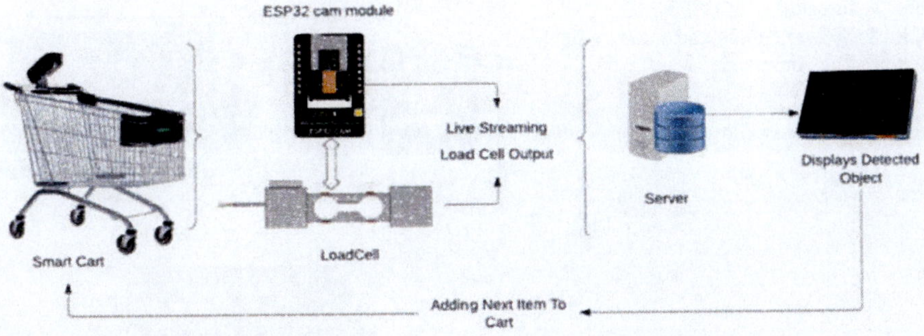

Fig. 1 System design of research work

Fig. 2 Flow chart of the overall system

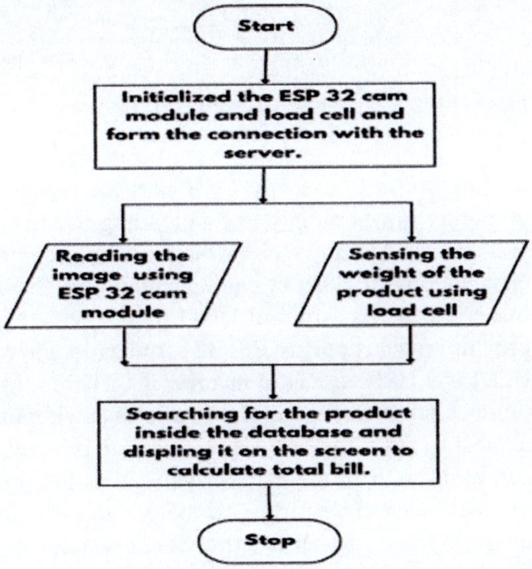

4 Flowchart

5 Method

Hardware Implementation

In this proposed system an ESP 32 camera module and a load cell of capacity 38 kg is mounted onto a shopping trolley. ESP 32 camera module has OV2640 camera with very low power consumption and small size. It also provides an on-board TF card slot. ESP 32 camera modules are used in the proposed system for capturing the frames to detect a product. ESP 32 camera module is used worldwide in IoT applications like QR identification, wireless video monitoring, etc. ESP32 camera

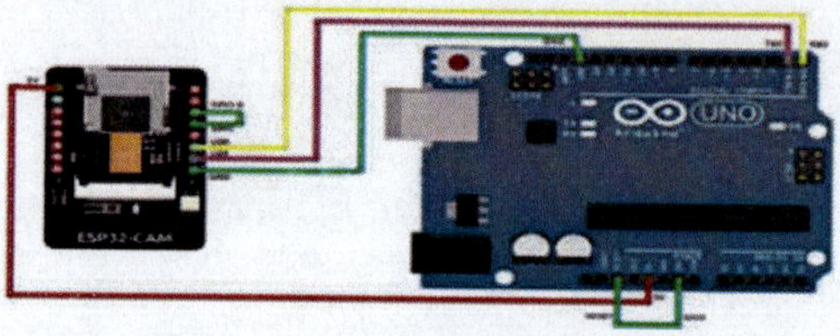

Fig. 3 Integrating of ESP32 Cam for power supply and sending the commands

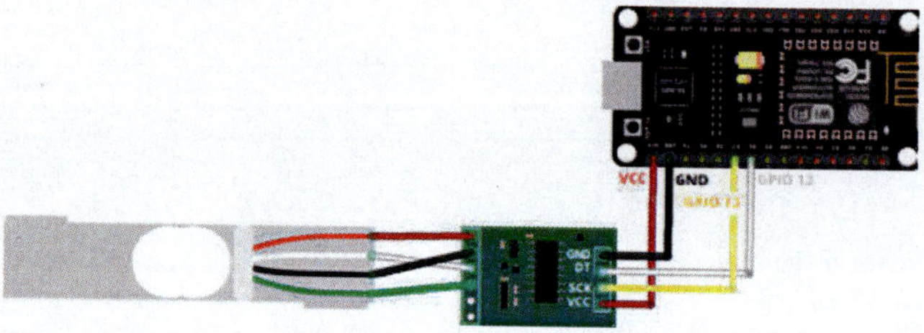

Fig. 4 Integration of load cell with HX 711

module doesn't have any USB port for power supply and data transmission. For providing power supply and uploading the program to ESP32 camera module the system uses Arduino UNO which is a microcontroller board based on ATMega328P. Connections of ESP 32 camera module with Arduino UNO are shown in Fig. 3. Instead of using Arduino UNO there is another option to upload a program and provide power supply to ESP32 camera module which is by using FTDI programmer. FTDI is a USB to TTL converter IC. There might be an error because of the same product but different weight. In this case without the load cell the system would not be able to fetch the correct price of the product, So to deal with this issue. A load cell is inserted in the cart for validating the product inserted and the price of the product inserted according to the weight. The load cell is a force transducer i.e. it converts force into an electric signal which can be measured and standardized. To detect the weight of the product, load cell is used along with HX 711 which is used as an amplifier as it is a 24 bit analog to digital converter (ADC). The connection of the Load cell with HX 711 is shown in Fig. 4.

Software Implementation

The system uses the ESP32 cam for capturing the live events happening in the cart, So for streaming this live events the ESP 32 cam needs to be connected to network which is accomplished using the Wi-Fi module present on the ESP32 cam. For accessing the live stream from the ESP32 cam the server uses the IP address of ESP 32 cam. The server takes help from python for accessing the frames from the live stream and providing that frame as input to the Image detection model. Furthermore, after the detection of the product the Load cell checks the weight of the product to give the accurate price of the product. Thus, the continuous output of load cell (in grams) along with the time stamp is sent to the server. To send the response from the

ESP32 camera board to server and from server to ESP32 camera board, client server communication is established. Further the server checks the time stamp and for the same time stamp it checks the output from the load cell and the machine learning model is matching with the detected product's weight in the database or not. In this case if the load cell's output and product's weight match in the database then the product name along with its price is added in the bill of the customer. The bill is also displayed on the LCD which is mounted on the cart. Also, if a customer wants to remove any product from the cart, the camera is going to detect the product again and the load cell is going to detect a deduction in the weight. Then the system cross checks that the weight of the product removed is equal to the weight deducted. If the load cell's output is deducted by product's weight, then that object is removed from the cart and simultaneously the product will be removed from the total bill of the customer. A similar cycle of adding or removing any product from the cart continues and later on the final bill is generated.

6 Results and Discussions

The digital shopping cart fetches images from the from the datasets and recognizes them using the ML algorithm. It's an application of modern AI technology. The user just has to put the items they want to purchase in the cart and the items in the cart are directly added to their bills. The proposed system of smart shopping cart is extremely beneficial for customers to steer away from the inconvenience that they typically face while shopping, mostly during the weekends. Also queue time is minimized for the customers at the time of payment to the cashier, which helps supermarkets by attracting more customers towards them.

Figure 5 shows the interface of the ESP32 cam live stream an this stream divided into frames. The frames also are given as input to the machine learning model which returns the name of the detected product with time stamp.

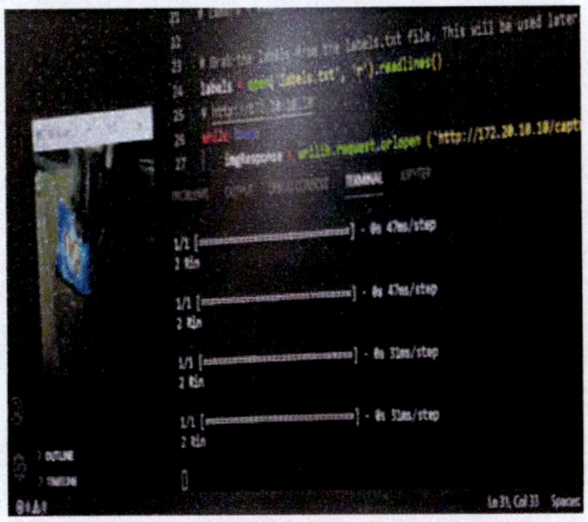

Fig. 5 Product Rin is detected

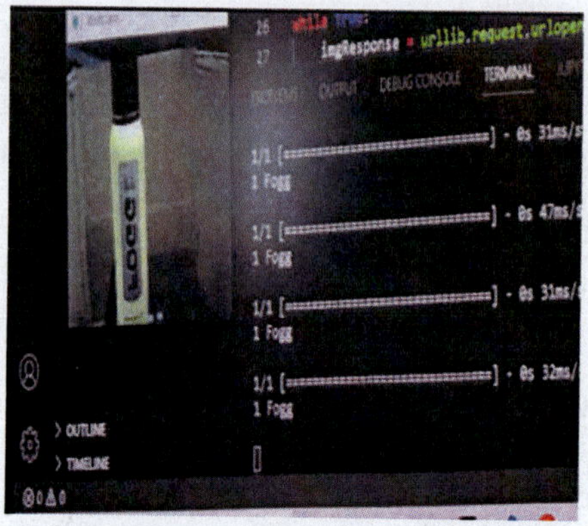

Fig. 6 Product Fogg is detected

Fig. 7 Product Colgate is detected

Figure 6 shows product Fogg is detected on the console, and it is printing the time stamp

Figure 7 shows product Colgate is detected on console and it is also printing time stamp.

Figure 8 shows the load cell output the system takes reading 5 times and the first and the average of 5 readings gets printed. The weight in our cart was 70 g and the output shows 72.5 which is quite accurate. It gives continuous output and as we add new products to it the weight gets added.

Fig. 8 Output of Load Cell

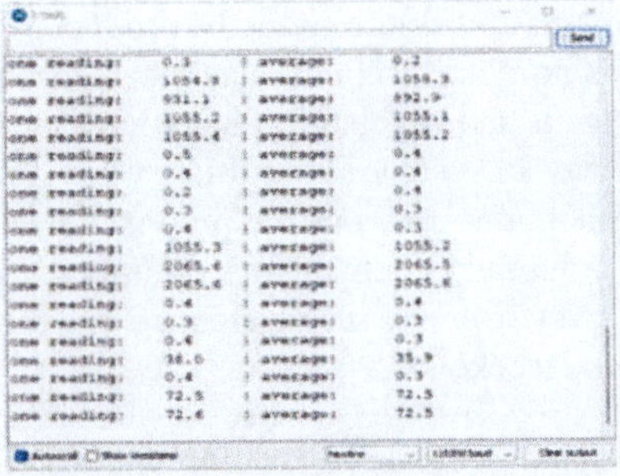

7 Future Scope

Few of the updates that the proposed system can have are displaying nutrition facts of the edible products. Checking of availability of the product in the store on the cart itself. Customers can make a shopping list for themselves before visiting the supermarket on the cart application. We can integrate an automatic payment machine on the cart itself where the user can swipe their card or pay through UPI. Putting an AR/VR map of the store on the cart itself will direct the user through the supermarket to the products they want. Adding screen touch display with a GUI to make the system user friendly.

8 Conclusion

The research aimed to bring a hassle-free and wait-free experience for the customers of the supermarket by providing a long queue solution in a single cart. It also helps the supermarket owners to attract people to their supermarkets. The overall research helps in solving the inefficiencies in the current supermarket system. This system helps to minimize the labor work in supermarkets which is beneficial to supermarket owners. This proposed system aims to implement the smart shopping cart system with low cost in an effective way. As discussed in the above Existing system section that using a high specification board is not feasible or cost-effective way to implement this system in day to day life. That's why in this proposed system a low-cost board ESP32 is used which is very low cost compared to Raspberry Pi and Beagle bone black board. In this system the maximum computation task is done on the server that's why ESP32 is used. As this system is providing cloud computing and storing solutions, our data is secured even after system failure. Maximum computational work is done on the server that's why power consumption for individual shopping cart is also low. For this system our machine learning model is present only on the server, that's why systems maintenance is also easier as compared to updating every system mounted on each different shopping cart. Use of multiple product recognition systems i.e. object detection using deep learning and weight detection using load cell helps to reassure about the product by which the exact product is recognized and there is no doubt in both customer and owner's mind while billing. Some products like fruits and vegetables do not have any bar code attached to them, so existing systems may face some difficulties while detecting the objects while adding them into the shopping cart, but the proposed system is going to overcome such problems. The proposed system of smart shopping cart is extremely beneficial for customers to steer away from the inconvenience that they typically face while shopping, mostly during the weekends.

Acknowledgements Guidance throughout the project, researching and development, was provided by Prof. Puja A. Chavan. Our sincere gratitude to Director VIT, Pune, Head of the department, Multidisciplinary Engineering and VIT, Pune for this opportunity.

References

1. Sarwar MA, Daraghmi Y-A, Liu K-W, Chi H-C, İk T-U, Li Y-L (2020) Smart shopping carts based on mobile computing and deep learning cloud services. IEEE Wirel Commun Netw Conf (WCNC) 2020:1–6. https://doi.org/10.1109/WCNC45663.2020.9120574
2. Shetty R, Pawar D, Poojari R, Patel R, Mali J, Student B (2022) Smart shopping cart using machine vision along with machine learning. IRJET 8:2395–0056
3. Meghana TK, Bedare RS, Ramakrishna M, Vignesh P, Pavithra M (2020) Smart shopping cart with automated billing system. Int. J. Eng. Res. Technol. (IRJET) IETE. 8(11):88–91
4. Namiki S, Yokoyama K, Yachida S, Shibata T, Miyano H, Ishikawa M (2021) Online object recognition using CNN-based algorithm on high-speed camera imaging: framework for fast and robust high-speed camera object recognition based on population data cleansing and data ensemble. In: 2020 25th International Conference on Pattern Recognition (ICPR), pp. 2025–2032. https://doi.org/10.1109/ICPR48806.2021.9413042
5. Oka T, Morimoto M (2015) An extraction and recognition method for partially hidden objects. In: 2015 International Conference on Informatics, Electronics & Vision (ICIEV), pp. 1–4. https://doi.org/10.1109/ICIEV.2015.7334055
6. Cruz JPN, Dimaala ML, Francisco LGL, Franco EJS, Bandala AA, Dadios EP (2013) Object recognition and detection by shape and color pattern recognition utilizing artificial neural networks. Int Conf Inf Commun Technol (ICoICT) 2013:140–144. https://doi.org/10.1109/ICoICT.2013.6574562
7. Anitha R, Subburam S, Keerthana G, Yoganandarajurs KH (2020) Smart trolley billing system. Eur J Mol Clin Med 7(11):2264–2270
8. Chong CB, Tan T, Lim FL (2006) A model-based approach for rigid object recognition. In: 18th International Conference on Pattern Recognition (ICPR 2006), pp. 116–120. https://doi.org/10.1109/ICPR.2006.103
9. Ragesh N, Giridhar B, Lingeshwaran D, Siddharth P, Peeyush KP (2019) Deep learning based automated billing cart. Int Conf Commun Signal Process (ICCSP) 2019:0779–0782. https://doi.org/10.1109/ICCSP.2019.8697995
10. Mekruksavanich, S (2019) Design and implementation of the smart shopping basket based on IoT technology. In: 2019 14th International Joint Symposium on Artificial Intelligence and Natural Language Processing (iSAI-NLP), pp. 1–6. https://doi.org/10.1109/iSAI-NLP48611.2019.9045336.

11. Prithvish VN, Agrawal S, Alex JSR (2018) An IoT-based smart shopping cart using the contemporary barcode scanner. In: DanielThalmann N, Subhashini KM, BalaMurugan MS (eds) Intelligent Embedded Systems: Select Proceedings of ICNETS2, Volume II. Springer, Singapore, pp 45–58. https://doi.org/10.1007/978-981-10-8575-8_6
12. Belibov D, Tudose DS (2019) Smart shopping cart. In: 2019 18th RoEduNet Conference: Networking in Education and Research (RoEduNet), pp. 1–6. https://doi.org/10.1109/ROEDUNET.2019.8909680
13. Subudhi SR, Ponnalagu RN (2019) An intelligent shopping cart with automatic product detection and secure payment system. In: 2019 IEEE 16th India Council International Conference (INDICON), pp. 1–4. https://doi.org/10.1109/INDICON47234.2019.9030331
14. Kowshika S, Madhu mitha SS, Madhu Varshini G, Megha V, Lakshmi K (2021) IoT based smart shopping trolley with mobile cart application. In: 2021 7th International Conference on Advanced Computing and Communication Systems (ICACCS), pp. 1186–1189. https://doi.org/10.1109/ICACCS51430.2021.9441866
15. Bhagya V, Kiranagoud SG, Lankesh HR, Harish SV (2021) An implementation of electronic shopping cart transaction and display system using wireless technology on IOT. Int. J. Eng. Res. Technol. (IJRET) NCCDS. 09(12):1–4
16. Viswanadha V, Pavan Kumar P, Chiranjeevi Reddy S (2018) Smart shopping cart. In: 2018 International Conference on Circuits and Systems in Digital Enterprise Technology (ICCSDET), pp. 1–4. https://doi.org/10.1109/ICCSDET.2018.8821103
17. Lekhaa TR, Rajeshwari S, Sequeira JA, Akshayaa S (2019) Intelligent shopping cart using bolt Esp8266 based on Internet of Things. In: 2019 5th International Conference on Advanced Computing & Communication Systems (ICACCS), pp. 758–761. https://doi.org/10.1109/ICACCS.2019.8728346
18. Vallabhuni RR, Lakshmanachari S, Avanthi G, Vijay V (2020) Smart cart shopping system with an RFID interface for human assistance. In: 2020 3rd International Conference on Intelligent Sustainable Systems (ICISS), pp. 165–169. https://doi.org/10.1109/ICISS49785.2020.9316102
19. Arciuolo T, Abuzneid A (2019) Simultaneously shop, bag, and checkout (2SBC-Cart): a smart cart for expedited supermarket shopping. In: 2019 International Conference on Computational Science and Computational Intelligence (CSCI), pp. 1162–1167. https://doi.org/10.1109/CSCI49370.2019.00219
20. Yusuf K, Abdurohman M, Putrada AG (2019) Increasing passive RFID-based smart shopping cart performance using decision tree. In: 2019 5th International Conference on Computing Engineering and Design (ICCED), pp. 1–5. https://doi.org/10.1109/ICCED46541.2019.9161139
21. Shaik S, Jabirullah M, Vishwakarma AK, Ranjan R (2020) Advancement of shopping handcart for supermarket. In: 2020 Sixth International Conference on Parallel, Distributed and Grid Computing (PDGC), pp. 526–531. https://doi.org/10.1109/PDGC50313.2020.9315742

Analysis of Automated Music Generation Systems Using RNN Generators

Ruchir Shiromani, Tanisha Mittal, Anju Mishra, and Anjali Kapoor

Abstract Deep Learning Algorithms has advanced drastically, finding their way in almost every field of where automation can be done or is required such as Computer vision, Natural Language Processing, Speech recognition, Robotics and many other. Most of these tasks are limited to unimaginative or less creative tasks in society. Generating Music is one of the challenging tasks for a Deep learning algorithm as it requires content generation, but it finds its application in co-creative tools. Thus, Artificial Music generation is one of the fields where Deep Learning is being applied, generating new compositions based on historical data. Our report "music generation using deep learning" is based on LSTMs – long short-term memory to generate music using Maestro dataset. Our main goal is to generate music using Deep learning algorithm i.e., LSTMs based on previous notes.

Keywords Music Generation · Deep Learning · LSTM

1 Introduction

Music is a part of our everyday life from listening to radios to YouTube music videos [12]. With everyone having a distinct music taste, the area is expanding day by day and new music is being generated every day. In recent years, deep learning can be seen useful in various fields but there is less research done on generation music. Music generation proves to be a greater challenge compared to images, such as Music is temporal multiple instruments are used together in the generation of music. Music is a combination of various chords and melodies grouped together so which produces multiple output each time.

It's a misconception that music can be generated by an expert only. In today's advanced world a person who loves music can also generate music, on the basis of his/her preference. Till a few years back music was generated using analogue

R. Shiromani (✉) · T. Mittal · A. Mishra · A. Kapoor
Amity University Uttar Pradesh, Noida, India
e-mail: ruchirshiromani3344@gmail.com

N. Hasteer et al. (eds.), *Decision Intelligence Solutions*, Lecture Notes in Electrical Engineering 1080,
https://doi.org/10.1007/978-981-99-5994-5_22

signals manually [6]. But in today's time it is generated by the use of advanced technology which are assisted by human beings. Deep learning can be applied to the field of Artificial music generation to generate new composition of music using the historical data present. The paper is based on the construction of generative neural network architecture that is capable of effectively portraying the complex details of harmony and melody without the need for humans' intervention.

In the paper music generation using deep learning algorithms has been displayed. The main motive was to devise an algorithm that can create music. This paper displays the use of LSTM (LONG-SHORT TERM MEMORY) to generate music based on previous notes using Maestra datasets. The instrument used to generate music in this paper is a Grand acoustic piano. The report comprises of the precise details of the music and the methodology that has been used for the generation of the music notes.

2 Literature Review

Kinbert chou [12] aims at whether music generation is achievable using various model architectures or not, their paper is based on generation of classical music dataset MAESTRO. This dataset contains 200 h of piano music recorded over the years. Their paper shows, long short-term memory (LSTM) layer that takes sequence of notes and thus generates or outputs tokens predicted according to the pre-processed vocabulary.

Shuqi Dai [1] explained various recent advancements in Deep Learning algorithms, they aimed at exploring automatic music generation using Deep learning, thus worked on a solution called Music Frameworks, based on the design of hierarchical music representations in music frameworks inspired by Hiller and Ames [6]. Music framework architecture is based on controllable melody generation systems that make use of Auto-regressive model based on Transformer and LSTMs on POP909 dataset.

Jean-pierre briot [7], surveyed and analysed different ways of generation of music using Deep Learning techniques and thus proposed methodology to generate musical content. They based their methodology based on five dimensions namely objective, representation, architecture, challenges, strategy.

Prafulla Dhariwal [11], introduced Jukebox that is a deep learning model generating music with singing raw input into it. Paper also shows the use various other factors such as vocals, artist to make singing more comfortable, for compressing the audio into a discrete space VQ-VAE architecture has been used in combination with Auto regressive transformers for generating or recreating the lost information while compression. It made use of LyricWiki dataset for consuming metadata into the networks and trained their first model over Maestro dataset for generation of classical music using their architecture.

Hasim sak [8], shows acoustic modelling in speech recognition using deep LSTM RNNs on TIMIT dataset. Deep LSTMs allows to successfully predict various sequences with lower loss, deep BLSTM RNN have outperformed traditional DNNs

Table 1 Summary of Selected Research Work for Literature Study

S.No	Year	Authors	Dataset	Methods
1	2022	Shuqi Dai, Huiran Yu, Roger B. Dannenberg [2]	POP909	Repetition and structure analysis
2	2021	Shuqi Dai, Zeyu Jin, Celso Gomes, Roger B. Dannenberg [1]	POP909	Auto-regressive model based on Transformer and LSTM
3	2019	Chou, K., Peng, R. [12]	Maestro	LSTM for music generation using NLP
4	2014	Hasim Sak, Andrew Senior, Francoise Beaufays [8]	TIMIT	Deep LSTM RNN
5	2020	Prafulla Dhariwal,Heewoo Jun, Christine Payne, Jong Wook Kim, Alec Radford, Ilya Sutskever [11]	Lyric Wiki, Maestro	Autoregressive Sparse Transformer

in speech recognition. Thus, the paper discusses various deep architectures for better speech recognition using RNNs (Table 1).

3 Methodology

3.1 *Artificial Intelligence*

Field of computer science and engineering that concerns with the understanding of intelligent behavior [13] or the ability of a machine to imitate human intelligence [14] is known to us as Artificial intelligence. It is emerging out to be a vital tool due to its versatility; it has proved to be tool essential in future. The study of various algorithms that improves with their own experience is called as Machine Learning-

"A computer Program is said to be learning from experience with respect to some task and performance measures, if its performance in task improves with experience". In modern Day Machine Learning covers various aspects of our day to day from web searches to content filtering on social media; scope of machine learning is very vast. Rise of machine learning is shaping artificial intelligence for a new future, dream of auto driven cars and development of mechanical men can be shaped through machine learning.

Representation learning uses a huge amount of raw data and turns it into useful information to automatically discover information for detection or classification, thus deep learning a form of representation learning make use of multilayer representation that is obtained from transforming representation at each module and turns it into information [16]. This report would further discuss multi-layer representation with deep learning algorithms.

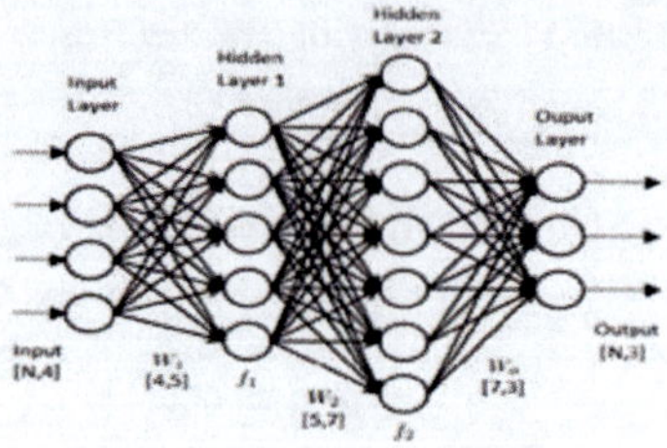

Fig. 1 [18] Artificial neural network with input, output, and hidden layers

3.2 *Deep Learning*

Deep Learning is a subset of Machine Learning. It helps us to perform various analytical and automation tasks without much human interference [15]. Computational models are mathematical models that need extensive computational resources to study behavior of simulations that are very complex [9]. These computational models consist of multiple layers of processing in deep learning which allows them to represent data after multiple layer abstraction. This multi layered structure can be achieved by ANNs (Artificial Neural Networks) [16].

Neural network is composed of connections known as processors (or neurons) each producing an activation in real values. It is a multi-level process in which input neurons get activated through sensors that in turn activates other neurons through weighted connections [17]. Learning is done through finding correct weights that ensures desired behavior in response to the given problem. In these there are multiple layers out of which the first layer also known as input layer and the last layer also known as output layer rest of the layers are known to be hidden layers.

Since many decades many shallow neural networks were with very less layers but as time passed a method called backpropagation was developed for supervised learning with depth inn neural networks, but supervised training of neural networks faced many difficulties. Although supervised learning won many contests in visual training, pattern recognition etc. The emergence of unsupervised learning that made deep learning practically feasible [17]. Since then, many neural networks have been developed (Fig. 1): -

In "Fig. 2" input networks get activated through sensors and thus activate other neurons through weighted signals each producing a real valued activation. It contains hidden layers which are interconnected with each other.

3.3 *Recurrent Neural Networks (RNN's)*

RNN has been developed in the 1980s, but at that time they were not into use. RNN came into more use with the development of LSTM which came into existence in 1990s because of the large amount of data that has to be worked upon and the increase computational power [3].

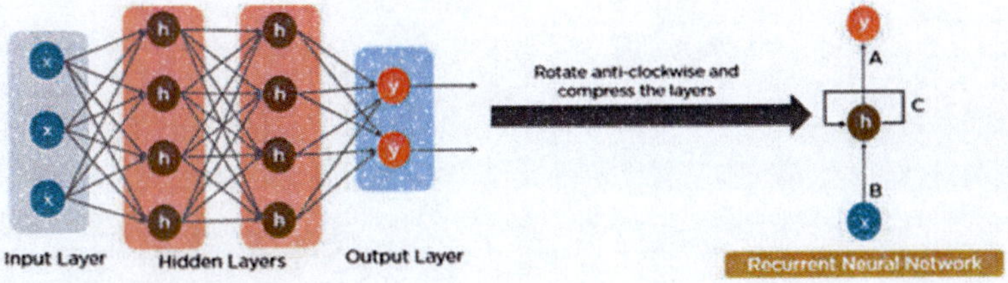

Fig. 2 [3] Recurrent Neural Network

Recurrent neural networks are important where prior information is important to take future decisions for e.g., text, speech analysis etc.... Recurrent neural networks store information from past results and use it to make decisions in the present time to make new results, and hence this helps them exploit information from both stored (at time t-1) and present (at time t). A variant of RNNs known as LSTMs i.e., long short-term memory network has been developed due to the problem of vanishing gradient faced by RNNs [10]. Unlike static networks like CNNs which carry different weights in their network RNNs carries same weights throughout. RNNs overrun many static networks but lags due to its gradient problem and they are not simple to be trained, hence backpropagation cannot be used [3].

RNN was developed to overcome the problem of standard neural networks. They consist of a hidden layer which can remember the previous information which is necessary for the next step. The hidden layer is the most important component of the RNN architecture.

RNN also has a memory. It remembers or maintains a record of all the information about the calculations that have been carried out. It maintains the same setting for performing tasks on all hidden layers or inputs because it provides the same output.

3.4 Long Short Term (LSTM)

Problems of sequence prediction have been faced for a long time. With recent breakthrough happening in data sciences, LSTM is found to be the most effective solution to these problems [3]. LSTM belongs to a complex area of Deep Learning. LSTM is a technique that makes use of several algorithm that try to perform human like actions and show their behavior to uncover the underlying relationships in the given sequence data [3]. It is a Recurrent Neural Networks (RNN) architecture designed to overcome the limitations of conventional RNN [4]. It tends to model sequences temporal in nature accurately, with their long -range dependencies. It is a sequential network. LSTMs are used in handling the vanishing gradient problem that arises in RNNs [5]. LSTM is capable of handling entire sequence of data apart from single data points such as images. Use of LSTM can be found in machine learning, speech recognition, handwriting recognition, sentiment analysis, language modelling, videos analysis, etc. LSTMs are widely used because they have an edge over the conventional

Fig. 3 LSTM

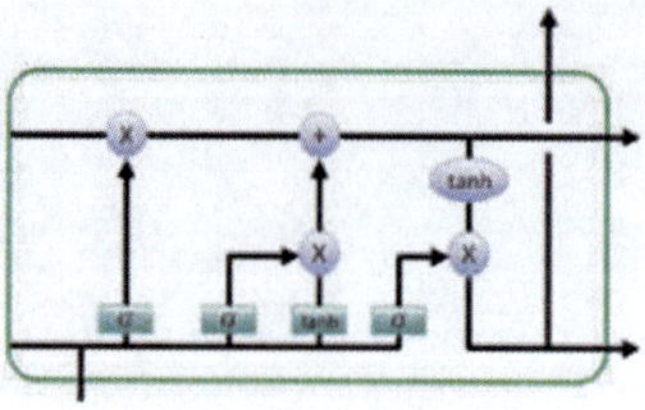

RNN and feed-forward networks. They have the property of selectively remembering patterns for long dutions of time. Time series data are used in LSTMs for making effective predictions because there is possibility of time lags of varying length between important events.

Various fields where use of LSTM can be seen:

Music Generation, Robot Control, Speech Recognition, Time Series Prediction, Sign Language Translation, Handwriting Recognition, Market Prediction, Semantic Parsing, Human Action Recognition, Short-Term Traffic Forecast, etc.

3.5 Architecture of LSTM

LSTM has a chain like architecture having different memory blocks which are known as cells and 4 neural networks. LSTMs works on both LONG-TERM MEMORY (LTM) AND SHORT-TERM MEMORY (STM) (Fig. 3).

It makes calculations simple and effective by making use of gates.

1. FORGET GATE: In this LTM forgets the information that is unnecessary [3].
2. LEARN GATE: In this necessary information that have been learned from LSTM is used by the current input by combining the current input and STM [3].
3. REMEMBER GATE: In this gate the information that has not forgotten in the LTM and the STM along with the current input are combined altogether and works as an updated LTM [3].
4. USER GATE: This gate predicts the output of the current event by making use of LTM, STM and the current input which further works as an updated STM [3].

3.6 Description

The motive of this paper is to generate music through Deep Learning algorithms on provided dataset, the MAESTRO data set has been used. MAESTRO stands for MIDI and Audio Edited for Synchronous Tracks and Organization. MAESTRO is a dataset which is composed of around 200 h of piano performances. Various libraries have been used for data visualization and data extraction; after data pre-processing has been done for converting various MIDI input files to notes for deep learning algorithms to train over it. Paper displays use of LSTM (Long Short-Term Memory) for generating output.

Libraries used: Fluidsynth, pretty_midi, Pandas, Tensorflow Processes:

- Importing data
- Data Pre-processing
- Training LSTM for generating output.

Pretty_midi library has been used for reading midi files from MAESTRO dataset. Pretty MIDI is a library that contains various utility functions for handling MIDI data. Thus, for analyzing, manipulating, and synthesizing data pretty_midi can be used.

Here, Fluid synth acts as virtual synthesiser that takes input or waveform of any form and transforms it into audio.

Data Pre-processing

In data pre-processing sequence of data are fed to LSTM network, for that various libraries have been used such as pretty midi to extract notes from MIDI data. From those notes various essential components such as pitch, duration, and step between each note has been extracted (Fig. 4).

This function is used to convert midi files into notes so that sequences can be formed that helps create sequences used to train our LSTM network. Create sequence function is used to generate sequences essential for our network.

Visualization of music data in the form of plot using pitch and time i.e., start, end and duration of a particular node.

Training Network:

Using our pre-processed data i.e., the sequences formed are going to fed to the network for it to learn on historical data, learning rate set to 0.005 for the network to learn at a suitable rate that is, it doesn't reach saturation too early, and overfitting occurs and not so slowly that underfitting occurs. The various input and output parameters have been set for the network to learn and provide outputs.

Model summary is provided below (Fig. 5):

```
array(['E4', 'G3', 'C4', 'C3', 'D4', 'B3', 'E4', 'C4', 'B3', 'D4'],
      dtype='<U3')
```

Fig. 4 Note names

Model: "model_1"

Layer (type)	Output Shape	Param #	Connected to
input_2 (InputLayer)	[(None, 25, 3)]	0	[]
lstm_1 (LSTM)	(None, 128)	67584	['input_2[0][0]']
duration (Dense)	(None, 1)	129	['lstm_1[0][0]']
pitch (Dense)	(None, 128)	16512	['lstm_1[0][0]']
step (Dense)	(None, 1)	129	['lstm_1[0][0]']

Total params: 84,354
Trainable params: 84,354
Non-trainable params: 0

Fig. 5 Model Summary

4 Results and Discussions

An LSTM network is trained at different Epochs to find the results respectively. At different epochs notes appeared at different durations of time, thus the output generated has different levels of tempos to it leading to different melodies. In the graphs provided of the Pitch vs time we can check the step, pitch, and time, that represent different notes generated at output providing us with various different melodies. To start generating notes have been provided to our network with sequence of notes that makes it possible to develop or generate a sequence from SoftMax distribution of notes generated by the model. Thus, picking the most probable note that i.e., the note with the highest possibility of recurring, this leads to formation of sequences (Fig. 6).

Different Epoch generates different distributions of notes; since at every epoch final loss function was different distributions obtained were different. Network is trained at 10 epochs, 20 epochs, and 50 epochs, at every epoch different result can be obtained, due to different SoftMax probability distributions. As you can see distribution of pitches at different epochs is different, thus the generated music or output has different melody to it. At 20 epochs the most soothing output was generated (Table 2).

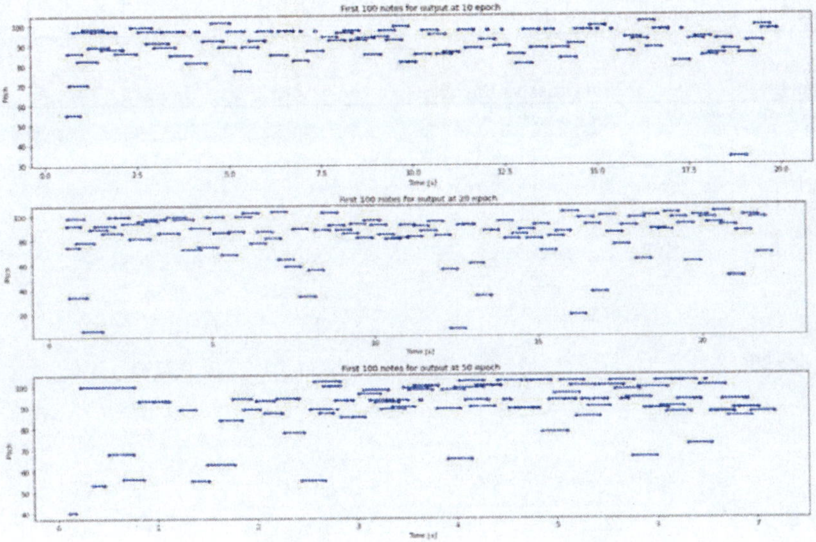

(a)

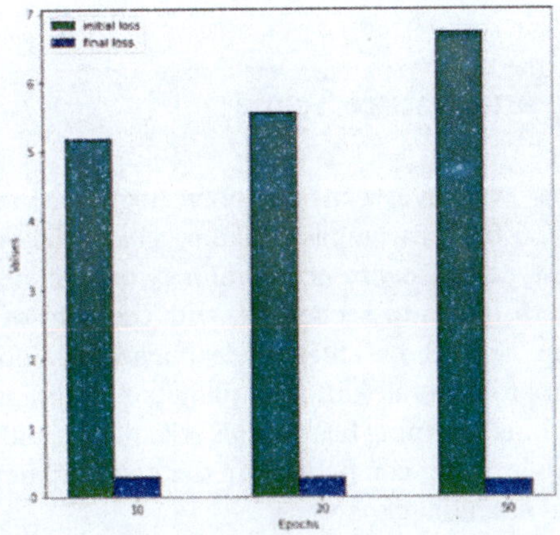

(b)

Fig. 6 **a.** outputs at different epochs **b.** Loss vs Epochs comparison.

Table 2 Loss comparisons at different epoch

Epochs	Plot (Loss vs Epochs)	Loss
Epoch 10	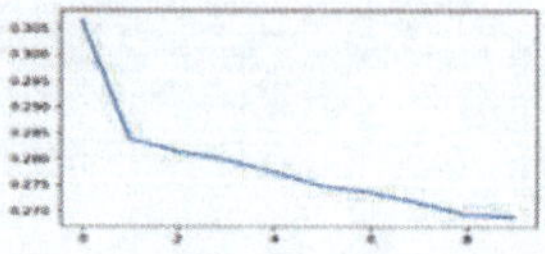	0.2643
Epoch 20	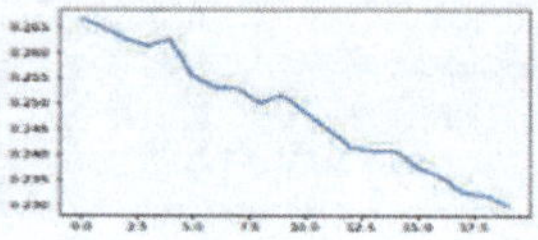	0.2273
Epoch 50	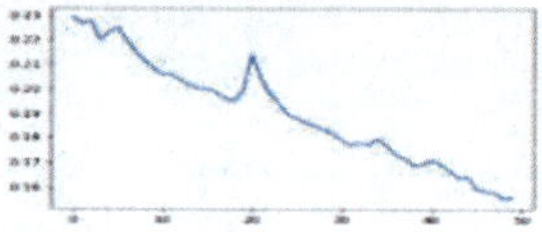	0.1613

5 Conclusion and Future Scope

Through this paper we have seen that how Artificial intelligence through deep learning can be used for creative tasks like music or melody generation. We have seen how LSTM a type of Recurrent Neural network can be implemented on music dataset after converting it into sequences, with the help of such data how we can train Deep Learning algorithms, and how deep learning algorithms can be trained at different parameters to generate different outputs for similar inputs. Deep Learning as a branch of Machine Learning has various advantages and with this report, we can conclude how deep learning can be used in the entertainment industry with respect to provide data and essential expertise.

References

1. Dai S, Jin Z, Gomes C, Dannenberg RB (2021) Controllable deep melody generationvia hierarchical music structure representation
2. Dai S, Yu H, Dannenberg RB (2022) What is missing in deep music generation? A study of repetition and structure in popular music. arXiv preprint arXiv:2209.00182
3. www.analyticsvidhya.com
4. www.intellipaat.com
5. www.wikipedia.org
6. Hiller L, Ames C, Franki R (1985) Automated composition: an installation at the 1985 international exposition in Tsukuba, Japan. Perspect New Music 23(2):196–215

7. Briot JP, Hadjeres G, Pachet FD (2017). Deep learning techniques for music generation--a survey. arXiv preprint arXiv:1709.01620
8. Long Short-Term Memory Recurrent Neural Network Architectures for Large Scale Acoustic Modeling Hasim Sak, Andrew Senior, Francoise Beaufays
9. Ravì D et al (2016) Deep learning for health informatics. IEEE J Biomed Health Inform 21(1):4–21
10. Bengio Y, Simard P, Frasconi P (1994) Learning long-term dependencies with gradient descent is difficult. IEEE Trans Neural Netw 5(2):157–166
11. Dhariwal P, Jun H, Payne C, Kim JW, Radford A, Sutskever I (2020) Jukebox: a generative model for music. arXiv preprint arXiv:2005.00341
12. Chou, K., Peng, R. (2019). Deep Learning Music Generation
13. Korb KB, Nicholson AE (2010) Bayesian Artificial Intelligence. CRC Press, Boca Raton. https://doi.org/10.1201/b10391
14. Aghion P, Jones BF, Jones CI (2017) Artificial Intelligence and Economic Growth (No. w23928). National Bureau of Economic Research
15. Mitchell TM (1997) Machine Learning. McGraw-Hill Inc, New York
16. LeCun Y, Bengio Y, Hinton G (2015) Deep learning. Nature 521(7553):436–444
17. Schmidhuber J (2015) Deep learning in neural networks: an overview. Neural Netw 61:85–117
18. www.medium.com
19. Rawat W, Wang Z (2017) Deep convolutional neural networks for image classification: a comprehensive review. Neural Comput 29(9):2352–2449

Differential Evolution Image Contrast Enhancement Using Clustering

Samiksha Yadav, Ankita, Shweta Singhal, and A. K. Mohapatra

Abstract Differential Evolution (DE) is a population-based totally metaheuristic search algorithm that optimizes problems via iteratively enhancing candidate solutions primarily based on an evolutionary technique. The current study aims to showcase the capability and success of finding a globally optimal solution in order to enhance contrast and detail in color images. This paper introduces a new technique for contrast enhancement of an image based on clustering using Differential Evolution Algorithm. DE's role is to optimize the turnaround function's metrics by optimizing the outcome fitness standard. Experimental results were compared with the firefly image contrast enhancement technique and concluded that pro-posed method is an improvement over preceding technique.

Keywords Contrast Enhancement · Differential Evolution Algorithm · Clustering · Firefly Algorithm

1 Introduction

The motive of image enhancement is to enhance image quality for later evaluation or display. Image enhancement refers to emphasizing or polishing image functions which includes counters, borders, edges, or contrast to facilitate in addition evaluation and display. It features low-level approaches that consists of noise reduction, contrast enhancement, and display adjustment to get images of desired quality.

The dispersion of the pixels along the entire dynamic range of an image, or overall difference among the lowest and highest values, can be used to quantify contrast in an image [1]. The contrast makes it possible to distinguish between an image's constituent parts. When most of the pixels are concentrated in a small area of an image, it can be claimed that the contrast in the image is low. When using a grayscale image as benchmark, the image may have a most of it.

S. Yadav · Ankita · S. Singhal (✉) · A. K. Mohapatra
Department of IT, IGDTUW, New Delhi, India
e-mail: miss.shweta.singhal@gmail.com

N. Hasteer et al. (eds.), *Decision Intelligence Solutions*, Lecture Notes in Electrical Engineering 1080,
https://doi.org/10.1007/978-981-99-5994-5_23

Contrast enhancement approaches try to better give our pixels around the range, allowing the values of the input image to be stretched to use the entire spectrum. This has been done specifically for electron microscopy, aerial and satellite photography, medical-biological images analysis, and archaeology. Log transmute, contrast stretching, thresholding, and transform of law of power are a few examples of contrast enhancement. These techniques are more effective in certain situations and call for knowledge to choose the optimal technique for a particular image; some of them can also need manually chosen parameters [1].

[2] Differential Evolution is a sort of metaheuristic algorithm which is utilized to achieve fast convergence and low computing time. It optimizes a prob via attempting to enhance a viable solution across generations utilizing a fitness function and employing a crossover technique that differs from that of the Genetic Algorithm.

We introduce a new contrast enhancement approach in this research that makes use of differential evolution and clustering. Clustering is a prominent statistical data analysis aspect in many fields, including image analysis, machine learning, bioinformatics and pattern recognition [2]. Clustering is the act of categorizing relevant ones into diverse groups, or, more specifically, reconfiguring a data number of clusters so that data in each cluster is sorted premised on a few distance measures. Various problems have been solved by the combination of DE and clustering, such as image segmentation, pixel grouping, and color quantization. This work is a tweak to the Differential evolution method to increase its performance for color images.

2 Differential Evolution Algorithm

Differential evolution (DE) [3] algorithms are efficient and effective. This evolutionary optimization method was introduced in 1995 by Stern and Price.

Differential Evolution (DE) is quiet a concept and framework that's simple to put into practice and has a high degree of effectiveness and fast convergence. Its use of floating point encoding rather than binary has caused it to garner a great deal of attention, and it has been widely used in a variety of areas. As a result, many modifications of DE have been presented [3]. A feasible nomenclature using DE/a/b/c Used to categorize distinct options. where a is how to choose a vector to transform, b stipulates a number of distinct vectors accustomed to perturb base Vectors a and c represent crossover mechanisms (Fig. 1).

Suppose the search space DE has X dimension, the number of parameters of the function being optimized is n, then the nth individual in population is a X-dimensional vector, , $P_n = (x_{n1}, x_{n2}, .., x_{nD}) \in R^X$ The basic operators of DE are:

- Initialization - For resolving optimization issues, differential evolution keeps track of a range of potential solutions correlated to individuals over a time interval [1]. Each population in generation t contains M individuals.

$$P_{n_j^0} = P_j^L + rand_j(0,1) * P_j^U + P_j^L \tag{1}$$

$$n = 1, 2, \ldots, M; \; j = 1, 2, \ldots, X; \}$$

where P_j^U, P_j^L are the upper and lower bound respectively, and $rand_j(0, 1)$ is an evenly distributed random value range between that [0, 1].

- Mutation - Three individuals are opted from the population. Arbitrarily, i.e., P_{r1}, , $P_{r2,}$ P_{r3} satisfies $r1 \neq r2 \neq r3$ and r1, r2, r3 ∈ [1, M]. [4] For DE variant rand/1/bin or rand/1/exp, we have

$$A_{nj}{}^{t+1} = P_{r3j}{}^{t} + Z \times P_{r1j}{}^{t} - P_{r2j}{}^{t} \tag{2}$$

where Z is a real-valued mutation factor and Z ∈ [0,2].

- Crossover - Crossover operators are prepared to increment Distinction of population. This action assures one aspect in $P_n{}^{(t+1)}$ provided through $P_n{}^{(t)}$.

$$B_{nj}{}^{(t+1)} = \begin{cases} A_{nj}{}^{(t+1)}, rand_j(0, 1) \le CRorj = int(H) \\ P_{nj}{}^{(t)}, rand_j(0, 1) \le CRorj \neq int(H) \end{cases} \tag{3}$$

where H = M* $rand_j(0, 1)$

- Selection - Selection is performed through collating the fitness of individuals to create the further generation population.

$$P_{nj}{}^{(t+1)} = \begin{cases} B_{nj}{}^{(t+1)}, f(B_{nj}{}^{t+1}) > f(P_{nj}{}^{t}) \\ P_{nj}{}^{(t)}, f(B_{nj}{}^{t+1}) \le f(P_{nj}{}^{t}) \end{cases} \tag{4}$$

For maximizing optimization problems, an individual possess greater fitness value is endured and vice-versa.

3 Related Work

Differential evolution (DE), a technique used in evolutionary computation, aims to continually enhance a candidate solution in relation to a predefined quality metric in order to optimize a problem. Such techniques are frequently referred to as metaheuristics since they can search extremely huge spaces of potential solutions and make little to no assumptions about the problem being optimized. Metaheuristics like DE, though, do not ensure that an ideal solution will ever be discovered.

Differential evolution is a population-reliant random search technique, was given by the Storn and Price [5]. It was discovered that this method is effective well when solving difficult optimization issues despite its initial purpose of being used to solve Chebyshev polynomial problems. Numerous engineering optimization disciplines have used the DE, including signal refining, resource allocating, industrialized design, automation, fabricated neural networks, and power structure optimization.

DE is based on a straightforward, doable idea. Sturdy robustness, lesser control parameters, and improved search proficiency are its benefits [6]. It can tackle complicated issues concurrently and solve multimodal and nonlinear functions.

It also performs better in terms of convergence. The DE is prone to premature maturity when tackling complex optimization problems, slipping within local optimums, and having poor precision at latter search stage as a result of the decrease in population variety. These difficulties have also become in importance in DE theory research. Many academics have worked hard over the last 20 years to enhance DE performance [5]. The DE has been improved primarily by the use of adaptive control parameters, the creation of novel mutation operators, the fusion of other optimization methods, and the implementation of multiple population parallel search, among other things. Various schemes using Differential evolution have been suggested for various image segmentation like [2] chaotic Differential evolution for image enhancement; Image segmentation using DE and OTSU.

Ha Che Ngoc introduced a novel method for enhancing contrast of an image predicated on a correlation between the DE algorithm and the sigmoid function. The DE method is used for determining the parameters of the sigmoid function in order to maximize the estimate of contrast. Moreover, Parihar, A.S Verma and O.P. Yadav [3] presented a differential evolution-based the best fuzzy system to improve image contrast. For undersaturated and oversaturated regions of an image, the algorithm employs distinct fuzzification functions to obtain an enhanced image, [7] fuzzy characteristics are being made clearer by reversing the membership functions.

The firefly algorithm is a hypothetical ongoing adaptation that is applied for concurrent exploration on local and global feature point. Incorporating firefly algorithm various image enhancement techniques have been proposed to get better results. The firefly algorithm is a recent heuristic approach for non-linear or non-continuous optimization [7]. The optimistic heuristic approach is applied to enhance image contrast by Majumder and Irani. Hassanzadeh, Mahmoudi and Vojodi created a technique of adaptive local enhancement which is based on the firefly algorithm to enhance the quality and tone of input images. Incorporating chaos sequence and levy flight method, Gopal et al. established two techniques to enhance the contrast across moderate images.

In 2004, Munteanu and Rosa demonstrated a technique for improving the contrast of an image and employed an Evolutionary Algorithm [4] to determine the most effective adjustment. Furthermore, they implemented a heuristic strategy to conduct the image enhancement [8], and they had to assess their technique through both qualitative and quantitative methods, underlining its superiority to Linear Stretching and Histogram Equalization.

4 Proposed Method

The proposed approach employs the DE algorithm's mutation, crossover, and selection tactics for clustering [5]. The inclusion of the DE algorithm's mutation, crossover, and selection strategies aids in the thorough examination of the entire set of potential solutions and fast attainment of the best outcome. An illustration of the proposed approach is shown in Fig. 2 (Fig. 3).

Steps involved in proposed approach are explained below:

1. Image Preprocessing

Convert input image into double two-dimensional matrix.
Convert input TrueColor image RGB to the grayscale image.
Reshaping grayscale image to vector

2. Start DE Clustering

Assign number of clusters(k)

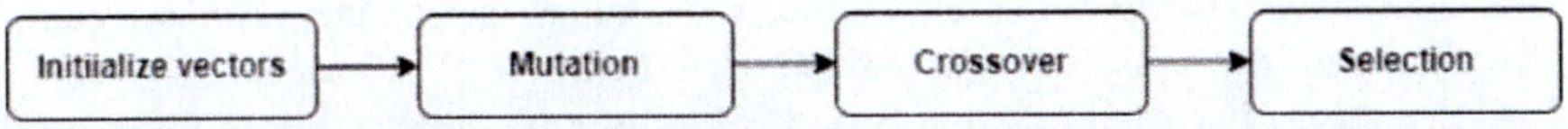

Fig. 1 Differential Evolution Algorithm

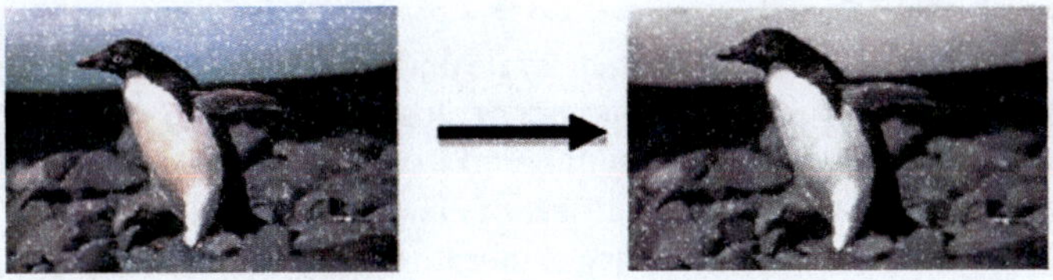

Fig. 2 Image Preprocessing

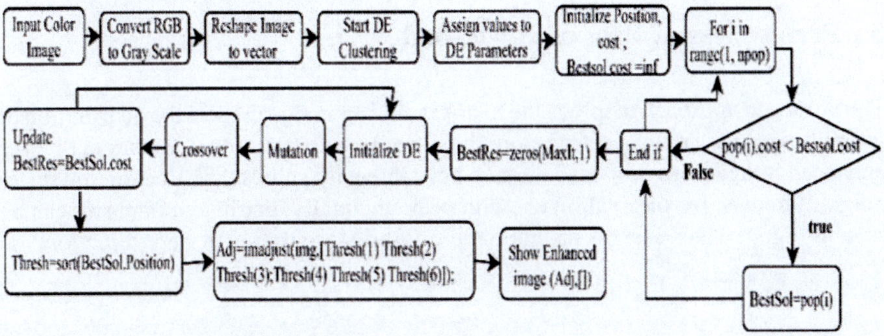

Fig. 3 Proposed Algorithm

Table 1 Values for Proposed DE parameters

Parameters	Value
Cluster size	6
Maximum number of iterations	50
Crossover Probability	0.2
Population size	6

Define Cost function and decision variables
Assign values to DE parameters (Population size, maximum iteration, scaling factor, crossover probability)
Start clustering with empty position. And get best Cost value for each cluster

3. Initialize Population
4 Mutation: Generates a random number of decision variable matrix size from continuous uniform distribution with the lower and upper bounds of scaling factor.

Calculate mutant vector using *DE/rand/1* scheme.

5. Crossover: Using Crossover scheme, trial solution is generated. And compared with recent best solution to get higher cost value.
6. Selection: For each iteration Based best solution is updated or remains same based on cost value returned.
7. Map intensity values in grayscale image to best cost values for each position.
8. Return enhanced image.

Some parameters were kept constant while performing proposed method. Population size was kept equivalent to the number of clusters in order to search efficiently in multi-dimensional area. Maximum number of iterations that return best cost values. Crossover Probability controls the number of elements that will be altered and it was kept low in order to avoid unnecessary exploration. The values of the DE parameter are shown in Table 1.

5 Preciseness of Proposed Method

The proposed approach employs the histogram of an image reflects the distribution of intensities across an image. And from the histogram distribution it is evident that the enhanced image from proposed method have intensity values that fill entire intensity range. Whereas, for original image with peak unequally distributed intensity can be seen in Fig. 4. Thus, assures us that the proposed algorithm is in right direction.

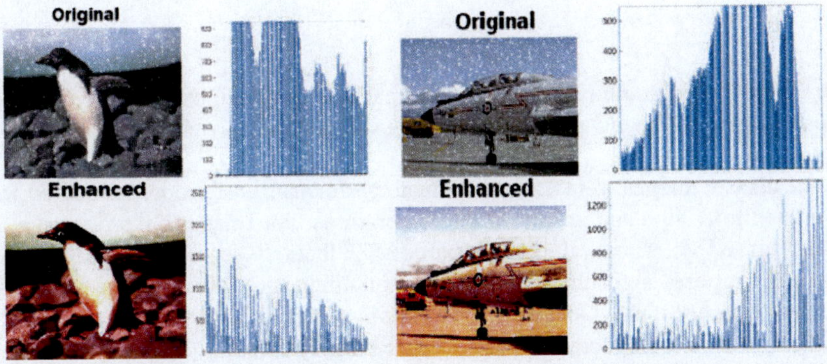

Fig. 4 Histogram distribution of original and enhanced image by proposed algorithm

6 Experiment and Results

This research utilized multiple color images for executing Image contrast enhancement experiments; however, the outcome of two images is presented. The proposed differential evolution-based image contrast enhancement using the [7] Clustering (DE-ICC) technique's results are compared to those of other techniques i.e., firefly-based image contrast enhancement [9].

The two algorithms were enacted and simulated using the MATLAB R2022b software on a desktop pc outfitted with an Intel® Core (TM) i3-7020U processor at 3.0 GHz, 8 GB of RAM, and a 64-bit operating system.

The proposed approach and the Firefly-based image contrast enhancement techniques were applied over some random RGB images chosen from Berkeley-dataset in JPG format with differing sizes and an 8-bit pixel density for every channel. Resulting images of the performed task are shown in Fig. 5.

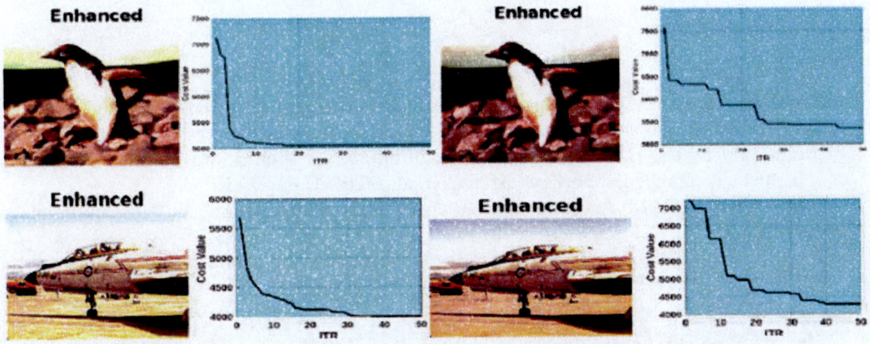

Fig. 5 Experimental results with Convergence graph for image (a) Modified Firefly Algorithm; (b) Proposed Method

6.1 Comparison

The Convergence graph (Fig. 5) for each image were drawn based on the best cost value returned for each iteration using proposed method as well as modified firefly method.

The dataset images were used to evaluate performance of proposed method with modified firefly algo and estimated on 50 iterations. For Images the proposed algorithm returned better cost value as compared MFA that reflects the competitive performance of proposed algorithm. For Image 1, too much variation can be seen. Considering collectively, it is evident from the convergence graphs. The proposed algorithm has been proven to be more efficient than the already existing firefly algorithm.

7 Conclusion

In this study, we proposed a Differential Evolution Image contrast enhancement algorithm driven by clustering. The algorithm takes color image as input and outputs the intended contrast enhanced image. The proposed algorithm advances the contrast and traits of the dataset images. This algorithm is evaluated using images taken from the Berkeley image dataset. The resulted images procured by original images and firefly image enhancement images are then compared with proposed images.

The histogram distribution (Fig. 4.) of images showed that the proposed algorithm images have much better intensity distribution as compared to the original images. Moreover, the resultant image has much better contrast as compared to original and firefly image enhancement images, this is clearly visible from Fig. 5. The convergence graphs revealed that the proposed method outperformed the prevailing method for the images.

References

1. Seema GB, Bansal G (2017) Image contrast enhancement approach using differential evolution and particle swarm optimization. Int Res J Eng Technol 4:1134–1138
2. Chakraborty S et al (2022) Differential evolution and its applications in image processing problems: a comprehensive review. Arch Comput Methods Eng 1–56
3. Kaur K, Vijay M (2018) Adaptive differential evolution-based lorenz chaotic system for image encryption. Arab J Sci Eng 43:8127–8144
4. Kumar N et al (2022) Color image contrast enhancement using modified firefly algorithm. Int J Inf Retrieval Res (IJIRR) 12:1–18
5. Nag S (2019) Vector quantization using the improved differential evolution algorithm for image compression. Genet Program Evolvable Mach 20:187–212
6. Solórzano-Espíndola CE, Anzueto-Ríos Á (2018) Automatic fuzzy contrast enhancement using gaussian mixture models clustering. In: International congress of telematics and computing. Springer, pp 120–134

7. Parihar AS et al (2018) Image contrast enhancement using differential evolution. In: Advances in communication, devices and networking. Springer, pp 517–526
8. Prakash SJ et al (2021) Contrast Enhancement of Images Using Meta-Heuristic Algorithm. Traitement du Signal 38(5)
9. Shih MY, Enríquez AC, Hsiao TY, Treviño LMT (2017) Enhanced differential evolution algorithm for coordination of directional overcurrent relays. Electric Power Syst Res 143:365–375
10. Wang XN, Feng YJ, Feng ZR (2005) Ant colony optimization for image segmentation. In: 2005 international conference on machine learning and cybernetics. IEEE, pp 5355–5360
11. Yang Q (2010) An adaptive image contrast enhancement based on differential evolution. In: 2010 3rd international congress on image and signal processing. IEEE, pp 631–634

Smart Traffic Monitoring System

A. Bragadeesh, S. Harish, and N. Sabiyath Fatima

Abstract This paper presents a prototype design for a smart traffic system to be used at junction roads. This project uses traffic cameras for vehicle detection where automatically the signal timing changes by sensing the traffic data. Here, the parameters of the traffic dataset are the number of vehicles, speed, and vehicle type. If there is no vehicle on the road with a green signal, the traffic keeps accumulating on other roads of the junction. The conventional traffic system uses a fixed time, and its performance is considered poor in efficiency. To provide an alternative to conventional traffic systems, this proposed system is designed to effectively manage traffic congestion. This proposed method uses computer vision and analyzer algorithms to detect the vehicle's speed and calculate whether or not the vehicle will be able to reach the stop line to pass through within the available green signal time in real-time. If there is no vehicle in the green zone of a traffic light that has the green light, then it will check the other camera which is planted at a particular distance away from the stop line to detect vehicles and if there is a vehicle crossing this camera, the computer vision and the analyzer will check the speed, of the vehicle and calculates whether the vehicle will pass the traffic light before the red light is ON. If the detected vehicle will not pass on time the traffic light will automatically change to Red and the next road will get a green light instantaneously. So, the road traffic can be cleared and the signal time can be used effectively, Thus it will address the problem which causes traffic congestion at the junctions and more importantly prioritize green signal changes for emergency vehicles in real-time scenarios.

Keywords Greenzone · Traffic · Analyzer algorithm · Emergency vehicles · Traffic camera

---

A. Bragadeesh · S. Harish · N. S. Fatima (✉)
Department of Computer Science and Engineering, B.S. Abdur Rahman Crescent Institute of Science and Technology, Chennai, India
e-mail: sabiyathfathima@crescent.education

N. Hasteer et al. (eds.), *Decision Intelligence Solutions*, Lecture Notes in Electrical Engineering 1080,
https://doi.org/10.1007/978-981-99-5994-5_24

1 Introduction

Traffic congestion leads to unnecessary consumption of fuel in the vehicles that are stuck in this traffic and also leads to sound pollution. And the emergency vehicles like fire engines and ambulances can't reach their destination quickly, potentially increasing the chance of human loss. So to address these problems, this proposed traffic control system automatically changes the traffic signal if there is no vehicle on a particular road and this traffic control system with the help of signal adjustment gives priority to emergency vehicles. This traffic system uses programmed cameras to detect vehicles on the roads. And the algorithm used in the analyzer camera also detects emergency vehicles like ambulances and fire engines. The algorithm will control priority over the roads and will automatically change the traffic signal in real-time for efficiency.

The traffic light system that already exists is a prefixed timer type, it set a constant time for every signal cycle leading to a waste of time during the empty road, and the other roads will be left with heavy traffic, and it will lead to noise pollution and the lots of fuel will be wasted in the vehicle on the other roads. Emergency vehicles like ambulances and fire engines cannot get to their destination in time. In the cross junctions, the system is programmed in such a way that it'll change the light in a particular interval of time. It's an annoying problem of congestion during heavy traffic hours. Here, the proposed traffic control system will generate traffic light signal accordingly, based on whether the vehicle is present on the road or not, and priority will be given to the emergency vehicles. This system will be used for urban cities and therefore, is a better solution to deal with traffic congestion.

The purpose of this paper is to coin out and model a system where the green signal time at the traffic light is not wasted. The analyzer near the traffic light will detect whether there are vehicles near it, if no vehicle is near the traffic light then the other analyzer that is at distance from the traffic light will start to check, whether a vehicle passed through the road on the time of the green signal, and if so, then the speed of the vehicle is calculated and the analyzer will check whether it can pass through the green signal with the remaining time, If not, the traffic light will change to red. In the case of an emergency vehicle, once a road is changed from green to red, it will check whether any other road has an emergency vehicle, if it does, the signal will be changed to green. If an emergency vehicle is detected by the analyzer that is at distance from the traffic signal, the particular road will remain green until the emergency vehicle passes the road.

This paper is organized by the proposed method's working workflow and its implementation in real life. Followed by practically generated data to test the model efficiency and compared using a Gantt chart, and then concluded.

2 Related Works

[1] Basavaraju et al., proposed a sensor system to detect vehicle density to manage traffic. It calculates the traffic density along the way by several sensors. This system can be controlled remotely and the timing of the traffic lights can be delayed through the microcontroller and the recorded data can be monitored. The recorded data can also be erased after it is analyzed. The limitation of this paper is it did not consider any emergency vehicles.

[2] R. Gosh et al., proposed an auto-density traffic system using AT89S52. This model controls the traffic using traffic density data by installing an IR transmitter at both ends of the road. If a road has a higher density, the AT89S52 will trigger the green light to glow. The limitation of this design is, no traffic light sequence is followed in this model, i.e. red, yellow, green. Only red and green lights are shown, and it also does not accommodate emergency vehicles in the traffic system.

[3] Udoahah et al., proposed a traffic light surveillance system which can be used in the cross junction to determine traffic density on each road using IR sensors. The flow of traffic on each road was determined based on the traffic density. The drawback of this design is it gives more importance to dense traffic roads but it causes more traffic on other roads. The emergency vehicles don't get priority. The overall working of this design is time consuming.

[4] Gayathri proposed a density-based traffic control system with an emergency vehicle alert and the system is powered by solar. This design uses IR sensors and photodiodes, these are placed in the line of sight of configuration across roads. An RF transmitter will be present in the emergency vehicle. The power for the circuits will be from solar energy to run the traffic light all the time. The limitation of this system is; the RF transmitter should be set in the emergency vehicles; it makes the design expensive and unachievable. And the system won't get solar energy every season, so the system will fail if it won't get enough power to run the traffic control system.

[5] K. Shekar et al., proposed an automatic traffic management system. To control the traffic, IR sensors are installed on the side of the road in the junction. Importance is given to the higher-density traffic road. A mechanism is installed to detect absence of sunlight and activate street-lights using light-dependent resistors (LDR). The limitation of this design is it doesn't give priority to the emergency vehicle.

[6] Dauda Elijah et al., proposed a traffic signal controlling system that uses an ultrasonic sensor and a sound sensor. The traffic density is caliberated using the ultrasonic sensor and the signal timing changes automatically. If an emergency vehicle is present in any of the roads, a sound sensor is used and importance is given for that road. Immediately green signal is given to that road and another road will be given the red signal.

[7] K. Swathi et al., proposed a video monitoring system to determine the traffic density and vehicle classification.This design uses IR and GPS to acquire the emergency vehicle alert and traffic density calculation.

[8] Dr..godese et al., proposed a traffic controlling system to detect density of the road. If the density is higher, the road will be changed from red to green. The component used in this design is a raspberry pie to detect the density of the traffic. This paper tries to reduce the traffic jam to a certain extent. This design is based on image processing based traffic light system, and also canny edge detection is used to detect the traffic density. The limitation of this paper is there is no priority given to emergency vehicles.

[9] Nur Shaaadah Nik Dzulkefli et al., proposed a traffic controlling and managing system to detect traffic densityc. The green light will be changed to red light if there's no vehicle on the road. The components used in this design are an Arduino UNO microcontroller and infrared sensor to detect density. The limitation of this paper is there's no importance given to emergency vehicles.

[10] Rotake et al., proposed a traffic controlling system using an embedded system and it has a microcontroller with in built 8-channel ADC for IR – input from the IR transmitter. This transmitter will be in an emergency vehicle. If the emergency vehicle is detected, the divider gate will be opened. The limitation of this paper is it cannot control traffic congestion.

[11] Anand Shah et al., proposed a traffic control system, this system is used to investigate whether the traffic light systems that are controlled by the traffic density are IR sensors, RFID tags, and thermal cameras also this system uses reinforcement learning, this system doesn't consider emergency vehicles to give priority.

[12] Usikala et al., proposed a traffic control system that uses Arduino integrated development environment (IDE) to control the traffic light. It will prioritize the lane which has the higher vehicle amount, the components used in this system are Arduino and infrared sensors. The disadvantage of this system is there's no importance given to emergency vehicles.

[13] Jyoshna et al., proposed a traffic controlling system in which the traffic light signal changes automatically sensing the traffic density at the junction. The components used in this system are Arduino and IR sensors. These sensors will be mounted on the side of the roads. But the disadvantage of this system is that this system doesn't give priority to emergency vehicles.

[14] T.E. Somefun et al., proposed a density-based traffic management system which uses infrared sensors as a counter to measure the traffic density. And also, the infrared sensor is used for detecting the speed of the vehicles. The components used in this system are infrared sensors, Arduino mega with Atmega 2560 chips, and the Bluetooth model, Bluetooth serial monitor. The disadvantage of this system is it doesn't prioritize emergency vehicles.

[15] Anand Gaikwad propsoed a traffic system where the green light's and red light's time period is based on traffic density present at that exact moment. This system has a number plate recognition system as well. The components used in this system are ARM7TDMI – S a general-purpose 32-bit microprocessor, piezo sensor, and web camera, but this system doesn't give priority to emergency vehicles.

3 Proposed Methodology

A prefixed timing system is pre-planned and fixed for a particular road junction by analyzing the traffic behavior by specifying how long should each signal last. This doesn't require any human intervention after installation. Let's assume the average speed within 400 m of a junction is 60Kmph, and usually, it is maintained until the end of the signal as it's the highest speed within the economic range for most vehicles. The techniques involved in this proposed system are green zone and analyzer algorithms (refer to Fig. 4).

The Green Zone: In this method, the first 20 m from the stop line from a traffic signal is referred to as the Green zone. This is an imaginary calculated area where generally all vehicles stop and hold their position for the RED signal and wait for the green signal.

The Analyzer: Its an additional camera that is placed 330 m away from the stop line on each road's traffic signal junction that is designed to capture data of passing vehicles for their type and speed. It is simply a tool used to collect necessary data for the algorithm. Generally, the analyzer will actively collect data for 15 s starting from the last 20 s of remaining green signal time. This is because by observation it's around 60 kmph the speed that most vehicles maintain on highways as it is the highest within the economic speed range in most vehicles. So any vehicle that is passing the analyzer at the initial of its 15 s will always make it to the green zone if the speed is maintained. All this collected data will be sent to the traffic signal algorithm for taking further decisions depending on traffic on other roads through cable wires underground.

4 Implementation

Generally green signal lasts 60 s for each road and when one road gets a green signal then all other three roads are switched to red signals instantaneously throughout the 60 s. In this method each road is referred to as **R1**, **R2**, **R3**, **R4** as visible in Fig. 1, and **EV**-Emergency vehicles, **NV-** normal vehicles, **Gzone**- Greenzone, and **@x**- where x refers to the green signal time used up.

CASE 1: When a green signal countdown reaches the last 20 s, it will trigger the Analyzer that is 330 m from the stop line as seen in Fig. 3 for each road in a traffic signal junction. The Analyzer will now start to monitor every single vehicle that passes by it for 15 s, and the information is sent to the traffic signal on its road which can add 5 more seconds to the green signal if more than 70%of vehicles that passed the analyzer can reach the Gzone as highlighted in Fig. 3.

If only less than 70% of passed vehicles can reach the green zone, then the analyzer won't send any information to the traffic signal and let the default pre-fixed timer run its routine.

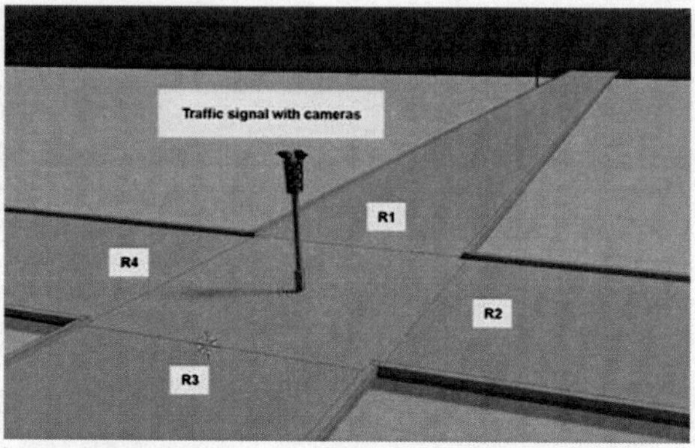

Fig. 1 The junction of traffic signals

CASE 2: Assuming the same scenario as in CASE1, among the 70% of passed vehicles, if it has Firetruck, Ambulance, or other emergency vehicles, the Analyzer will add 8 s which can be seen in Fig. 4 to the green signal to ensure these vehicles don't wait for long.

CASE 3: Suppose at the start of the Green signal countdown if there are no vehicles in the gzone captured in traffic cameras from Fig. 2, then a trigger will be sent to Analyzer for 5 s to start collecting data. If no vehicles pass or vehicles pass with less than 60Kmph during this 5 s then vehicles won't make it to the zone in time, then Analyzer will quickly send a request to the traffic signal to RED as seen in Fig. 4 so that the remaining 55 s of the wait time is saved for other roads where traffic is accumulating.

During the Clash of Cases: Generally, pre-fixed traffic signals run in either a clockwise or anti-clockwise direction strictly. If four roads are clockwise labeled as R1, R2, R3, and R4 and the green signal always follows the same sequence, not all highways will have a reserved road. But in this proposed methodology if R1 holds the

Fig. 2 The traffic cameras

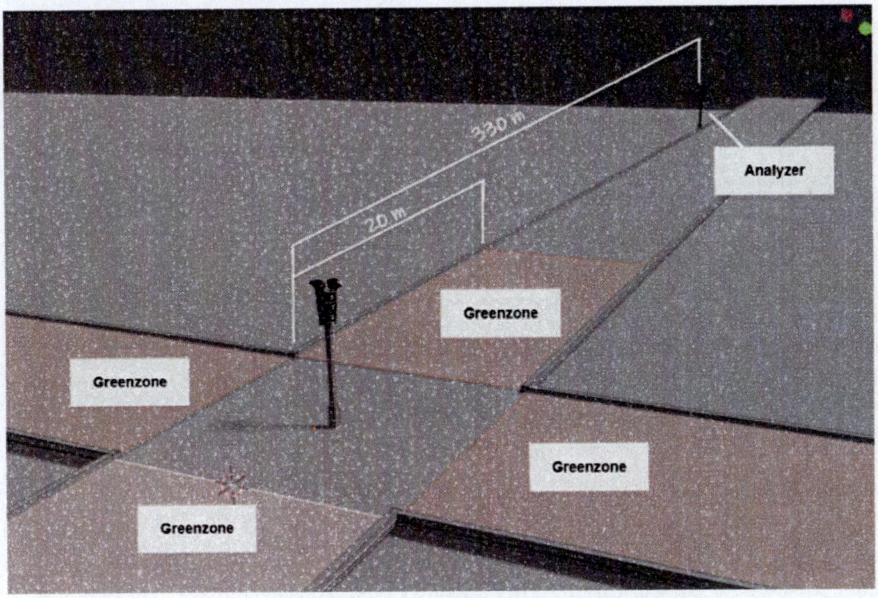

Fig. 3. 3D View where the analyzer is placed 330 m away from the stop line and gzone (Greenzone) of 20 m is highlighted in the orange shade (not in actual proportions)

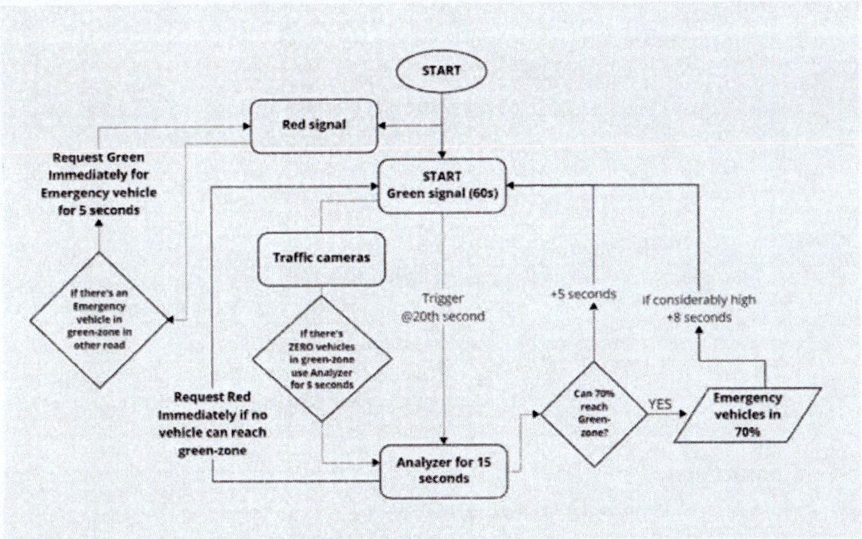

Fig. 4 The Workflow of the Analyzer algorithm

green signal and simultaneously R3 has an emergency vehicle within Gzone, then after the green signal in R1 is timed out, instead of switching the green signal to R2 by order, the analyzer in R3 would have requested (generally for 5 s), for emergency vehicle and will be granted with green signal until the emergency vehicle is detected to disappear from zone from a traffic camera. The moment the emergency vehicle passes through, the green signal is given back to R2 for default timing and direction and continues as normal.

Analyzer Workflow

Figure 4 below gives brief information as an alternative for the Prefixed timer type of traffic signal that already exists. Also, this flow diagram is represented how to develop the analyzer algorithm with only one traffic light facing one road, and there exist a total of four in a junction facing all four directions working interdependently as seen in Fig. 1.

The Pseudo-Code

The following code provides a clear image of the structure of the analyzer algorithm to be embedded into the traffic system for one road out of four, so each road will have the same code working interdependently based on the signal.

```
//START OF CODE
GreenSignal()
            TrafficCam()
                  If( VehiclesDetected ==0)
                        Analyzer(5 seconds)
                              If(VehiclesSpeed< 60 Kmph)

                                      RedSignal();

            Analyzer(@45th second)
                  ForEachVehicle(Speed,type)
                        If (GreenZone()>70% && Type==Emergency)
                              Green(+8 seconds)
                        else if (GreenZone()>70% && Type!=Emergency)
                              Green(+5 seconds)
RedSignal()
      trafficCam()
            if(vehiclesDetected !=0)
                  if(EmergencyVehicleCount > 0)
                        GreenSignal(+5 seconds)

//END OF CODE
```

5 Result and Analysis

Assuming it takes 10 s for 20 vehicles to pass through the stop line (this varies depending on locations, so has to be assessed regularly by local traffic authorities).

R1-Road 1, **R2**- Road 2, **R3**- Road 3, and **R4**- Road 4 are the defined roads of the junction. And the Gantt chart at top of the diagram is referring to the existing prefixed timer system of 60 s green, while the bottom chart represents the proposed analyzer algorithm. The green bar indicates a green signal and the red indicates a red signal.

Scenario 1:
R1-100 nv, R2-130 nv, R3- 140 nv, R4- 80 nv

In scenario 1 looking at the Gantt chart from Fig. 5, the prefixed timer system at end of one cycle still holds 10 vehicles in R2 and 20 vehicles in R3 in traffic at 240th seconds, while this proposed method resulted in only 10 vehicles in R3 and all other roads cleared by 230rd second. This is a 10 s improvement for regular traffic that involves no emergency vehicles.

Scenario 2:
R1-100NV, R2-70NV, R3- 130NV with EV@55, R4- 150NV.

In scenario 2, Fig. 6, the prefixed timer system at end of one cycle still holds 10 vehicles in R3, with emergency vehicles not cleared yet and 30 vehicles in R4 at end of 240th seconds, while the proposed method resulted in only 20 vehicles in R4 and no emergency vehicle is stuck in R3 by end of 228th seconds. This is 12 s improvement in realistic peak-hour traffic in urban areas with no Traffic police to handle it manually. In R2 it takes only 35 s for 70 vehicles to pass through, but it took 40 s of the green signal to ensure there are actually no other vehicles following up immediately to use the green signal for 5 s.

Scenario 3:
R1-100NV, R2-0NV, R3- 60NV with EV@Gzone, R4- 80NV with EV @Gzone.

In scenario 3, in Fig. 7, it took 240 s for the prefixed timer system to clear all roads but held one emergency vehicle in R3 until the 130th second and another in R4 until the 190th second wasting too much emergency time. But the proposed method not

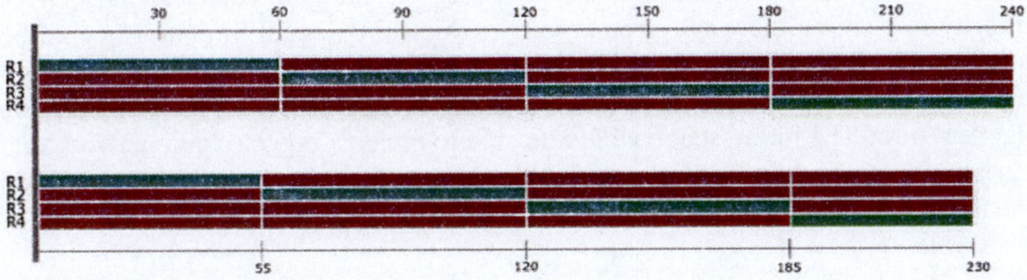

Fig. 5 Gantt chart for scenario 1

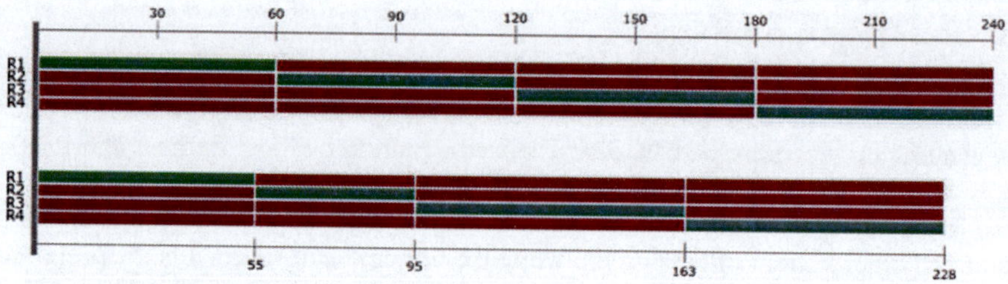

Fig. 6 Gantt chart for scenario 2

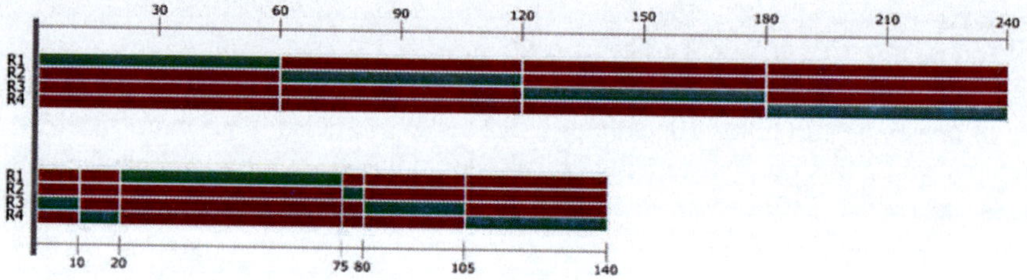

Fig. 7 Gantt chart for scenario 3

only took just 140 s to clear the whole traffic with an improvement of significantly 100 s but also cleared emergency vehicles in R3 within 10th seconds and R4 within 20th seconds by overriding the green signal cycle, for exceptional cases like these, that provide priority for emergencies.

Gantt charts used above aren't generated using real data, but out of pure mathematics with practical generation data for different scenarios.

6 Conclusion

This proposed traffic monitoring system is better than the existing conventional system, by managing traffic light's green signal time for the road that has no vehicle, with emergency vehicles, and accumulating traffic. It spends the traffic time for the vehicles on all roads of junction wisely by using real-time traffic data and also gives priority in clearing the way for emergency vehicles. Potential limitations could be costly as it involves additional cameras, and winds may carry plastic covers that can cover the camera's lens making it unusable, and short circuits may occur disrupting the data flow. The future scope would be able to create a bigger cloud network that can actively monitor traffic density using data from all neighbouring junctions within the same area using 5G and manage the traffic even better.

References

1. Basavaraju A, Doddigarla S, Naidu N, Malgatti S (2014) Vehicle density sensor system to manage traffic. Int J Res Eng Technol 3:1–4
2. Ghosh R, Rasaily D, Dey I (2016) Auto density sensing traffic control system using AT89S52. Int J Eng Trends Technol 32(5):208–212
3. Udoakah YN, Okure IG (2017) Design and implementation of a density-based TrafficLight control with surveillance system. Niger J Technol 36(4):1239–1248
4. Gayathri R et al (2016) Solar powered traffic control system based on traffic density with emergency vehicle alert. Middle-East J Sci Res 24(2):234–238
5. Aruna KS (2016) Implementation of power saver street lighting and automatic traffic management system. Int J Comput Sci Mob Comput 5(1):15–24
6. Mshelia DE, Alkali AH, Dada EG, Ismail K (2019) Design and development of traffic density detection and signal adjustment system. Asian J. Appl. Sci. Technol. (AJAST) 3(1):86–98
7. Swathi K, Sivanagaraju V, Manikanta AKS, Dileep Kumar S (2016) Traffic density control and accident indicator using WSN. Int J Mod Trends Sci Technol 2(4):1-4
8. Godse SP, More N, Surana A, Patil P, Kamble S (2019) Traffic density detection with vehicle identification for smart traffic monitoring. Int. J. Res. Eng. Technol. Sci. 8:1–6
9. Dzulkefli NNSN et al (2019) Density Based Traffic System via IR Sensor. J Phys Conf Ser 1529:022061
10. Rotake D, Karmore S (2012) Intelligent traffic signal control system using embedded system. Innov Syst Des Eng 3(5):11–20
11. Shah A, Prof JB, Kulkarni RP, Gogave S, Sisode V (2020) Traffic control system and technologies: a survey. Int J Eng Res Technol V9(01):1–5. https://doi.org/10.17577/IJERTV9IS010246
12. Usikalu MR, Okere A, Ayanbisi O, Adagunodo TA, Babarimisa IO (2019) Design and construction of density based traffic control system. IOP Conf Ser Earth Environ Sci 331(1):012047. https://doi.org/10.1088/1755-1315/331/1/012047
13. Chanda J (2021) Density based traffic control system using Arduino. SSRN
14. Somefun TE, Awosope COA, Abdulkareem A, Okpon E, Alayande AS, Somefun CT (2020) Design and implementation of density-based traffic management system. Int. J. Eng. Res. Technol. 13(9):2157–2164
15. Gaikwad A, Shreya S, Patil S (2018) Vehicle density based traffic control system. Int J Trend Sci Res Dev 2(3):511–514. https://doi.org/10.31142/ijtsrd10938

Intelligent Waste Bot Using Internet of Things and Deep Learning for a Smart City

Rudresh Shirwaikar, Yogini Lamgaonkar, Lyzandra D'souza, and Diksha Prabhu Khorjuvenkar

Abstract The buildup of solid garbage in metropolitan areas is causing environmental contamination and may be harmful to human health if not adequately controlled. A sophisticated/intelligent waste management system is required to handle various wastes. Separating garbage into its component parts is a crucial stage in waste management and is generally done manually by hand- picking. We propose a bot-based waste management system with nodes consisting of mechanical robotic arms to pick up waste material. Deep learning is a kind of machine learning that harvests information by layering data representation and feature extraction. The robotic arm has been designed to move in the same way as a human arm does. This is accomplished by using an Arduino Nano microcontroller with servo motor control which is programmed in the Java language. Convolution neural network, a deep, feed- forward artificial neural network, was used to accurately assess the image. This technique is offered as a part of a solid waste sorter. The combination of robotic waste management with a deep learning system allows trash to be classified and disposed off automatically. As a result, trash segregation with deep learning includes image acquisition from a camera, object recognition, prediction, and categorization into biodegradable and non-biodegradable categories.

Keywords Waste Segregation · Convolution Neural Network · Smart City · IoT · Deep Learning

1 Introduction

Every year, millions of tons of waste is produced globally, with India generating roughly 100,000 tones every day. With a population of 1.38 billion, waste disposal in India is still a laborious procedure fraught with health risks [1]. With increasing

R. Shirwaikar (✉) · Y. Lamgaonkar · L. D'souza · D. P. Khorjuvenkar
Department of Computer Engineering, Agnel Institute of Technology and Design, Goa University, Taleigão, India
e-mail: rs@aitdgoa.edu.in

N. Hasteer et al. (eds.), *Decision Intelligence Solutions*, Lecture Notes in Electrical Engineering 1080,
https://doi.org/10.1007/978-981-99-5994-5_25

population and environmental concerns, garbage management and waste segregation (biodegradable and non-biodegradable) issues must be addressed [2]. Due to the rapid rise in population and consumption, major cities around the world are currently facing a variety of waste management issues. Waste management, in general, entails a variety of processes, involving garbage collection, transport, processing, disposal, and monitoring. The difficulties in this area include inadequate garbage collection infrastructure, a lack of funding, health concerns, and proper waste management planning.

Currently, garbage management is done manually utilizing auto tippers that go to each community/society to collect garbage. In some areas of the city, garbage is separated into biodegradable and nonbiodegradable categories. In Some Countries, garbage is collected from house-to-house by the government and automated waste collection is also available [3–5]. In India, no automated waste segregation technology is available at the household level, and low-cost, user-friendly and a compact, waste segregation system for urban houses is urgently needed to streamline the waste management process.

To properly manage waste, it must be collected, sorted, transported, and disposed off in a way that causes the fewest health and environmental concerns. Automation can decrease the need for human contact while maintaining good sanitary standards [6]. Artificial intelligence (AI) and the Internet of Things (IoT) are two technologies that can be used to develop an efficient system and automate waste management processes. They provide innovative methods for lowering the expense and complexity of trash management.

Our proposed method, which employs a deep learning system and a robotic arm, utilizes completely autonomous garbage sorting. Deep learning waste segregation includes obtaining pictures from a camera, object identification, prediction, and categorization into biodegradable and non-biodegradable categories [7]. The robotic arm attached to the waste truck will include a gripper that will automatically sort biodegradable and non-biodegradable waste using Convolutional Neural Network as a segregation technique and deposit it in predetermined storage sections of the vehicle [8]. The objective is to replace human labor with automated waste segregation and management robot system that uses deep learning to improve efficiency.

2 Method

The task of image recognition is carried out by three units, data processing, training and classification. The system architecture as shown in Fig. 1, depicts how the system differ and accomplish various functions. In data processing, raw image data is captured by an external or internal camera, and classifications are performed using the input. The model is trained using CNN in Python programming. The robotic arm functions through microcontroller.

As the program starts (Fig. 2), all of the necessary packages are imported, including numpy, tensorflow, imutils, and a trained CNN model. The reading is done using an infrared model that determines whether or not there is an item present. It returns to sensing the entity, if the object is not present. When an item is detected, it takes a picture of it using a lens and sends signals to a USB port. The captured image is then loaded, in the imutils library for OpenCV.

Basic tasks like translation, rotation, scaling, skeletonization, and displaying Matplotlib pictures in image processing are made easier in Python 2.7 and Python 3 technology. The data is then given to the trained model, which employs deep learning techniques and the CNN algorithm to predict the final output. After the forecast is made, based on bio or non-bio, the robotic arm picks up the garbage and separates it into bio and non-bio bins.

We have included the necessary components in Fig. 3, the ESP32 is used to identify an object or garbage that is written in Python using the Python IDLE library. We have implemented the LENET Architecture for image training, and we have used the ESP32 microcontroller for which we wrote the code in C++ , where the serial data supplied from the image processing for the output has been classified here for the robotic arm for a certain dust bin. The ESP32 Microcontroller functions as a slave to a host MCU, lowering communication stack overhead on the most important software application. Microcontroller is used to identify an item or trash, with Python IDLE which is used to write the code.

Arduino comprises of a programmable circuit board and Arduino IDE which is ready-made software is used to develop and upload computer program to the physical board. A motor can be started, lights can be turned on and off, a connection to the cloud can be made and many other operations may be performed using an Arduino board to receive analogue or digital input data from various sensors [9] A CNN is a Deep Learning technique that takes an input image and assign priority to different aspects in the image while also identifying them. Pre-processing requirements for classification approaches are much higher than for ConvNets. With enough training, ConvNets can pick up these filters and attributes [10]

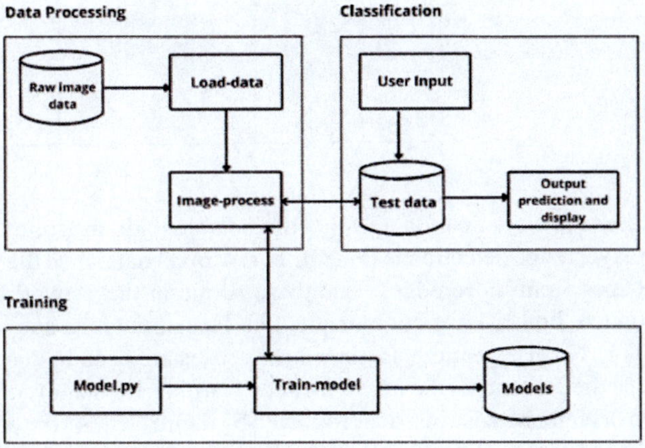

Fig. 1 System architecture design

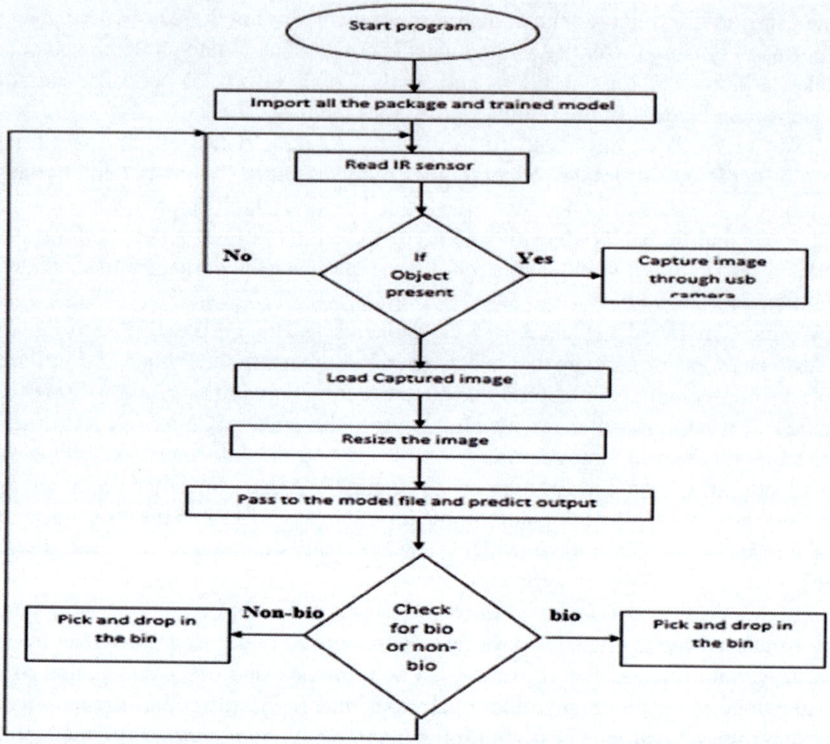

Fig. 2 Flowchart of the proposed model

Fig. 3 General components

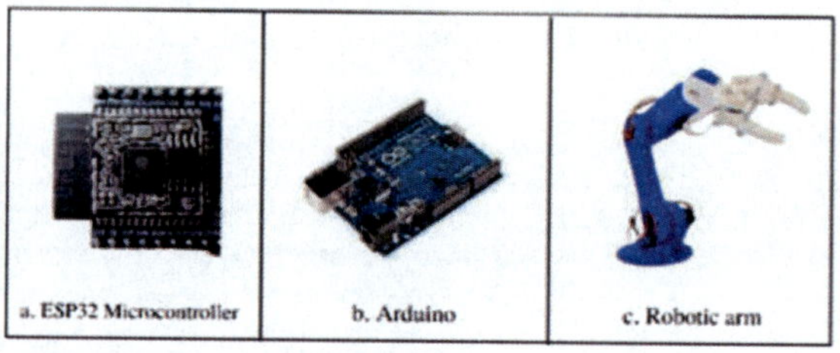

In terms of image classification, CNN could be able to do the following things:

In the first layer, edge detection is done from raw pixel data, then the second layer recognizes shapes from those edges. Use these forms in the network's uppermost layers to recognize higher-level components like face structures, automotive parts, and so forth [11, 12]. These more advanced properties are used by the CNN's final layer to forecast the contents of the image. In deep learning, a (image) convolution is a multiplication of element-wise matrices followed by a sum [13]. As the program starts (with reference to Fig. 1), all the necessary packages are imported, including NumPy, TensorFlow, imutils, and a trained CNN model. The reading is done using an infrared model that determines whether or not there is an item present. It returns to sensing the entity, if the object is not present. When an item is detected, it sends signals to USB to take a picture of it through the lens. After that, the image is loaded and the imutils library for Python and OpenCV is used to make basic image processing operations like scaling, skeletonization, translation, rotation, and presenting Matplotlib images easier. The data is then transferred to the Trained model, which utilizes the CNN algorithm to predict the final return using the deep learning approach [14, 15]

After the prediction is made based on the type of garbage (biodegradable or non-biodegradable), the robotic arm picks up the garbage and separates it into designated bins. The first stage of practice is training, which occurs after the dataset has been acquired. It is critical to train the dataset in order to acquire correct findings. A large number of inputs must be provided to train the network to make its own decisions. To recognize pictures, train a layer of features that accept input from the image's pixels with the purpose of image recognition [16, 17].

Fine-tuning begins with a layer-by-layer pre-training, despite the fact that the training process has previously been defined. The term "pre-training" refers to the process of manually training data by specifying the inputs. Because the user is initially in charge of the outputs of the inputs, this is known as supervised training. Deep Learning is used to determine whether or not the waste photos are biodegradable [18].

Edge detection and matching methods, as well as grayscale and gradient matching approaches, are utilized in the picture identification phases, and are analogous to those used in image processing. The input is provided via the camera, which might be external or internal. The TensorFlow library and imutils are used to process images on the computer, while the camera lens is used to detect the objects. The image is identified by combining the findings of numerous algorithms. Every layer in the CNN network employs a different set of filters, typically hundreds or thousands, and then aggregates the results before passing the output to the following layer [19, 20] (Fig. 4).

During training phase, automatically a CNN will learn the values for these filters. Following the detection of an item in the software, it is compared to the other alternatives which are offered during data training, and a probability list, for each option is generated.

As a consequence, the picture with the greatest probability index is selected as the item, resulting in the detection of the object in the trash, and the garbage is sorted into various mounds using the robotic arm. The system can learn and train on its own due to the self-learning algorithm, which eliminates the need for human intervention [21].

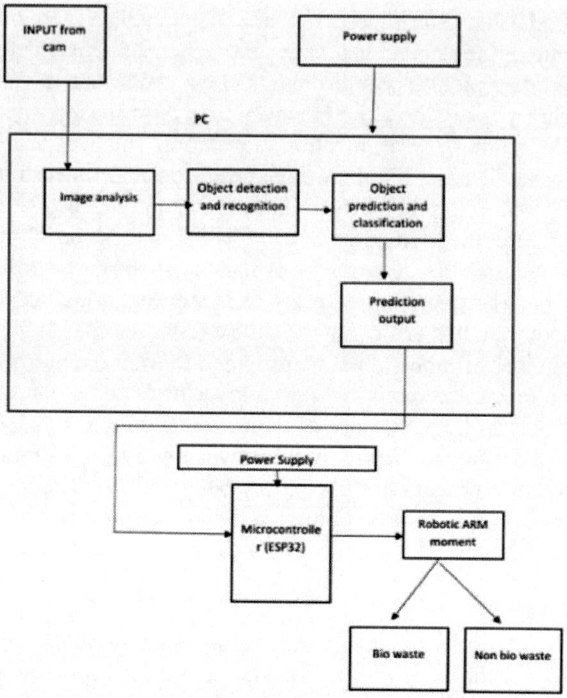

Fig. 4 Proposed system class diagram

3 Results and Discussion

The experimental results demonstrate successful trash segregation utilizing an auto waste segregation system. This system segregates metal and other wastes using an inductive proximity sensor. The robotic arm scoops up the waste as soon as it is identified.

3.1 Training Dataset

3.1.1 Video Capture

The photos are taken from the webcam, and the video must be displayed from the webcam according to the user's preferences. The input would be high-resolution video via a webcam. The expected output is to show video from the webcam at the user-specified resolution, with reference to Table 1 (Test case 1).

Table 1 Test Cases

Test Case	Name of Test	Input	Expected Output	Actual Output	Result
1	Capturing Video	Resolution, Webcam (width, height)	Display video from a camera with resolution set by the user	User selected video is shown	Successful
2	Loading Trained Model	File model.h5	No errors during the loading of the model	No errors when the model was loaded	Successful
3	Classifying Waste	Webcam input has been resized	Waste from frame input must be classified	Waste is categorized, from the frame input	Successful
4	ARM Commanding	Waste production forecast	The ARM movement	ARM moves to the left or right	Successful

3.1.2 Loading Trained Model

The models that have been trained are loaded to see if they contain any errors. Video and the model file model.h5 would be the input. The anticipated output is for the model to load without any errors. This test case has been completed and passed, as stated in Table 1 (Test case 2)., with the expected results.

3.1.3 Classify Waste

The webcam images are compressed or preprocessed before being used to classify the garbage in the image. The input would be a webcam image that has been compressed. The anticipated output is to classify the gesture from the user- specified input image. This test case was run and passed with the expected results, as indicated in Table 1 (Test case 3).

3.1.4 Movement of ARM

After identifying the waste output, ARM advances to the left or right, based on the object input and categorization. This test case has been completed and passed, as stated in Table 1 (Test case 4), with the expected results.

3.2 Data Estimation of the Sensor

The system utilizes an inductive proximity sensor to segregate metal and other wastes, and the experimental findings show that the wastes are successfully separated using an automated waste segregation system. The robotic arm picks up the waste once it has been spotted. When plotted against the period, the average training accuracy was 94.5%, which is practically perfect (see Fig. 5). The "epoch" serves as a benchmark for calculating time.

The loss function, which gives us the training and validation loss, is plotted against the epoch in Fig. 6; if the values of the loss function are low, the evaluation produces meaningful findings. It is reasonable that a small number of data points can be incorrectly classified in a functional classification task. Hence, the loss functional model agrees with our results. Figure 7displays the training dataset images and Fig. 8 shows the working model of designed robotic arm.

Fig. 5 Accuracy comparison

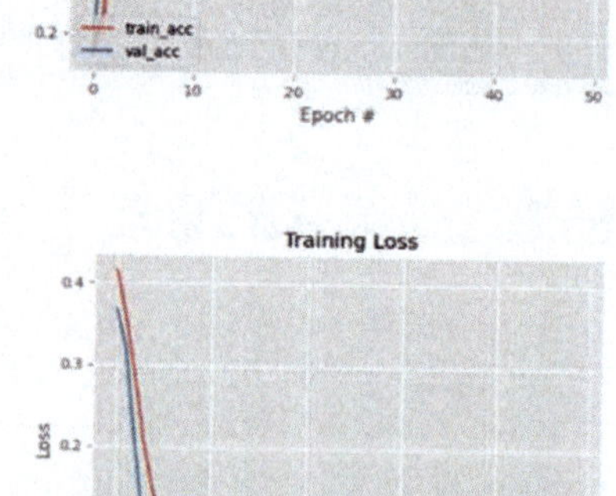

Fig. 6 Training loss and Validation accuracy

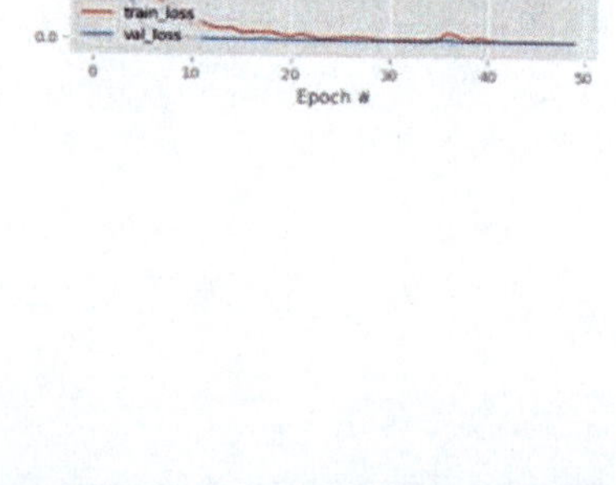

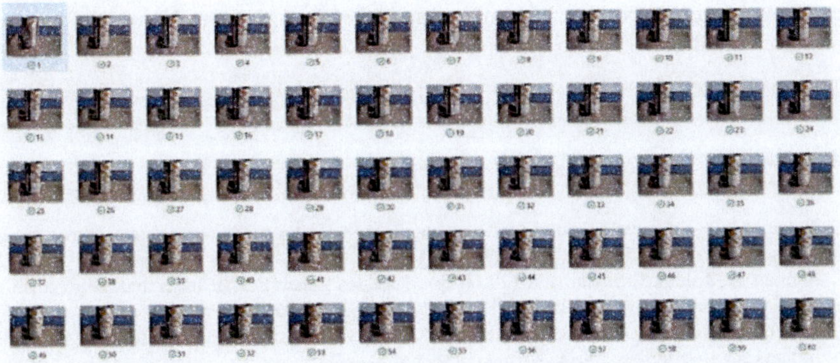

Fig. 7 Training Dataset

Fig. 8 Working model of a robotic arm

4 Conclusion

With the aid of deep learning algorithms, we proposed a waste segregation system that can separate various waste components. This technology can be used to classify garbage automatically, helping to reduce human intervention and guard against illness and pollution. When data was tested against the trash dataset, we got an accuracy of 87%. The smart garbage separating robot arm can separate between degradable and non-degradable rubbish. When the obstacle sensor is triggered, the camera is activated, and garbage is recognized using image processing. The robot is then informed to drop the waste in the appropriate waste bin. This system goes a step further in contributing to the cleanliness of our society, hence supporting our modest Prime Minister's notion of "SWACHH BHARAT ABHIYAN". The future scope of our system is to be able to categories more garbage item, by turning some of the used parameters, also crusher and advanced Artificial Intelligence could be added to this system to make it more advanced and efficient. With the help of our suggested methodology, the trash will be separated more quickly and intelligently, requiring less human intervention, making it a practical model for the Waste management agencies.

References

1. Tapan N (2009) Municipal solid waste management in India: from waste disposal to recovery of resources? Waste Manage 29(3):1163–1166
2. Asokan P, Mohini S, Shyam RA (2007) Solid wastes generation in India and their recycling potential in building materials. Build Environ 42(6):2311–2320
3. Mufeed S, Kafeel A, Gauhar M, Trivedi RC (2008) Municipal solid waste management in Indian cities – a review. Waste Manage 28(2):459–467
4. Yinghao C, Chen H, Xiaodan X, Bohai T, Shyam K, Xiaogang X (2018) Multilayer hybrid deep-learning method for waste classification and recycling. Comput Intell Neurosci 2018:1–9
5. Shuchi G, Krishna M, Rajkumar P, Sujata G, Arun K (1998) Solid waste management in India: options and opportunities. Resour Conserv Recycl 24(2):137–154
6. Kellow P, Joel R, Sergei AK, Neeraj K, Vasco F (2019) IoT-based solid waste management solutions: a survey. J Sens Actuator Netw 8(5):1–25
7. Alex K, Ilya S, Geoffrey EH (2012) ImageNet classification with deep convolutional neural networks. In: NIPS 2012 Proceedings of the 25th International Conference on Neural Information Processing Systems, vol. 1, pp. 1097–1105
8. Sindhu R, Vidhya S, Rajat K, Rohit S, Rachana S (2019) Waste segregation using artificial intelligence. Int J Sci Technol Res 8(12):903–905
9. Joseph J (2019) Smart waste management using deep learning with IoT. Int J Netw Syst 8(3):37–40. https://doi.org/10.30534/ijns/2019/10832019
10. Sandhya D, Vijaykumar VR, Muthumeena M (2018) Waste segregation using deep learning algorithm. Int J Innov Technol Explor Eng 8(2S):401–403
11. Dharmana MM, Aiswarya MS (2020) A low cost solution for automatic plastic segregation. Int J Eng Adv Technol 9(3):784–788. https://doi.org/10.35940/ijeat.C5275.029320
12. Anjali PA, Monisha BV, Manasa P, Ankit Kumar P, Shekhar R, Riyaz S (2019) Drone based solid waste detection using deep learning & image processing. In: Alliance International Conference on Artificial Intelligence and Machine Learning, pp. 357–364 (2019)
13. Adedeji O, Wang Z (2019) Intelligent waste classification system using deep learning convolutional neural network. Proc Manuf 35:607–612. https://doi.org/10.1016/j.promfg.2019.05.086
14. Myra GF, Jose B, Tan J (2020) Literature review of automated waste segregation system using machine learning a comprehensive analysis. Int. J. Simul. Sci. Technol. 20:15.1-15.7
15. Bansal S, Patel S, Shah I, Patel P, Makwana P, Thakker DR (2019) GDC: Automatic Garbage Detection and Collection. arXiv:1908.05849, pp. 1–12 (2019)
16. Pradeep K, Saravanan M, Andm A (2017) Hybrid network intrusion detection system based on Gann models. Int J Pure Appl Math 116(11):31–39
17. Tran Anh K et al (2020) Waste management system using IoT-based machine learning in university. Wirel Commun Mob Comput 2020:6138637.1-6138637.13
18. Bueno-Delgado M-V, Romero-Gázquez J-L, Jiménez P, Pavón-Mariño P (2019) Optimal path planning for selective waste collection in smart cities. Sensors 19(9):1973. https://doi.org/10.3390/s19091973
19. Cicerone LP, George C, Costel Emil C, Nicoleta LC, Tiberiu D (2017) Smart city platform development for an automated waste collection system. J Sustainab. 9(11):2064

20. Souptik P, Kolkata Sayan B, Srutayu B (2019) Smart garbage monitoring using IoT. In: IEEE 9th Annual Information Technology, Electronics and Mobile Communication Conference (IEMCON)
21. Bircanoğlu C, Meltem A, Fuat B, Özgün G, Merve Ayyüce K (2018) Recycle net: intelligent waste sorting using deep neural networks. In: 2018 Innovations in Intelligent Systems and Applications (INISTA), pp. 1–7. IEEE (2018)

Effect of Climate Change on Soil Quality Using a Supervised Machine Learning Algorithm

Ramdas D. Gore and Bharti W. Gawali

Abstract Climate change influences the characteristics and processes of the soil. More research is needed to understand how climatic change affects erosion processes and how these factors affect soil N, P, K, EC, pH, and OC storage. The response of soil organic matter levels to changes in the N, P, K, EC, pH, and OC cycles determines how well the soil is supported crop development. The impact on food security is significant. This article illustrates how climate change affects the mechanisms that impact soil fertility, including temperature increases, changes to precipitation patterns, and fluctuations in humidity and wind. The chemical lab analysis of 300 soil samples from the Marathwada region, coupled with meteorological data, has been compared to the ASD FieldSpec data. To conduct our climatic analysis, we merged the soil data from five Marathwada districts, Aurangabad, Beed, Parbhani, Osmanabad, and Nanded, with the IMD climate dataset. We have used preprocessing techniques like median imputation, outlier, and data normalization and machine learning algorithms such as KNN, SVM, Logistic Regression, Random Forest, and Decision Tree to the analysis of results respectively, 76, 89, 92, 99, and 98% accuracy. The accuracy of the ASD FieldSpec dataset has given good results, and the Decision Tree has provided better results (99%).

Keywords Climate Smart Agriculture · Machine Learning · Supervised Classification · KNN · SVM · Random Forest · Decision Tree

1 Introduction

Climate change refers to any significant alterations in climatic phenomena that last a long time. Global warming is a phenomenon frequently associated with greenhouse gases created by human activity and the burning of fossil fuels such as coal, oil,

R. D. Gore (✉) · B. W. Gawali
Department of Computer Science and Information Technology, Dr. Babasaheb Ambedkar, Marathwada University, Aurangabad , Maharashtra 431004 , India
e-mail: ramdasgore1888@gmail.com

N. Hasteer et al. (eds.), *Decision Intelligence Solutions*, Lecture Notes in Electrical Engineering 1080,
https://doi.org/10.1007/978-981-99-5994-5_26

Table 1 Soil Textures of Marathwada Region Soil Samples

Name of Village/City	Sand %	Clay %	Silt %	Texture class
Aurangabad	23	37	40	clay loam
Jalna	20	29	51	clay loam
Rithi	9	28	63	silt clay loam
Chikhalthan	13	26	61	silt clay loam
Osmnabad	15	27	58	silt clay loam
Parbhani	18	26	56	Silt clay loam
Nanded	17	52	31	clay
Beed	19	27	54	Silt clay

and natural gas. The ecosystem, particularly the soil, is impacted by climate change [1]. The soil has given globally satisfied the growing population's demands for food and fiber. Still, its effects on soil properties and processes can damage the safety of the world's food supply. Between 2030 and 2052, global warming will reach 1.5 °C if it keeps increasing at the present rate [2]. The expected global climate change is crucial for recovering soil fertility and production, which includes rising temperatures, increased atmospheric CO2 levels, altered rainfall patterns, and increased atmospheric nitrogen deposition. The presence of organic matter, the soil temperature pattern, the hydrology of the soil, and salt are some of the factors that climate change impacts. The impact of climate change on soils has been researched in light of all these factors [3].

There is much anxiety worldwide about the speed and scope of climate change, such as greenhouse gases and human activities [4]. It has increased heat retention, temperature, and high spatial and temporal variability. The precipitation characteristics include the volume and intensity of rain and snow. The geographical distribution pattern is also noticeably affected by changes in the temperature pattern [5].

According to reports, soil moisture stress impairs soil activities, lowering plant yield. Important soil physical characteristics have an impact on soil fertility as a result of climate change.

Aurangabad and Jalna District's soil texture is under the clay loam, and Rithi, chikhalthan, Osmanabad and Parbhani are falling in the silt clay loam, as shown in Table 1. The Beed district soil texture is silt clay, and the Nanded soil texture is clay. Overall Marathwada region's soil texture is silt clay loam.

2 Climate and Soil Database

The Indian Metrological Department (IMD), Pune, India's Ministry of Earth Science, has provided the climate database from 1952 to 2018. We have used five districts of Marathwada region climate databases such as Aurangabad (1952–2018 – 66 years),

Beed (1960–1996 – 35 years), Osmanabad (1976–2011 – 29 years), Nanded (1960–2007 – 42 year) and Parbhani (1944–2018 – 74 year). We have included meteorological variables, including temperature (in degrees Celsius), rainfall millimeters), relative humidity, the number of rainy days (RD), and mean wind speed (MWS, in kilometers per hour) in our forecast models. We have collected 300 soil samples from the Marathwada region. Soil samples are used for chemical analysis and ASD FieldSpec Spectroradiometer analysis.

2.1 *Methods and Techniques*

Historical data is utilized as inputs in the forecasting process to provide informed estimations that are predictive in determining the direction of future trends. Weather forecasting uses science and technology to predict atmospheric conditions at a given location and time. Since the eighteenth century, people have made informal and scientific attempts to predict the weather [7]. The gathering of quantitative information on the current locations of the atmosphere, land, and water, as well as the use of meteorology to forecast how the atmosphere will change in a particular area, provide the basis for weather predictions. Choosing the model that will match the data from a scatter plot may be difficult.

2.1.1 Preprocessing

The raw data (soil and climate) is converted into an Excel file. We separated the database into districts and years.

2.1.2 Missing Values

The mean and median imputers are used for missing numeric data. For missing values broken down by month, we utilized a median imputer.

2.1.3 Exploratory Data Analysis (EDA)

Two graphical methods are used to find outliers when the distribution is normal: Two different kinds of plots are box plots and scatter plots. For better visual comprehension of data behavior in a distribution's middle and tails (Fig. 1).

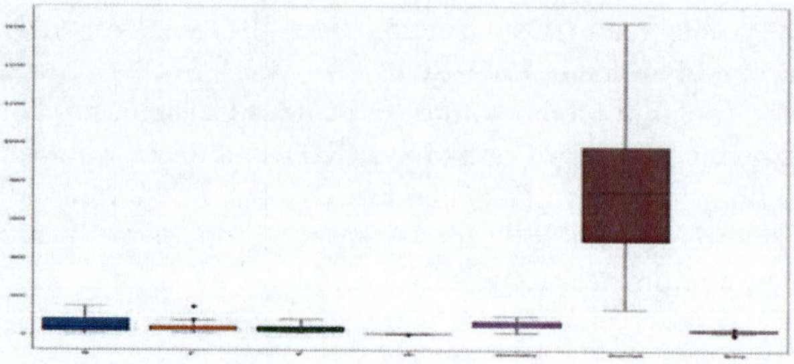

Fig. 1 Data Visualization-Outliers

2.1.4 Correlation

The statistical measures of the mean, standard deviation, variance, and covariance are all intimately connected to correlation. Relationships between two or more dataset variables (or features) are commonly explored in statistics and data science. Every data point in the collection is an observation, and the features are the characteristics or properties of those observations. For every dataset, we have used variables and observations. The dataset presents the factors or variables (climate parameters).

The following pair diagram depicts the pairwise association between these closely related attributes (Fig. 2).

3 Implementation of Techniques/Experiment Work

All of the districts in the Marathwada area have experienced variations in the rainfall pattern. The trend for temperature, rainfall, and heavy rainfall is rising, whereas the trend for wet days is decreasing. The amount of rainfall is rising, but the number of wet days is not rising, which impacts the soil's quality.

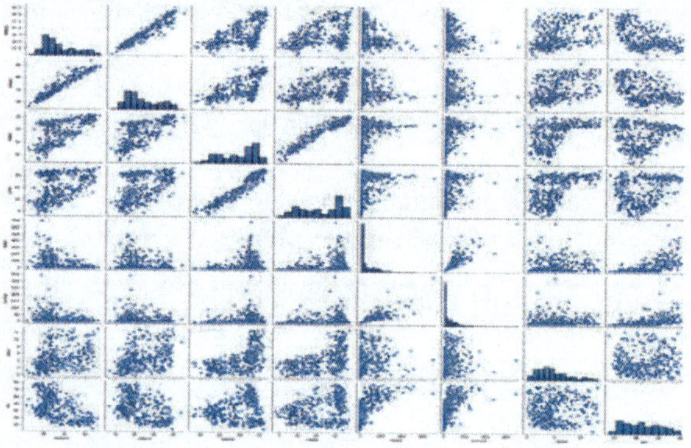

Fig. 2 Correlation of Climate and Soil Parameters

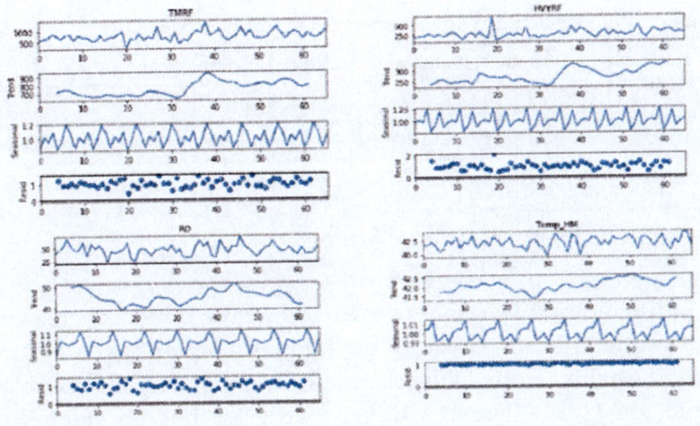

Fig. 3 Seasonality and Trend of Climate Data

3.1 *Holt Winter's Method*

Demand data from any nation frequently shows significant seasonality and trends in the real world. Agriculture uses the trend and seasonality in climatic data. Holt-Exponential One of the earliest time series analytical forecasting techniques that considered trend and seasonality was Winter's Smoothing. Based on trend and seasonality, this approach is used to create forecasts in three distinct ways, and it has an average value (Fig. 3).

Winter's approach is also known as triple exponential smoothing for analysis. There are three variants of exponential smoothing for each of the three attributes. The fundamental tenet of the additive model is that the anticipated value for each data point is equal to the sum of its baseline, trend, and seasonality components. Climate trends, seasonality, and forecast inaccuracy are shown positive results.

3.2 *Mean Absolute Percentage Error (MAPE)*

The MAPE metric measures the forecasting system's accuracy. The formula below is used to determine MAPE.

$$MAPE = \frac{1}{n}\sum_{t=1}^{n}\left|\frac{e_t}{Y_t}\right| \tag{1}$$

where n denotes the number of data points, and e_t denotes the forecast error from Y_t - $\hat{Y}_t$. Y_t is shown as the actual value, and $\hat{Y}_t$ is demonstrated as the predicted value (Table 2).

Temperature and precipitation have increased in the most recent decade (2010–2020) and the next decade (2020–2030). Between 2010 and 2020, there were fewer

Table 2 Result of Methods and Moving Average for the Time Series Data

Parameter of database	Holts winter with additive seasonality and trend	MAPE
TMRF	17.12	13.88
HVYRF	18.21	10.83
RD	10.47	9.14
Temp	1.93	1.52

rainy days. In the monsoon season, particularly heavy rainfall increases, and there were 3–4 days of heavy rain, covering the normal monthly rainfall. It is a warning sign for rainy droughts. Holts Winter's additive seasonality approach yields good temperature results (1.52). The MAPE values for the monthly rainfall total, heavy rainfall amount, and wet days are 13.88, 10.83, and 9.14, respectively.

3.3 *First Derivative of Soil Sample*

It evaluates the slope of the spectral curve at each place. The baseline offsets of the spectral signature have no impact on the slope of the curve. It's an effective method for eliminating baseline offsets. The first derivative is viewed as a real-time rate of change. Alternately, the tangent line's slope is considered the first derivative (Fig. 4).

Due to recurrent cropping, the intensive use of chemical fertilizers and pesticides, and soil erosion in the Marathwada region due to heavy rains and high temperatures,

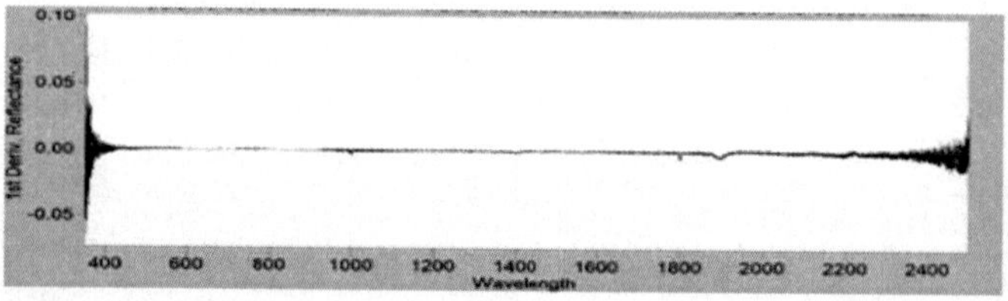

Fig. 4. 1st Derivatives of Soil Samples

Table 3 Marathwada Region Status of Soil Quality

Soil Parameters	Range From	To	2020	2021	2022	Remark	Unit
N	281.00	420.00	163.07	149.86	136.87	Low(Very Low)	kg/ha
P	15.00	21.00	49.61	40.12	39.13	High	kg/ha
K	151.00	240.00	298.37	282.24	293.10	High	kg/ha
OC	1.01	2.00	0.72	0.69	0.58	Low	%
pH	6.51	7.50	7.78	7.87	8.12	Moderately	Number
EC	0.01	1.00	0.81	0.74	0.72	Normal	mS /CM

N and OC are found to be low (less than 281 and 1.01, respectively). Water used for irrigation is contaminated by chemical pesticides and fertilizers, which raises the pH value (to a high of 7.5 > 8.5) (Table 3).

3.4 *Machine Learning Algorithms for Soil Analysis*

The key to creating a machine learning model is choosing the best algorithm among the many available in machine learning for the given dataset and problem. Its approaches include supervised and unsupervised machine learning. The advantage of supervised machine learning is that it makes it possible to gather data, output data from earlier experiences, optimize performance criteria, and solve several computation-related problems that arise in real-world settings. Supervised learning use of two categories of algorithms, such as regression and classification.

3.4.1 Logistic Regression

It is used to address classification problems. It employs discrete or categorical dependent values to predict outcomes like yes or no, 0 or 1, true or false, etc. We applied the equation below to it [8].

$$\frac{y}{(1-y)} \tag{2}$$

0 for Y = 0 and Y = 1 for Infinity (∞).

3.4.2 K-Nearest Neighbor (KNN)

Classifying newly discovered information as soon as practicable into the proper category is possible. Since it did not instantly learn from the training dataset, store the data, and then act on it at the time of classification. Here is a formula for calculating the Euclidean distance between two data points [9].

$$(A_1\&B_2) = \sqrt{((X_2 - X_1)^2 + (Y_2 - Y_1)^2)} \tag{3}$$

We determined the nearest neighbors and a new data point within category A by computing the Euclidean distance.

3.4.3 Support Vector Machine (SVM)

SVM is used to address classification problems. Finding the best restrictions for a decision is beneficial. A hyperplane is a perfect boundary or region. The SVM algorithm identifies the nearest line from each class. The points are referred to as support vectors. The margin, or the space between the vectors and the hyperplane, is what SVM seeks to maximize. The hyperplane with the biggest margin is the optimum one [10].

3.4.4 Decision Tree

It helps consider every option for resolving a problem and is very beneficial for difficult decision-making situations. Compared to other methods, it requires less data cleansing [11].

3.4.5 Random Forest

It took less training time than earlier algorithms and successfully predicted outcomes, even for large datasets. It has maintained its accuracy even without a major portion of the data. It can handle large datasets with several dimensions, enhance model correctness, and prevent overfitting [12, 13] (Fig. 5).

Using ASD data and Chemical soil analysis, we have implemented machine learning classification algorithms such as Logistic Regression, KNN, SVM, Decision Tree, and Random Forest. We got 92%, 85%, 91%, 99%, and 94% accuracy for Logistic Regression, KNN, SVM, Decision Tree, and Random Forest, respectively, for ASD FieldSpec data. For Chemical analysis, we have got 90%, 76%, 89%, 98%, and 93% accuracy for Logistic Regression, KNN, SVM, Decision Tree, and Random Forest, respectively. We got the highest accuracy, 99% and 98%, for decision tree algorithms for the ASD Fieldspec data and chemical data, respectively. Every

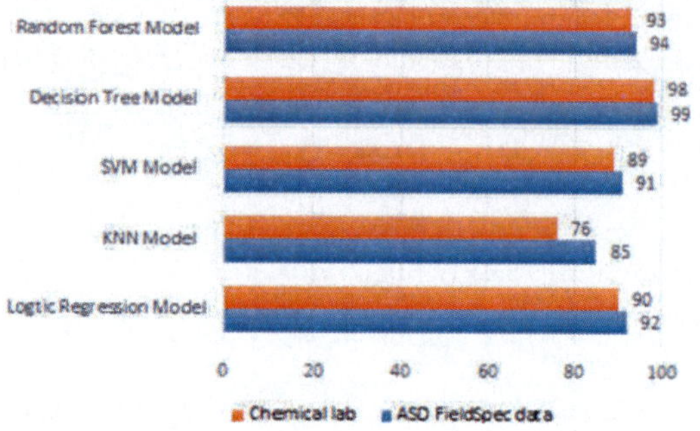

Fig. 5 Machine Learning Algorithm Results of ASD FieldSpec Data & Chemical Lab

supervised machine learning algorithm has given more accuracy to ASD FieldSpec data.

4 Conclusion

Climate change is expected to have an impact on soil characteristics, including texture, structure, N, P, K, pH, and EC, as well as soil fertility, which leads to salinization of the soil, a decrease in nutrient availability, a reduction in C and N dynamics, and a decrease in soil biodiversity. Several adaptations and mitigation approaches, including integrated nutrient management, residue management, conservation agriculture, etc., lessen climate change's detrimental effects on soil fertility. For data categorization, we employed supervised machine learning algorithms, and the decision tree approaches produced results with high accuracies, such as 99% for ASD FieldSpec data and 98% for chemical lab tests. Holt's winter with additive seasonality and trend is provided 1.52, 13.88, 10.83, and 9.14 MAPE for Temperature, the entire month of rainfall, heavy rainfall, and the number of rainy days, respectively. ASD FieldSpec and Chemical Lab analysis produce the same result for the soil quality analysis.

References

1. Suyal S (2022) Meena, nelofar tanveer; climate change and soil fertility: an issue of food security concern. Himalayan J Soc Sci Humanities 17:41–45
2. IPCC. Summary for Policymakers. In Climate Change 2007: The Physical Science Basis; Contribution of Working Group I to the Fourth Assessment Report of the Intergovernmental Panel on Climate Change; Solomon, S., Qin, D., Manning, M., Chen, Z., Marquis, M., Averyt, K.B., Tignor, M., Miller, H.L., Eds.; Cambridge University Press: Cambridge, UK. 2007, 1–18
3. Brevik EC (2012) Soils and climate change: Gas fluxes and soil processes. Soil Horiz 53. https://doi.org/10.2136/sh12-04-0012.
4. Pimentel D (2006) Soil erosion: a food and environmental threat. Environ Dev Sustain 8:119–137
5. Lal R (2010) Managing soils and ecosystems for mitigating anthropogenic carbon emissions and advancing global food security. Bioscience 60:708–721
6. Blum WEH, Nortcliff S (2013) Soils and food security. In: Brevik EC, Burgess LC (eds) Soils and Human Health. CRC Press, Boca Raton, FL, USA, pp 299–321
7. Brevik EC (2013) Climate change, soils, and human health. In: Brevik EC, Burgess LC (eds) Soils and Human Health. CRC Press, Boca Raton, FL, USA, pp 345–383
8. Rustad LE, Huntington TG, Boone RD (2000) Controls on soil respiration: Implications for climate change. Biogeochemistry 48:1–6
9. Lal R, Kimble J, Follett RF (1998) Pedospheric processes and the carbon cycle. In: Lal R, Kimble JM, Follett RF, Stewart BA (eds) Soil Processes and the Carbon Cycle. CRC Press, Boca Raton, FL, USA, pp 1–8
10. Mosier AR (1998) Soil processes and global change. Biol Fertil Soils 27:221–229
11. Brevik EC, Homburg JA (2004) A 5000-year record of carbon sequestration from a coastal lagoon and wetland complex, Southern California, USA. CATENA 57:221–232

12. Schlesinger WH (2013) An Overview of the Carbon Cycle. In: Soils and Global Change; Lal, R., Kimble, J., Levine, E., Stewart, B.A., Eds.; CRC Press: Boca Raton, FL, USA, 1995; pp. 9–25. Agriculture 2013, 3 411.
13. Mondal S (2021) Impact of climate change on soil fertility. In: Choudhary DK, Mishra A, Varma A (eds) Climate Change and the Microbiome. Soil Biology, vol 63. Springer, Cham. https://doi.org/10.1007/978-3-030-76863-8_28

Human Eye Fixations Prediction for Visual Attention Using CNN - A Survey

Judy K. George and **Elizabeth Sherly**

Abstract Human attention is closely related to the visual information perceived. Saliency prediction models help to concentrate on the core pixels of an image contributing to attention. Trends and advancements in Artificial Intelligence resulted in many prediction models but not reached the level of prediction as human eye selection. Currently available prediction models concentrate on reflexive or exogenous attention. Published works in the year 2014–2022 are considered for this work and a study on improvements in the path made by deep learning prediction models using Convolutional Neural networks (CNN) in unveiling the pixels which achieve the high visual conspicuousness. This paper shows the avenues which are possible for the saliency models in the future by incorporating voluntary attention to the current models which are beneficial for researchers in the visual attention field like activity recognition, visual question answering etc.

Keywords Saliency Prediction · Saliency Map · Eye Tracking Computer Vision · Visual Attention

1 Introduction

A torrent of visual information burst in our eyes when we open it to see this beautiful world. Around 10^8 - 10^9 bits of visual data enter into our perception area every second. This perceived information is too much to process all at once, as the processing speed of our brain is 40 bits per second [1]. A decision making happens inside our human visual perceptual system to select the most prominent information from the scene. The area of research which indulge in neuro-physiological support and modeling of processing the perceived data selectively is known as visual attention.

J. K. George (✉) · E. Sherly
Digital University Kerala, Thiruvananthapuram 695317, India
e-mail: judy.res21@duk.ac.in

E. Sherly
e-mail: sherly@duk.ac.in

N. Hasteer et al. (eds.), *Decision Intelligence Solutions*, Lecture Notes in Electrical Engineering 1080,
https://doi.org/10.1007/978-981-99-5994-5_27

Different researchers from diverse scientific backgrounds presented different theories to analyze the aspects of visual attention (VA) [2]. In the aspect of signal processing, Pre-attentive stage of VA concentrates on color, edge information, orientation or motion that are the basic features in an image and in Attention stage, areas with most "relevant" information is chosen. The bottom-up component occurs during the pre-attention stage and it is stimulus based and task independent. The top-down component is task dependent. Parallel search mechanism is used when the salient object is different from distractors considering a single feature. For example, if a scene differs in terms of color only then this mechanism can be used. Contrarily, if the target is different from the distractors in more than one feature, then serial searching is prescribed. It takes much time with number of distractors. If a scene differs in terms of both color and shape serial mechanism can be used.

The attention that involves the direct focus or shift in gaze which requires movements of eye to the next attracting locations is the overt attention. It requires the working of brain. Covert attention is the capability to attend to areas in a scene without accurate movements of eyes. While driving, the driver is able to understand the signals without focusing on the signals is an example for covert attention.

1.1 Visual Saliency

Visual saliency in a scene is defined as the tendency of an area in a scene to receive attentional focus. The saliency prediction algorithms concentrated on identifying the fixation points. It means the locations human observers would focus at first glance. In the computational modeling of visual attention, saliency map concept was introduced. Saliency map can be stated as a 2D topographic map that can exhibit each pixel's unique quality. Saliency map makes an image simple so that it becomes more meaningful and can be analyzed easily (Fig. 1).

Fig. 1 Illustrative image with saliency map [3]

1.2 *Significance*

Researchers have widely exploited the idea of where people look in images or videos to study the neural and behavioral response. It has got applications in fields of advertisement [4], image captioning [5], patient diagnosis [6], surveillance [7], activity recognition [8], object segmentation [9], Digitization [10], visual question answering [11], etc.

2 Related Works

Fixation prediction and detection of relevant object in the scene were the two areas focused by the researchers interested in Visual Attention. Both areas concentrate on the relevant information by filtering out irrelevant information but the application scenarios are different. Eye fixation prediction concentrates on information concerning cognition identified from the eye ball movements [12] which is used in visual search, scene analysis and perception. Contrarily, the salient object detection outputs the striking object present in a scene and it produces a binary map indicating salient objects [13].

2.1 *Saliency- Heuristic and Learning Based Models*

Classic models of saliency used the heuristic features like contrast, location, texture etc. for saliency [12]. Heuristic features act as a visual cue as the contrast, location and texture helps to identify the object and its background. After the extraction of features center-surround and normalization operations are performed. Traditional techniques contributed to saliency includes frequency domain analysis, random walks etc. These techniques help to identify the salient object by concentrating on the object by suppressing the background details.

Koch and Ullman [14] presented a model which combines the features. Saliency map was proposed and the winner-take-all feed-forward neural net-work concept got introduced. This selects the most prominent regions in the scene. The influential work by Itti [15] presented a biologically-plausible architecture to detect saliency. Color, Intensity and Orientation were considered by these early models but these heuristics combinations cannot play a good part in prediction.

The Itti model was expanded by the GBVS model [16] which uses jumbled walks on a graph structure. This model consists of the formation of activation maps and normalization but it produces Low resolution. Another classical approach is based on the concept Information maximization, which quantifies salient regions of the image. Bruce and Tsotsos [17] AIM model relies on Shannon's self-information measure. Frequency response of a picture [18] is used in constructing the saliency map by analyzing the log spectrum. Based on figure-ground segregation proposed by the Gestalt principle, Boolean map-based saliency model [19] evaluated the prominent pixels. This was the best among unsupervised learning models. By analyzing the Boolean maps in a topological structure saliency map got computed.

A comparison study of classical models depicted that the models produce low performance in terms of resolution. Smaller salient regions are highlighted than the larger ones and it failed to detect multiple salient objects. Classical models produced poor performance in noisy background and lightening variations. In some models, detection is possible at the center of the image and most of them are not well suited for task dependent applications.

2.2 *Deep Learning-Based Saliency Models*

A new wave on saliency came after the success of CNN. CNN extract the features of an image at pixel level and the FCN models learned or pretrained on ImageNet dataset can contribute much to the Saliency prediction. Though many classical models are available, this study focuses on the deep learning models which outperform the prediction of classical models. Figure 2 shows the illustration of various deep learning architectures extracting features.

In early models, which used plain CNNs, context is automatically captured.

Some models process the images at different scales for prediction. Due to pooling and convolution layers some information may be lost and it was retained in some models with the help of skip connections. Scanning the image vertically and horizontally with RNN or using the pre-trained classification models were used by some models to incorporate context in the prediction. Inception module with various receptive field sizes helps in feature extractions at multiple scales. Dilated Inception modules were used by some models to extract the dense features and to preserve the spatial information. However deep learning models failed to predict images with low level saliency features like feature contrast.

3 Visual Saliency Prediction Datasets

Saliency Prediction databases have been obtained by manual annotations by mouse click and eye-tracking devices. Eye tracking devices can disclose meaningful insights about behavior and performance of humans to an extent. Quantitative evaluation of saliency models is done with various metrics provided by datasets. Table 1 shows the commonly available saliency datasets.

In the saliency area, MIT300 [20] dataset was greatly accepted and used by the models. CAT2000 [22] contain 2000 images belonging to 20 categories. SALICON [25] is the one of the largest public datasets for predicting visual saliency and it use mouse click for annotation. The EMOd dataset [24] contains annotations which

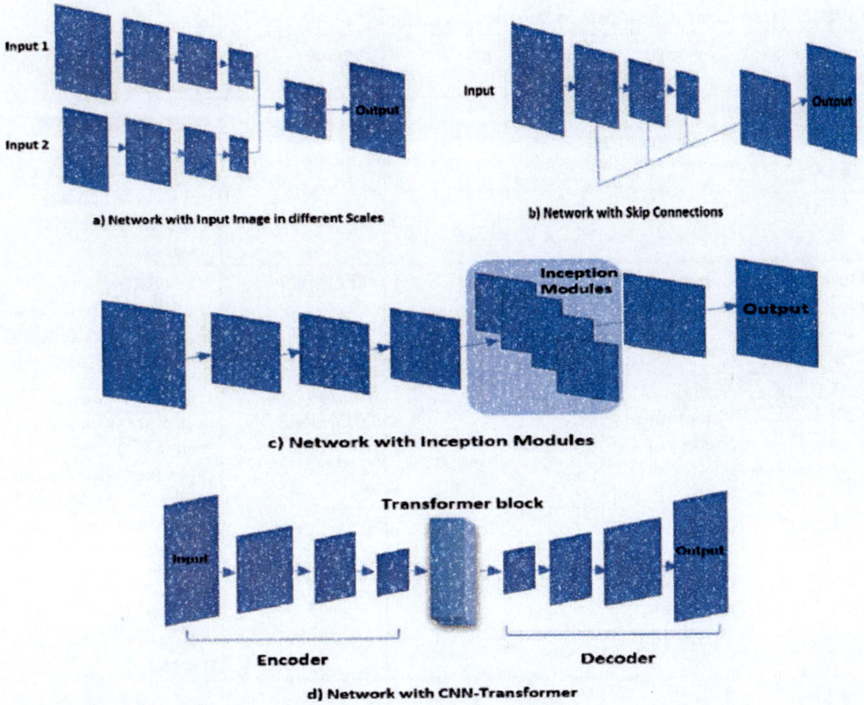

Fig. 2 Illustration of deep learning architectures extracting features

Table 1 Saliency Datasets

Name	Images	Observers
Tornoto [17]	120	20
MIT300 [20]	300	39
MIT1003 [21]	1003	15
CAT2000 [22]	2000	24
DUT-OMRON [23]	5168	5
EMOd dataset [24]	1019	16
SalECI dataset [4]	972	25
SALICON [25]	20,000	-

concentrate on image emotions. The SalECI dataset [4] is the primary dataset available for e-commerce purposes. Table 2 shows the analysis of Deep Learning models in predicting the saliency.

Table 2 Deep Learning Models in Saliency prediction

Reference	Objective	Methodology Used	Dataset Used	Remarks
eDN [26]	Saliency Prediction	CNN	MIT1003	First attempt to use CNN Pretrained models are not used. Accuracy was less than classic models
DeepGaze I [27]	Saliency Prediction	CNN- AlexNet	MIT1003	Use of AlexNet boost Prediction. High Level Features like Faces and Text are not Captured
SALICON [28]	Visual Attention Prediction	CNN- VGG16, AlexNet, GoogleNet	Tornoto, DUT-OMRON, MIT1003	Deep CNN which combines information from CNN that are pretrained on different image Scales. Failed to predict Synthetic Images
Mr.CNN [29]	Fixation Prediction	CNN- AlexNet	MIT	Low, mid, high features taken separately and integrated
DeepFix [30]	Saliency prediction	VGG-16	MIT300, CAT2000	First FCNN in Saliency Prediction. Inception Module is used. Location Biased Convolutional filters are used to learn location dependent patterns
ML-Net [31]	Saliency Prediction	VGG-16	SALICON	Combines feature maps extracted from VGG. Upsampling and Concatenation bring uncertainty to prediction
MxSalNet [32]	Saliency Prediction	CNN-VGG-16	CAT2000	Gating mechanism used to know the scene context
PDP [33]	Visual attention prediction	VGGNet	SALICON, MIT1003, MIT300	Probability distribution for prediction. Quickly learns the regions having high contrast
DeepGazeII [34]	Reading Fixations	CNN- VGG19	SALICON	A probabilistic model. Performance greater than the previous models

(continued)

Table 2 (continued)

Reference	Objective	Methodology Used	Dataset Used	Remarks
DSCLRCN [35]	Eye fixation prediction	VGG - LSTM	SALICON, MIT300	Learn local Context using CNN and Global Context using LSTM
SalGAN [36]	Visual Saliency Prediction	GAN	SALICON	Utilized GAN to build model Generator and Discriminator modules to learn and identify the predictions
iSEEL [37]	Fixation Prediction	CNN-VGG16	CAT 2000	Inter image similarities considered Ensemble of Extreme Learning Machine trained on fixation prediction is used
DVA [38]	Visual Attention Prediction	CNN -VGG16	SALICON	Encoder - Decoder Architecture Learning features by integrating direct supervision into different layers
PSM [39]	Personalized Prediction	CNN-PIEF	PSM	First personalized saliency dataset Effectiveness of person-specific information is studied
DiNET [3]	Saliency Prediction	CNN - ResNet	SALICON	Dilated Inception Module used
EML-Net [40]	Saliency Prediction	CNN-DenseNet-161, NasNet	SALICON	Scalable Combined evaluation metrics to get new loss functions
TransalNet [41]	Saliency Prediction	CNN-Transformer	SALICON	Long range visual information using self-attention

4 Evaluation Metrics of Visual Saliency

For the evaluation of visual attention models many metrics were proposed. On the basis of similarity measure, evaluation metrics can be divided into three major classes as value related, distribution related and location related. Eye-tracked data and the predicted saliency map can be compared by the similarity measure evaluation. Values at fixation points of predicted saliency map is done for evaluation in value related metrics. Correspondence of salient pixel location in predicted saliency map with original fixation point is considered in location related metrics. Dissimilarities in the statistical analysis of the fixation points and predicted saliency maps is considered in distribution related metrics. Table 3 depicts the set of metrics used by famous benchmarks for saliency model evaluation.

Let *SM* depicts the saliency map produced by the deep learning models and *GT* shows ground truth of an image provided by the dataset. Evaluation metrics relies on similarities and dissimilarities between *GT* and *SM*. Commonly used evaluation metrics in saliency includes:

Linear Correlation Coefficient (CC): Correlation between two random variables can be expressed linearly by Linear Correlation Coefficient (CC). Score of + 1 evinces a perfect linear relationship between *SM* and *GT* [30]. The CC metric between the SM and GT is given by:

$$CC = \frac{cov(GT, SM)}{\sigma GT * \sigma SM} \tag{1}$$

Table 3 Saliency Evaluation Metrics

Metrics	Class	Ground-truth
CC	Distribution related	Saliency Map
sAUC	Location related	Fixation Map
AUC	Location related	Fixation Map
SIM	Distribution related	Saliency Map
NSS	Value related	Fixation Map
EMD	Distribution related	Saliency Map
KL	Distribution related	Saliency Map

Table 4 SALICON Test Dataset Results. Bold values depict the top model

Model	CC	sAUC	AUC-J	NSS
DeepGazeII [34]	0.479	**0.787**	0.867	1.271
PDP [33]	0.765	0.781	0.882	-
ML-Net [5]	0.743	0.768	0.866	2.789
SalGAN [36]	0.781	0.772	0.781	2.459
SAM-ResNet [48]	0.842	0.779	0.883	3.204
SAM-VGG [48]	0.825	0.774	0.881	3.143
MxSalNet [32]	0.730	0.771	0.861	2.767
DSCLRCN [35]	0.831	0.776	**0.884**	3.157
DiNet [3]	0.860	0.782	**0.884**	**3.249**
EMLNeT [40]	0.886	0.746	0.866	2.050
TransalNet [41]	**0.907**	0.747	0.868	2.014

Area Under Curve (AUC): AUC metric acts as a standard for binary class pattern recognition. On the *SM*, which is a binary map, AUC metric assesses the classification performance. By setting the threshold between 0 and 1 *SM* can be grouped into relevant and non-relevant regions. By utilizing the ROC curve of *SM*, AUC metric can be measured. Variants of AUC are Shuffled Area Under Curve (sAUC) [42], AUC-Judd [43], AUC Borji [44].

Normalized Scanpath Saliency (NSS): Mean score evaluation of unit normalized saliency map MF (unit standard deviation and zero mean) at human eye fixations [45] is calculated.

$$\text{Normalized Scanpath Saliency} = \frac{1}{F}\sum_{i=1}^{F} MF(i) \tag{2}$$

where, F implies the number of human eye positions.

Similarity Metric (SIM): Similarity between *GT* and *SM* can be measured by SIM. Aggregate sum of minimum values at every pixel is calculated after the normalization of *SM*. A value of 1 indicates the distribution as same and 0 denotes no overlap.

$$SIM(SM, GT) = \sum_{i} \min(SMi, GTi) \tag{3}$$

Earth Movers Distance (EMD): EMD [46] denotes the value of probability distribution conversion from *SM* to *GT* and a low score indicates a high-quality prediction.

Kullback–Leibler (KL) Divergence: An asymmetric dissimilarity measure depicts the difference between probability distributions of *SM* and *GT* and low score indicates a high-quality saliency map [47].

$$KL(SM, GT) = \sum_{i} GTilog(\epsilon + \frac{GT}{\in + SMi}) \tag{4}$$

5 Comparison of Different Models in Prediction

SALICON: SALICON is the largest shared dataset for saliency prediction. Table 5 depicts the performance of selected deep learning models using the dataset. In the SALICON benchmark, human model scores are not fixed. For ranking the models, SALICON uses only four metrics namely NSS, CC, AUC and sAUC. So, the top scores are marked in bold. It denotes that none of the models perform well in all the metrics. Figure 3 shows the graphical analysis of CC, AUC, sAUC and NSS on the top performed models.

Table 5 MIT300 Test Dataset Results. Bold values depict the top model

Model	CC	sAUC	AUC-J	NSS	SIM	AUC-B	EMD↓	KL↓
	1	0.81	0.92	3.29	1	0.88	0	0
eDN [26]	0.45	0.62	0.82	1.14	0.41	0.81	4.56	1.14
DeepGaze I [27]	0.48	0.66	0.84	1.22	0.39	0.83	4.97	1.23
SALICON [28]	0.74	0.74	0.87	2.12	0.6	0.85	2.62	0.54
Mr.CNN [29]	0.48	0.69	0.79	1.37	0.48	0.75	3.71	1.08
DeepFix [30]	0.78	0.71	0.87	2.26	0.67	0.8	2.04	0.63
DeepGazeII [34]	0.52	0.72	0.88	1.29	0.46	**0.86**	3.98	0.96
PDP [33]	0.7	0.73	0.85	2.05	0.6	0.8	2.58	0.92
SalGAN [36]	0.73	0.72	0.86	2.04	0.63	0.81	2.29	1.07
iSEEL [37]	0.65	0.6	0.84	1.78	0.57	0.81	2.72	0.65
DVA [38]	0.68	0.71	0.85	1.98	0.58	0.78	3.06	**0.64**
SAM-ResNet [48]	0.78	0.7	0.87	2.34	0.68	-	2.15	-
SAM-VGG [48]	0.77	0.71	0.87	2.30	0.68	-	2.14	-
ML-Net [5]	0.67	0.7	0.85	2.05	0.59	0.75	2.63	1.1
DSCLRCN [35]	0.8	0.72	0.87	2.35	0.68	0.79	2.17	0.95
DiNet [3]	0.79	0.71	0.86	2.35	-	-	-	-
EMLNeT [40]	0.79	0.7	**0.88**	**2.47**	0.68	0.77	**1.84**	0.84
TransalNet-ResNet [41]	0.80	**0.75**	0.87	2.378	0.685	-	-	0.90
TransalNet-DenseNet [41]	**0.807**	0.746	0.87	2.41	**0.689**	-	-	1.01

MIT300: MIT300 is an unpublished dataset. In this dataset 8 scores CC, NSS, KL, SIM, EMD, AUC-J, AUC-B, sAUC is computed. A human model score is available in the dataset so that comparison of the model can be done. Top row in Table 6 depicts the human model score. From the table it is clear that none of the model attained performed well in all evaluation metrics (Fig. 4).

CAT2000: CAT2000 dataset comprises 4000 images of 20 categories. Each category with 200 images namely Art, Indoor, Outdoor, Line Drawings etc. A human model score is available in the dataset and top row in Table 6 depicts the human model score (Fig. 5).

6 Qualitative Analysis on Deep Learning Models

The Saliency map evaluation on some top performing models depicts that the models are not still good in predicting the pixels which are salient compared to the human score. Figure 6 shows that the top performing models are not predicting correctly

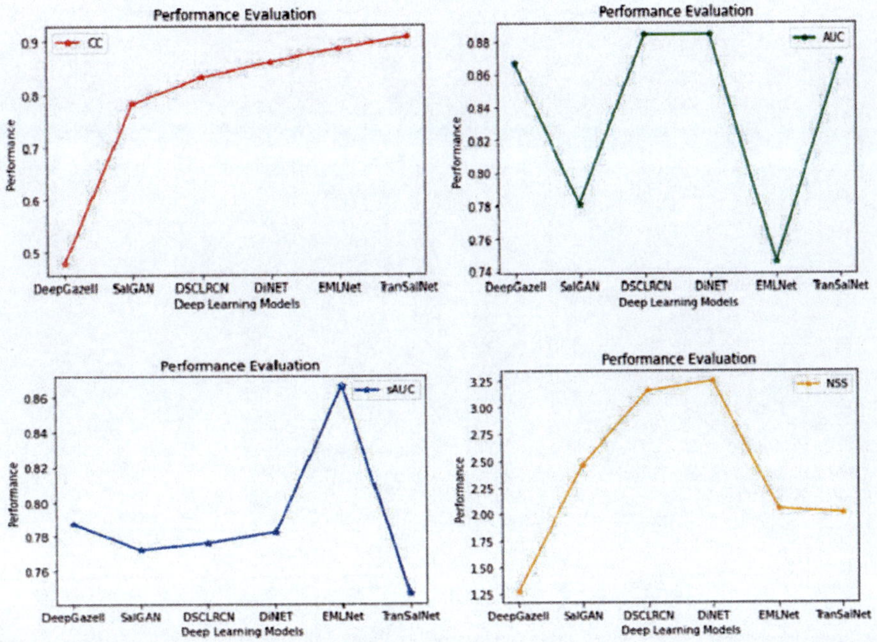

Fig. 3 CC, AUC, sAUC, NSS analysis on SALICON dataset. X-axis indicates the top performing models and Y-axis indicates the performance

Table 6 CAT2000 Dataset Results. Bold values depict the top model

Model	CC	sAUC	AUC-J	NSS	SIM	AUC-B	EMD↓	KL↓
	1	0.81	0.92	3.29	1	0.88	0	0
eDN [26]	0.54	0.55	0.85	1.3	0.52	**0.84**	2.64	0.97
DeepFix [30]	0.87	0.58	0.87	2.28	0.74	0.81	1.15	**0.37**
iSEEL [37]	0.66	**0.59**	0.84	1.67	0.62	0.81	1.78	0.92
SAM-ResNet [48]	**0.89**	0.58	**0.88**	**2.38**	**0.77**	0.8	**1.04**	0.56
SAM-VGG [48]	**0.89**	0.58	**0.88**	**2.38**	0.76	0.79	1.07	0.54
MxSalNet [32]	0.76	0.58	0.86	1.92	0.66	0.82	1.63	0.62
EMLNeT [40]	0.88	**0.59**	0.87	**2.38**	0.75	0.79	1.05	0.96

when compared to ground truth. Studies show that the gazing points varies with gender and emotional situation. A comprehensive study is required to find the cognitive factors which contribute in saliency and a model to be developed based on it.

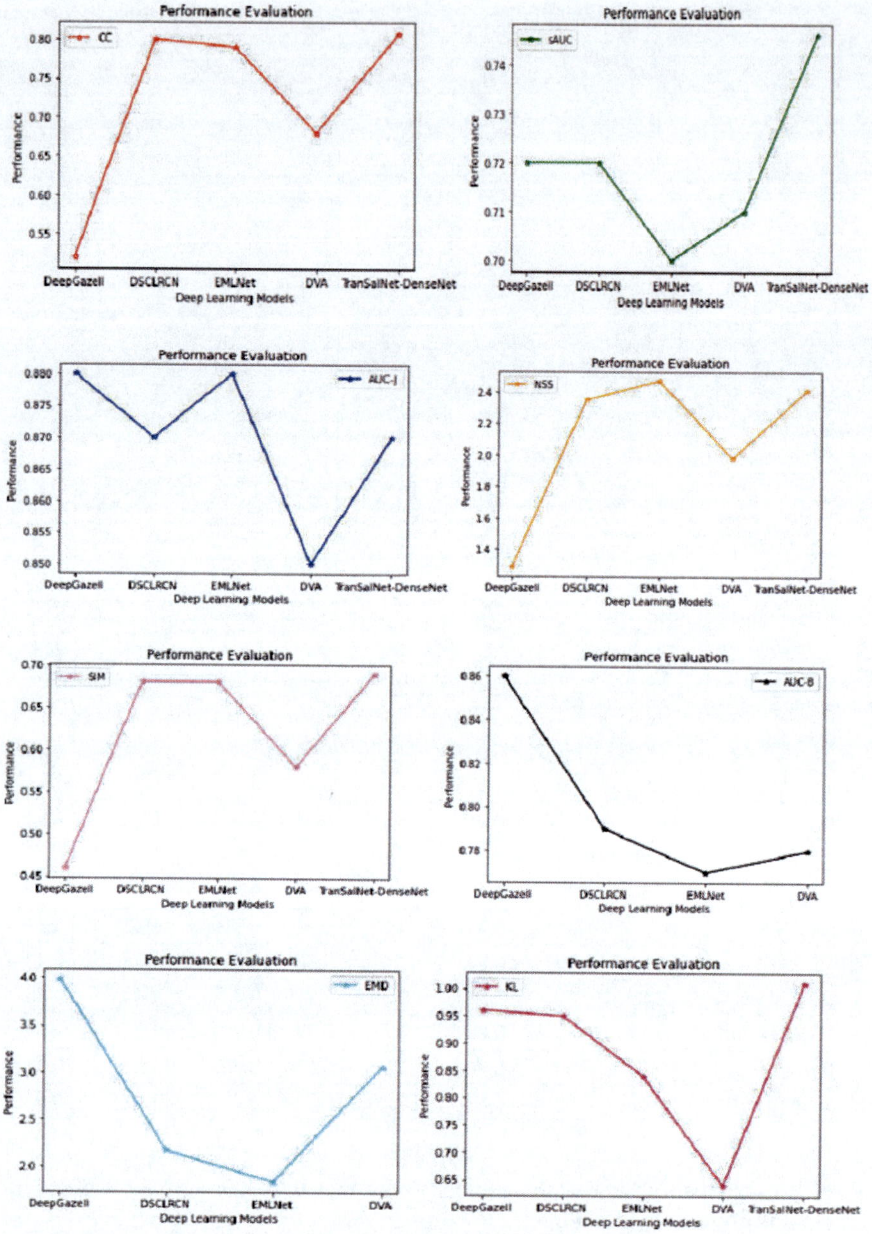

Fig. 4 CC, sAUC, AUC-J, NSS, SIM, AUC-B, EMD, KL analysis on MIT300 dataset. X-axis indicates the top performing models and y-axis indicates the performance

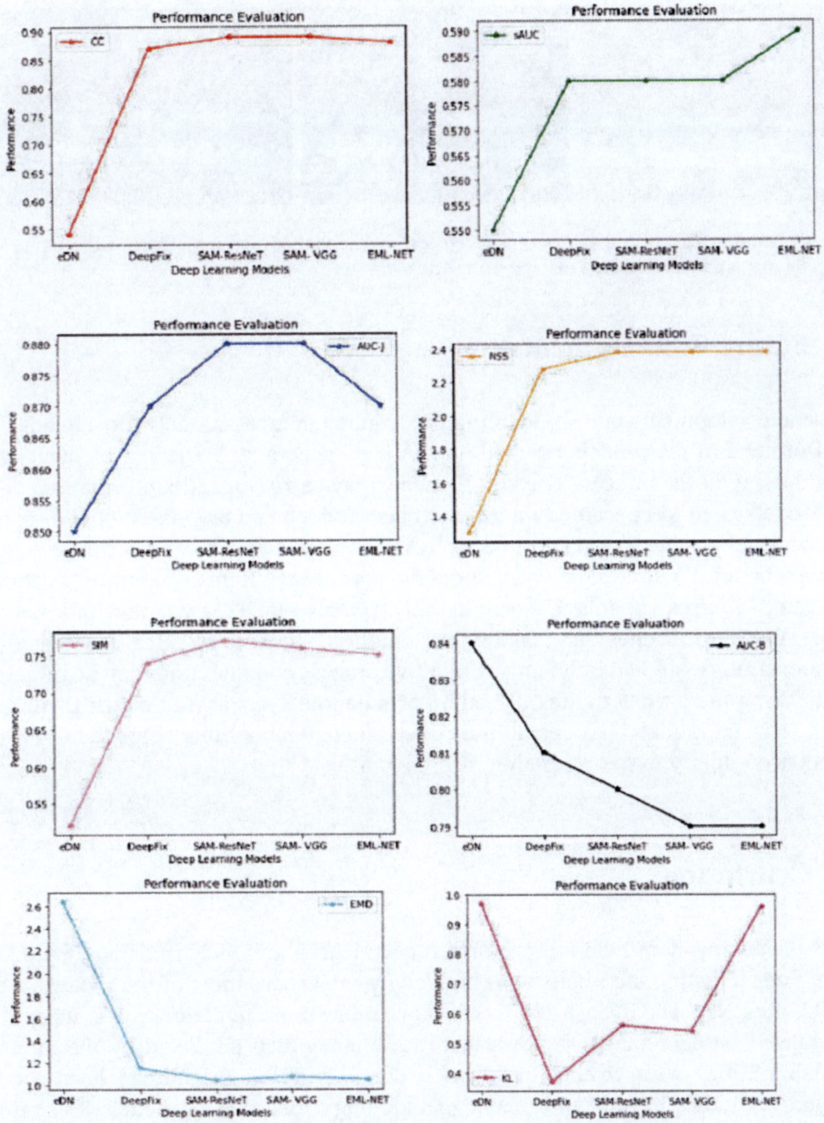

Fig. 5 CC, sAUC, AUC-J, NSS, SIM, AUC-B, EMD, KL analysis on CAT2000 dataset. X-axis indicates the top performing models and y-axis indicates the performance

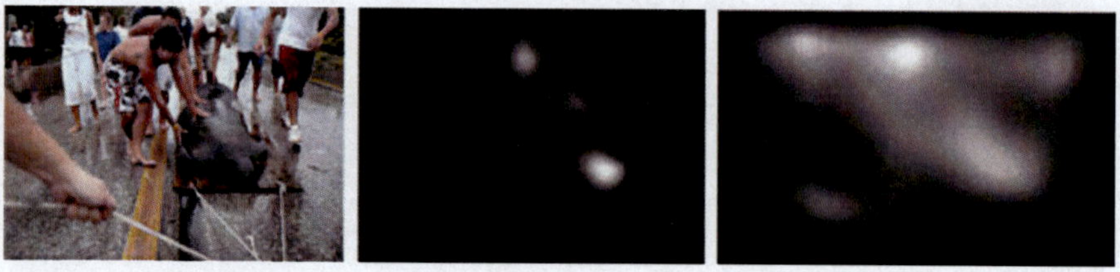

Fig. 6 Failure cases of top performing deep learning models. Columns indicates the input, the original map and the predicted map respectively

7 Future Enhancements

Recent developments in deep learning has resulted in many models for saliency.

But none of the models reached human level accuracy. Some key elements are still missing in the images. The key elements may carry semantic information.

Models need to be trained on different tasks which can help the model to reach human level accuracy. A dataset needs to be developed which can contribute to top down saliency. Existing saliency methods did not consider human cognition. Features extracted in current models cannot represent all pixels which carry salient information especially in the scenes that contain text, emotion, symbols etc. Reasoning quality of humans to be studied efficiently. Cognitive process and intentions can be revealed by tracking the gaze activities. Blending of potential eye tracking metrics with the deep learning models may result in a better saliency model that helps the observer to act accordingly to the prediction.

8 Conclusion

This paper considered the top-performed deep learning models in Saliency prediction. For predicting the pixels with high conspicuousness most of the models used VGG, Res-Net, and Dense-Net with skip connections, Dilations, and inception modules. The data analysis shows that the saliency map predicted by the models has similarities with ground truth but it does not reach the human level score. Cognitive studies have been a better methodology for saliency models. Fusion of cognitive features to saliency, making it possible to discover higher-level concepts in images. The higher-level concepts help to fill the semantic gap present in the currently available models.

References

1. Sziklai G (1956) Some studies in the speed of visual perception. IRE Trans Inf Theory 2(3):125–128
2. Gide MS, Karam LJ (2017) Computational visual attention models. Found Trends® Signal Process 10(4):347–427. https://doi.org/10.1561/2000000055
3. Yang S, Lin G, Jiang Q, Lin W (2019) A dilated inception network for visual saliency prediction. IEEE Trans Multimed 22(8):2163–2176
4. Jiang L, et al.: Does text attract attention on e-commerce images: a novel saliency prediction dataset and method. In: Proceedings of the IEEE/CVF Conference on Computer Vision and Pattern Recognition, pp. 2088–2097 (2022)
5. Cornia M, Baraldi L, Serra G, Cucchiara R (2018) Paying more attention to saliency: image captioning with saliency and context attention. ACM Trans Multimed Comput Commun Appl 14(2):1–21. https://doi.org/10.1145/3177745
6. Arun N et al (2021) Assessing the trustworthiness of saliency maps for localizing abnormalities in medical imaging. Radiol Artif Intell 3(6):e200267. https://doi.org/10.1148/ryai.2021200267
7. Borji A, Frintrop S, Sihite DN, Itti L (2012) Adaptive object tracking by learning background context. In: 2012 IEEE Computer Society Conference on Computer Vision and Pattern Recognition Workshops, pp. 23–30. IEEE
8. Yan Z, Younes R, Forsyth J (2022) Resnet-like CNN architecture and saliency map for human activity recognition. In: Deng S, Zomaya A, Li N (eds) Mobile Computing, Applications, and Services LNICSSITE, vol 434. Springer, Cham, pp 129–143. https://doi.org/10.1007/978-3-030-99203-3_9
9. Ren Z, Gao S, Chia L-T, Tsang IW-H (2013) Region-based saliency detection and its application in object recognition. IEEE Trans Circuits Syst Video Technol 24(5):769–779
10. Nousias S et al (2020) A saliency aware cnn-based 3d model simplification and compression framework for remote inspection of heritage sites. IEEE Access 8:169982–170001
11. Lin Y, Pang Z, Wang D, Zhuang Y (2017) Task-driven visual saliency and attention-based visual question answering. arXiv preprint arXiv:1702.06700
12. Ullah I et al (2020) A brief survey of visual saliency detection. Multimed Tools Appl 79(45):34605–34645
13. Kruthiventi SS, Gudisa V, Dholakiya JH, Babu RV (2016) Saliency unified: a deep architecture for simultaneous eye fixation prediction and salient object segmentation. In: Proceedings of the IEEE Conference on Computer Vision and Pattern Recognition, pp. 5781–5790
14. Koch C, Ullman S.: Shifts in selective visual attention: towards the underlying neural circuitry. In: Vaina LM (eds) Matters of Intelligence. Synthese Library, vol. 188, pp. 115–141. Springer, Dordrecht (1987). https://doi.org/10.1007/978-94-009-3833-5_5
15. Itti L, Koch C, Niebur E (1998) A model of saliency-based visual attention for rapid scene analysis. IEEE Trans Pattern Anal Mach Intell 20(11):1254–1259
16. Harel J, Koch C, Perona P (2007) Graph-based visual saliency. In: Schölkopf B, Platt J, Hofmann T (eds) Advances in Neural Information Processing Systems 19: Proceedings of the 2006 Conference. The MIT Press, pp 545–552. https://doi.org/10.7551/mitpress/7503.003.0073
17. Bruce N, Tsotsos J (2005) Saliency based on information maximization. In: Advances in Neural Information Processing Systems, vol. 18
18. Hou X, Zhang L (2007) Saliency detection: a spectral residual approach. In: 2007 IEEE Conference on Computer Vision and Pattern Recognition, pp. 1–8. IEEE
19. Zhang J, Sclaroff S (2013) Saliency detection: a boolean map approach. In: Proceedings of the IEEE International Conference on Computer Vision, pp. 153–160
20. Azam S, Gilani SO, Jeon M, Yousaf R, Kim J-B (2016) A benchmark of computational models of saliency to predict human fixations in videos. In: VISIGRAPP (4: VISAPP), pp. 134–142
21. Judd T, Ehinger K, Durand F, Torralba A (2009) Learning to predict where humans look. In: 2009 IEEE 12th International Conference on ComputerVision, pp. 2106–2113. IEEE

22. Borji A, Itti L (2015) Cat2000: a large scale fixation dataset for boosting saliency research. arXiv preprint arXiv:1505.03581
23. Yang C, Zhang L, Lu, H, Ruan X, Yang M-H (2013) Saliency detection via graph-based manifold ranking. In: Proceedings of the IEEE Conference on Computer Vision and Pattern Recognition, pp. 3166–3173
24. Fan S, et al (2018) Emotional attention: a study of image sentiment and visual attention. In: Proceedings of the IEEE Conference on Computer Vision and Pattern Recognition, pp. 7521–7531
25. Jiang M, Huang S, Duan J, Zhao Q (2015) Salicon: aaliency in context. In: Proceedings of the IEEE Conference on Computer Vision and Pattern Recognition, pp. 1072–1080
26. Vig E, Dorr M, Cox D (2014) Large-scale optimization of hierarchical features for saliency prediction in natural images. In: Proceedings of the IEEE Conference on Computer Vision and Pattern Recognition, pp. 2798–2805
27. Kummerer M, Theis L, Bethge M (2014) Deep gaze i: Boosting saliency prediction with feature maps trained on imagenet. arXiv preprint arXiv:1411.1045
28. Huang X, Shen C, Boix X, Zhao Q (2015) Salicon: reducing the semantic gap in saliency prediction by adapting deep neural networks. In: Proceedings of the IEEE International Conference on Computer Vision, pp. 262–270
29. Liu N, Han J, Liu T, Li X (2016) Learning to predict eye fixations via multiresolution convolutional neural networks. IEEE Trans Neural Netw Learn Syst 29(2):392–404
30. Kruthiventi SS, Ayush K, Babu RV (2017) Deepfix: a fully convolutional neural network for predicting human eye fixations. IEEE Trans Image Process 26(9):4446–4456
31. Cornia M, Baraldi L, Serra G, Cucchiara R (2016) A deep multi-level network for saliency prediction. In: 2016 23rd International Conference on Pattern Recognition (ICPR), pp. 3488–3493. IEEE
32. Dodge SF, Karam LJ (2018) Visual saliency prediction using a mixture of deep neural networks. IEEE Trans Image Process 27(8):4080–4090
33. Jetley S, Murray N, Vig E (2017) End-to-end saliency mapping via probability distribution prediction. Google Patents. US Patent 9,830,529
34. Kummerer M, Wallis TS, Bethge M (2016) Deepgaze ii: Reading fixations from deep features trained on object recognition. arXiv preprint arXiv:1610.01563
35. Liu N, Han J (2018) A deep spatial contextual long-term recurrent convolutional network for saliency detection. IEEE Trans Image Process 27(7):3264–3274
36. Pan J, et al. (2017) Salgan: Visual saliency prediction with generative adversarial networks. arXiv preprint arXiv:1701.01081
37. Tavakoli HR, Borji A, Laaksonen J, Rahtu E (2017) Exploiting inter-image similarity and ensemble of extreme learners for fixation prediction using deep features. Neurocomputing 244:10–18
38. Wang W, Shen J (2017) Deep visual attention prediction. IEEE Trans Image Process 27(5):2368–2378
39. Xu Y, Gao S, Wu J, Li N, Yu J (2018) Personalized saliency and its prediction. IEEE Trans Pattern Anal Mach Intell 41(12):2975–2989
40. Jia S, Bruce ND (2020) EML-Net: an expandable multi-layer network for saliency prediction. Image Vis Comput 95:103887
41. Lou J, Lin H, Marshall D, Saupe D, Liu H (2022) Transalnet: towards perceptually relevant visual saliency prediction. Neurocomputing 494:455–467
42. Bylinskii Z, Judd T, Oliva A, Torralba A, Durand F (2019) What do different evaluation metrics tell us about saliency models? IEEE Trans Pattern Anal Mach Intell 41(3):740–757. https://doi.org/10.1109/TPAMI.2018.2815601
43. Judd T, Durand F, Torralba A (2012) A benchmark of computational models of saliency to predict human fixations
44. Borji A, Sihite DN, Itti L (2012) Quantitative analysis of human-model agreement in visual saliency modeling: a comparative study. IEEE Trans Image Process 22(1):55–69

45. Peters RJ, Iyer A, Itti L, Koch C (2005) Components of bottom-up gaze allocation in natural images. Vision Res 45(18):2397–2416
46. Rubner Y, Tomasi C, Guibas LJ (2000) The earth mover's distance as a metric for image retrieval. Int J Comput Vision 40(2):99–121
47. Yan F, Chen C, Xiao P, Qi S, Wang Z, Xiao R (2021) Review of visual saliency prediction: development process from neurobiological basis to deep models. Appl Sci 12(1):309
48. Cornia M, Baraldi L, Serra G, Cucchiara R (2018) Predicting human eye fixations via an LSTM-based saliency attentive model. IEEE Trans Image Process 27(10):5142–5154

Comparative Analysis of Control Schemes for Fuel Cell System

Sadia Saman and Vijay Kumar Tayal

Abstract The fuel cell systems are highly suitable for trucks, boats and other electric vehicle applications. The nonlinear behavior of a fuel cell adversely affects its output. In this paper, three control schemes namely proportional integral (PI), integral and proportional integral derivative (PID) control are proposed for output improvement of a fuel cell system. In order to supply dynamically varying power demand, a DC-DC boost converter is developed to achieve the preferred DC output. The proposed PEMFC-based energy system's MATLAB/SIMULINK model is designed to judge the PEMFC's output regulation. The best settling time and least overshoot are achieved with PI control scheme.

Keywords MATLAB/Simulink · Proton Exchange Membrane Fuel Cell (PEMFC) · Modeling · PID control · DC-DC boost converter

1 Introduction

Renewable energy comprises of biomass, tidal, wind, solar, fuel cell, etc. Renewable energy is growing significantly and becoming more popular. Renewable energy sources are rising at a faster rate than conventional sources including coal and oil.

An effective and healthy alternative source of electricity is the fuel cell [1]. Hydrogen is utilized as input of Proton Exchange Membrane Fuel Cell (PEMFC) [1]. When used as a source of electricity in electric vehicles, PEMFC has almost no emissions, a higher efficiency with high power density at a low operating temperature [2]. It is essential to enhance fuel cell technology to cope the demand for high fuel efficiency, performance, and dependability as well as lower emissions.

The PEMFC runs at low voltage and its output voltage varies greatly depending on the loading and operating conditions. There are various control techniques presented in the literature for switching of converters [3], time delay control [4], and fractional

S. Saman (✉) · V. K. Tayal
Amity University Uttar Pradesh, Noida, UP, India
e-mail: sadiasaman1@gmail.com

N. Hasteer et al. (eds.), *Decision Intelligence Solutions*, Lecture Notes in Electrical Engineering 1080,
https://doi.org/10.1007/978-981-99-5994-5_28

order PI control [5]. The authors proposed closed-loop controller design by K factor [6] concept and standardized feedback amplifiers. A correct model of fuel cell with suitable power electronic interfaces is needed for a clear understanding of its response and characteristics.

With the aim of analyzing the dynamic behavior of PEM fuel cells with variations in temperature, a PSPICE-based parametric model is developed. This model is far superior in comparison with electrochemical equations-based models [8]. The DC-to-DC converter of a power unit based on fuel cells is an essential component. For better power regulation, the DC-to-DC converter design with proposed controllers plays a vital role [3].

The boost converter improves efficiency. Due to a fewer components and various advantages, a number of topologies of DC to DC converters are examined and contrasted. A battery, capacitors, and ultra-capacitors must be used in addition to the fuel cell generation capacity if the need for a sustainable load cannot be met. Energy storage devices increase the system cost with decreased reliability and life [8]. Further, a comparison of ultracapacitor and battery is carried out [8]. The cathode and anode receive air and hydrogen, respectively. An anode catalyst converts the hydrogen into electrons and protons. These electrons flow travel via a closed circuit result in current flow.

Now a day, fuel cell-based systems are preferred as the fuel cells offered clean energy with higher efficiency. Thus, it may be a good alternative to conventional sources of energy. Fuel cells are having a extensive variety of applications, especially in vehicles. Due to the need for better efficiency, higher reliability, and performance fuel cell-based systems are under research [5]. The fuel cell nonlinear behavior and variations in system characteristics with time present complex control. The approximate linear or nonlinear system descriptions are applicable in a narrow range. These approximate models are utilized for the determination of controllability, observability, and stability. Nevertheless, the system error can be reduced by the application of time-dependent parametric models in real-time. In order to accomplish the control objectives in these situations, it is desirable to employ adaptive controllers. These controllers are more adapted to fulfill the intended control objectives in presence of loading disturbances. In literature, to avoid the oxygen deprivation problem during loading, the observer-based feedback controller has been presented.

These reports also propose some related issues and methods of analysis. The models presented are one-dimensional to three-dimensional models. The time-varying conduct of the fuel cell in the former is modeled by lumped-parameter ordinary differential equations, meanwhile in the later it is modelled by distributed-parameter partial differential equations. Due to the accuracy of theory that is applicable to these systems, lumped-parameter system models are generally favored for the controller design applications.

These models describe each component by means of its stack voltage, exchange of heat, compressor, etc.

The vehicle's inertia is the only dynamic taken into account. The authors typically presumed optimal cathode conditions, such as the humidity of stack and temperature etc. In literature, by means of these studies, the integration of fuel cell with vehicle is well explained. There is a need for a precise mathematical model to understand the time-varying behavior of a nonlinear fuel cell accurately. The fuel cell model can be separated from the air feed scheme while maintaining the dynamic behavior.

This paper is organized as follows: Sect. 2 discusses the PEM fuel cell model, Sect. 3 elaborates on the boost converter design, Sect. 4 shows the PID controller design, Sect. 5 presents the simulation results, and Sect. 6 elaborates the conclusion.

2 PEM Fuel Cell Model

The fuel cell is a static means that generates electricity by means of a chemical reaction of fuels. The byproducts involve water and heat energy. In order to continue the chemical reactions, oxidant supply, and chemical reactions are to be maintained. The following equations summarize the chemical process that happens in the electrolyte membranes, cathode and anode to produce electricity: -

$$\mathrm{H2} \Rightarrow 2\mathrm{H}^{+} + 2\mathrm{e}^{-} \ \text{Anode reaction} \tag{1}$$

$$\frac{1}{2}\mathrm{O}_2 + 2\mathrm{H}_2 + 2\mathrm{e}^{-} \Rightarrow \mathrm{H}_2\mathrm{O} \ \text{Cathode reaction} \tag{2}$$

$$\mathrm{H2} + \frac{1}{2}\mathrm{O}_2 \Rightarrow \mathrm{H}_2\mathrm{O} \ \text{Overall reaction} \tag{3}$$

Both electrical and thermal efficiency serve as standards for fuel cell performance. The system's temperature control, water management, and fuel processing together affect the thermodynamic efficiency. The efficiency of a fuel cell depends on concentration and activation, as opposed to ohmic loss. Thus, the output voltage from a fuel cell model relies on ohmic loss, concentration and activation losses [10]. It has been noted that a slow chemical reaction causes the ohmic loss to decrease for low current values and the output voltage to rise [7].

The voltage decreases significantly at very high concentrations of current due to the reduction in efficiency of gas exchange. This is because of the availability of additional catalyst water. The relation between concentration zone and active zone is linear (Fig. 1).

It is essential to design a mathematical model for the analysis of varying behaviors of nonlinear PEMFC.

The ohmic and activation losses decide the stack voltage of fuel cell [9].

$$\mathrm{V}_{\mathrm{dc-stack}} = \mathrm{V}_{\mathrm{open}} - \mathrm{V}_{\mathrm{ohmic}} - \mathrm{V}_{\mathrm{act}} - \mathrm{V}_{\mathrm{con}} \tag{4}$$

$$\mathrm{V}_{\mathrm{open}} = \mathrm{N}_{\mathrm{o}}\,[\mathrm{V}_{\mathrm{o}} + \frac{RT}{2F}\ln(\frac{PH2\sqrt{\mathrm{PO2}}}{PH2O\sqrt{\mathrm{PO}}})] \tag{5}$$

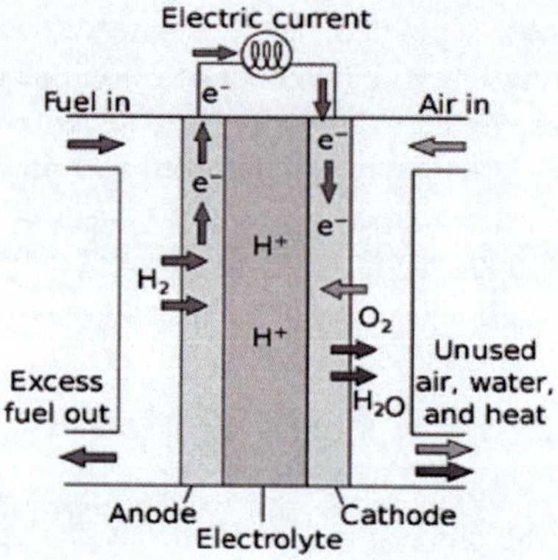

Fig. 1 PEM fuel cell operation diagram

$$V_{ohmic} = I_{dc} * R_{FC} \tag{6}$$

$$V_{act} = N_o * \frac{RT}{2aF} * \ln(\frac{Idc}{Io}) \tag{7}$$

$$V_{con} = -c * \ln(I - \frac{Idc}{Ilim}) \tag{8}$$

In a fuel cell, the ohmic loss becomes less significant at low current levels, and the acceleration of chemical reactions is primarily responsible for the rise in output voltage. There is a reduction in gas exchange efficiency, especially at high current density. This causes the fall in voltage level [6].

3 Boost Converter Design

The primary benefit of the boost converter is less components and higher efficiency. The converter regulates the voltage by duty cycle variation. The choice of converter components, such as inductance and capacitor value, is of utmost importance in order to minimize the generation of ripples for a known switching frequency. But, the small value of inductance results in higher coil current whereas the large inductance value increase the start time [11].

The fuel cell system voltage is increased by using a boost converter due to its superior efficiency and ease of control. It can be seen that DC bus cannot be interfaced with uncontrolled terminal voltage. As a result, the converter design only considers a linear region operating because of the resistance provided by the internal components of a fuel cell. A PWM DC to DC boost converter is depicted in Fig. 2. It's gain is given by: -

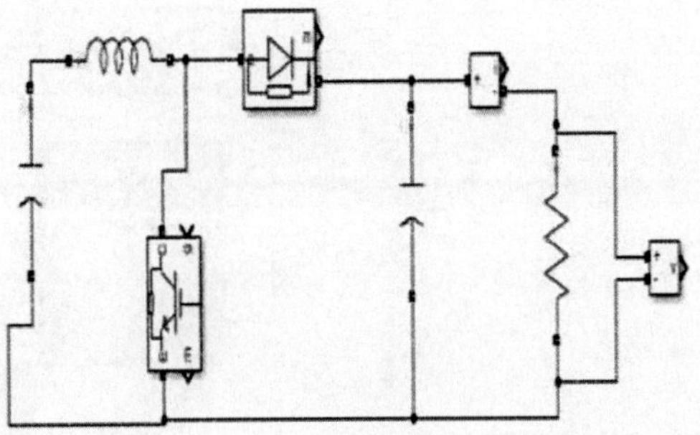

Fig. 2 DC to DC boost converter

Table 1 Boost Converter Parameters

Parameter	Values
L	0.0012
C	0.0051
Fs	10 kHz

$$\frac{\mathrm{Vo}}{Vin} = \frac{\mathrm{I}}{(I-d) + \frac{\mathrm{Rfc}}{(I-d)Rl}} \tag{9}$$

$$d = \frac{\mathrm{Vin}}{Vout} \tag{10}$$

$$\text{current ripple} = \frac{\mathrm{d(I-d)^2 * RTs}}{L} \tag{11}$$

$$\text{voltage ripple} = \frac{\mathrm{dTs}}{RC} \tag{12}$$

where;

duty cycle = d, output voltage = Vo, R = resistance, L = inductance, input voltage = Vin, C = capacitance,

Fs = switching frequency.

From the value of rated voltage, current ripple, switching frequency and voltage ripple, the boost converter size can be find out.

The components such as capacitor and inductor values for dc/dc converter are enlisted in Table 1.

4 PID Controller Design

The PID controller block is placed in a feedback loop's feedforward path. Input is the difference among the system output and the reference signal known as error signal. In spite of fluctuations in the fuel cell current, the pressure variation of oxygen and

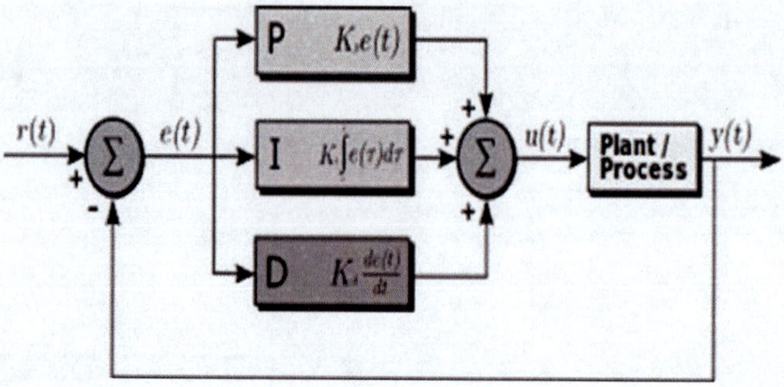

Fig. 3 PID controller

hydrogen is altered by means of Proportional plus Integral plus Derivative (PID) controllers (Fig. 3).

The basic operation of a PID control scheme is based on the information provided by output sensor [12–14]. Then it determines the proportional, integral, and derivative gains to obtain the optimal output.

$$G(s) = K_p + (K_i/s) + (K_d * s) \tag{13}$$

The PI control scheme is mathematically represented by Eq. (14).

$$G(s) = K_p + (K_i/s) \tag{14}$$

5 Simulation Results

The accurate plant model availability crucial prior to controller development. The PEMFC model parameters' values are derived from the PEMFC system. In a single cell, for fixed levels of input fuel pressures, the PEM fuel cell voltage, current and power without controller are depicted in Fig. 4

The variable voltage available at terminals can not be supplied connected to the DC terminals for grid purposes or residential purpose. A linear region of the fuel cell is considered for designing of converter. For controlling the voltage, the duty cycle of converter is varied.

Minimizing ripple generation for a given switching frequency is crucial for converter design. The choice of converter components, such as the capacitor and inductance values, is crucial in this regard. The output DC voltage of this converter is provided. To keep the constant DC bus voltage, a suitable controller is needed. Figure 5 shows Simulink model of fuel cell with closed loop boost converter with PID controller.

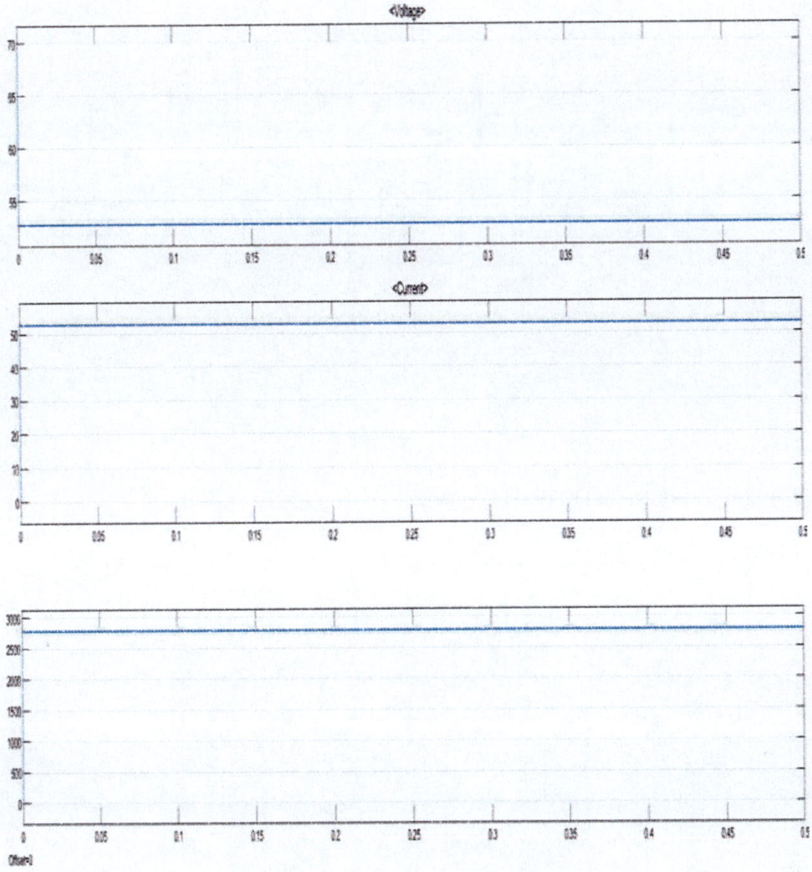

Fig. 4 Voltage, Current and Power of Fuel cell without controller

In this work PID, Integral and PI controllers have been used. These controllers are utilized to sustain a constant DC voltage in presence of load uncertainty. The Simulink model of the PEMFC system with the different tuning of the PID controller is used.

Figures 6, 7, and 8 depicts the output voltage, current, and power of the PEMFC DC to DC boost converter with PID, Integral, and PI controllers. The steady state error is zero because of PI controller action. This is observed from the different response curves that the PI controller is giving superior performance in terms of overshoot and settling time. Figure 8 depicts that fuel cell current and terminal voltage take a smaller time to achieve steady state and the load changes instantaneously. The performance of PI controller and DC to DC boost converter combination is significantly improved for load variations without any storage devices. In low power applications better performance for standalone connected applications is provided by the simple boost converter's implementation.

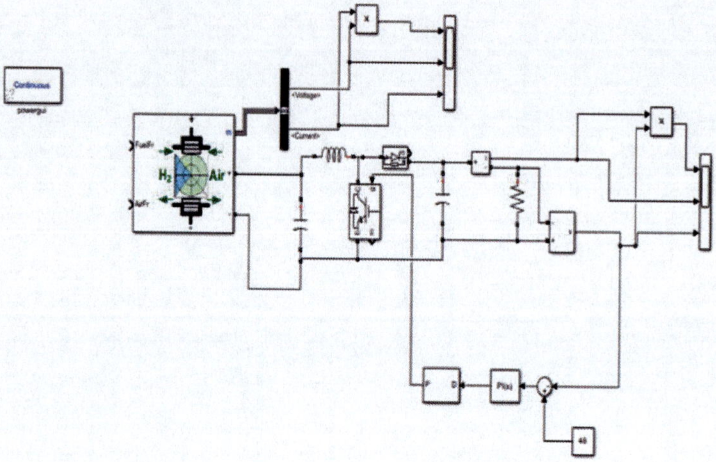

Fig. 5 PEMFC with DC to DC Boost Converter

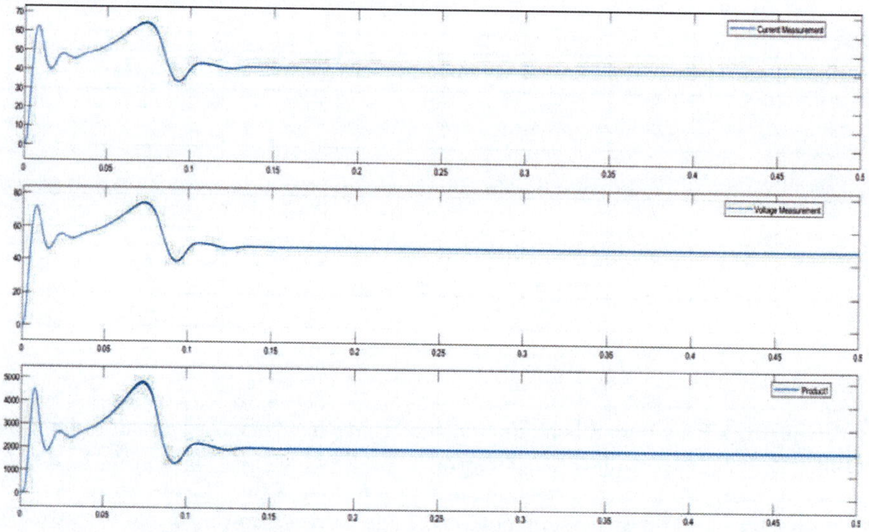

Fig. 6 Current, voltage and power measurement across load with PID controller

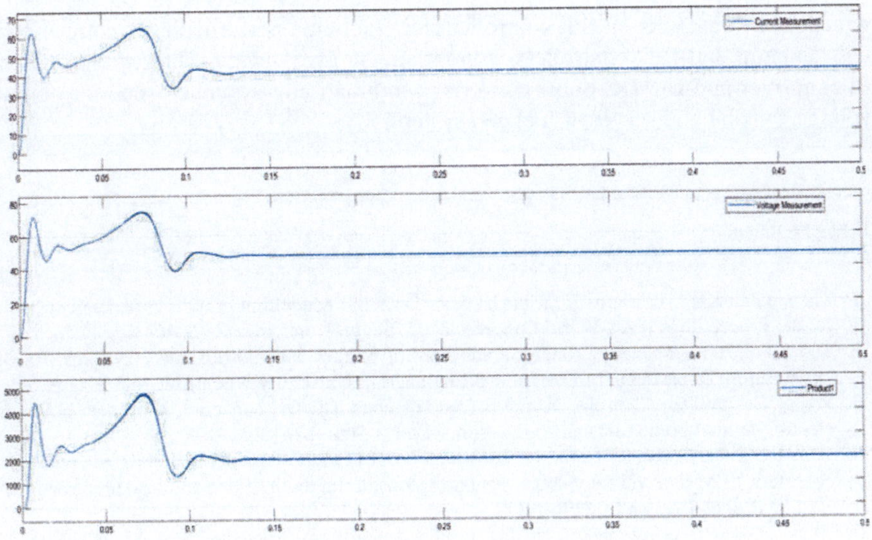

Fig. 7 current, voltage and power measurement across load with Integral controller

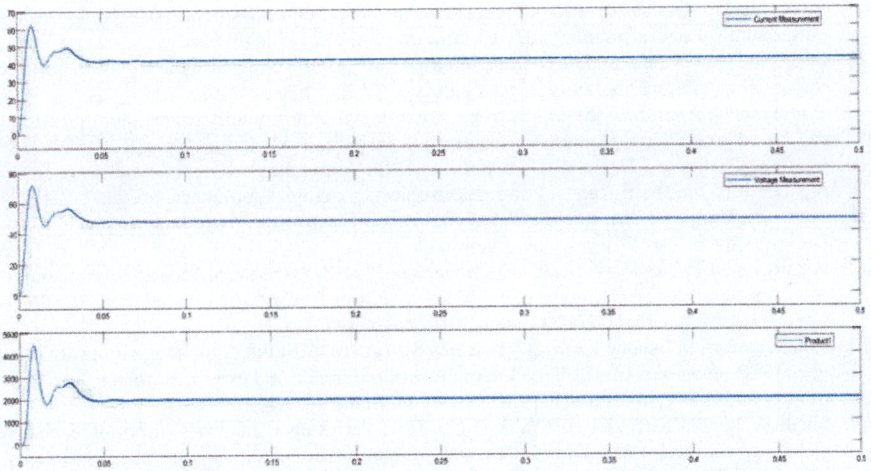

Fig. 8 Current, voltage and power measurement across load with PI controller

6 Conclusion

In order to regulate the fuel cell output voltage current and power, the PEM fuel cell with boost converter is simulated in MATLAB. The boost converter and feedback controller in PEM fuel cell system act as the energy source. This well-designed feedback controller effectively regulates the DC voltage. The PEM fuel cell system with proportional integral (PI) controller gives far better performance in comparison with Integral and PID controllers without any storage device. Thus, the proposed PI controller and DC-DC boost converter combination gives much superior output voltage, current and power of fuel cell system.

References

1. Thanapalan KKT, Williams JG, Liu GP, Rees D (2008) Modelling of a PEM fuel cell system. In: the Proceedings IFAC World Congress 2008, Seoul, Korea, pp. 4636–4641
2. Suh KW (2006) Modeling, analysis and control of fuel cell hybrid power systems. Ph.D. Dissertation, Department of Mechanical Engineering University, Michigan, Ann Arbor
3. Wang YX, Yu DH, Chen SA, Kim YB (2014) Robust DC/DC converter control for polymer electrolyte membrane fuel cell application. J Power Sourc. 261:292–305
4. Bankupalli PT, Ghosh S, Kumar L, Samanta S (2018) Fractional order modeling and two loop control of PEM fuel cell for voltage regulation considering both source and load perturbations. Int J Hydrogen Energy 43:6294–6309
5. Chiu L, Diong B, Gemmen R (2004) An improved small signal model of the dynamic behaviour of PEM fuel cells. IEEE Trans. Indus. Appl. 40(4):970–977
6. Arsov, G.L.: Improved parametric Pspice model of a PEM Fuel Cell. In: 11th International Conference on optimization of Electrical and Electronics Equipment, pp. 203–208. (2008).
7. Yu X, Starke MR, Tolbert LM, Ozpineci B (2007) Fuel cell power conditioning for electric power applications: a summary. IET Electr. Power Appli. 1(5):643–656
8. Fuel Cell Hand Book (2000) U.S. Department of Energy, Office of Fossil Fuel, 7th ed., National Energy Technology Laboratory, West Virginia
9. Tanrioven M, Alam MS (2006) Modeling, control, and power quality evaluation of PEM fuel cell-based power supply system for residential use. IEEE Trans Indus Appl 42(6):1582–1589
10. ChengKWE, Sutanto D, Ho YL, Law KK (2001) Exploring the power conditioning system for fuel cell In: 32nd IEEE Annual Power Electronics Specialists Conference, pp. 2197–2202
11. Mohan N, Undeland TM, Robbins WP (2001) Power Electronics Converters, Applications and Design, 3rd edn. Jon Wiley & Sons, New York
12. Palaniyappan TK, Yadav V, Tayal VK, Choudekar P (2018) PID control design for a temperature control system. In: International Conference on Power Energy, Environment and Intelligent Control (PEEIC), Greater Noida, India, 2018, pp. 632–637
13. Khubchandani V, Pandey K, Tayal VK, Sinha SK (2016) PEM Fuel Cell integration with using Fuzzy PID technique. In: IEEE 1st International Conference on Power Electronics, Intelligent Control and Energy Systems (ICPEICES), Delhi, India, pp. 1–4
14. Singh S, Tayal VK, Singh HP, Yadav VK (2021) Performance analysis of proton exchange membrane fuel cell (PEMFC) with PI and FOPI controllers. In: Vadhera S, Umre BS, Kalam A (eds) Latest Trends in Renewable Energy Technologies. LNEE, vol 760. Springer, Singapore, pp 211–219. https://doi.org/10.1007/978-981-16-1186-5_17

A Comparative Review on Channel Allocation and Data Aggregation Techniques for Convergecast in Multichannel Wireless Sensor Networks

Vishav Kapoor and Daljeet Singh

Abstract Consumers in today's fast-paced world have little patience for lengthy information-gathering processes. This has made it difficult for researchers to collect data in a timely fashion. Convergecast transmission is a promising method for quick messaging. In convergecast, each sensor node sends its packet directly to the sink node, bypassing the intermediary node. By reducing unnecessary transmissions and sending combined data to the network's anchor nodes, data aggregation extends the useful life of power-starved WSNs and increases their efficiency. Moreover, the role of multichannel communication has emerged as a need in WSNs. This paper describes and evaluates the different channel allocation and data gathering algorithms used to achieve efficient convergecast in WSN. Comparative evaluation of the techniques for data aggregation and channel allocation has been done on various metrics such as energy, QoS, robustness, channel type, scalability, mobility, and nature of techniques.

Keywords Wireless Sensor Networks · Convergecast · Multichannel · Aggregation · QoS · MAC

1 Introduction

Wireless sensor networks (WSNs) are a collection of low-power, low-capacity nodes that are connected by radio waves. Sensor nodes are devices that work together with others in a network to carry out a task, such as environmental monitoring. The number of connected devices in use today has surged during the last decade [1]. These days, it's hard to go somewhere without encountering at least one technological device that's linked to the internet [2]. One element of this trend that has several uses is

V. Kapoor (✉) · D. Singh
School of Electronics and Electrical Engineering, Lovely Professional University, Phagwara, Punjab, India
e-mail: Vishavkapoor55@gmail.com

N. Hasteer et al. (eds.), *Decision Intelligence Solutions*, Lecture Notes in Electrical Engineering 1080,
https://doi.org/10.1007/978-981-99-5994-5_29

WSNs [3]. In order to satisfy the demands of these networks in terms of power consumption and latency, various MAC protocols play a critical role. The amount of information created by the Internet of Things (IoT) grows exponentially as the number of devices that can be linked grows. These terminals get a lot of traffic, which may be difficult to manage. In particular for commercial processes, power consumption, and delay are major concerns in WSN. Wireless communications face the challenges of minimizing power usage while maintaining high throughput and zero data loss. Since sensors often need to run for weeks or even decades without human involvement or battery replacement, power consumption represents a significant challenge. Indeed, customers often opt to put such sensor nodes arbitrarily in a designated region, making replacement either very hard or prohibitively costly. In particular, these configurations are all about the cell optimization techniques that govern node operation throughout each time frame.

WSNs have come a long way in recent years, with smaller, more powerful sensors that can cover greater distances for less cost. Devices in single-channel WSNs (SWSNs) utilize the same frequency band for both data transmission and reception. When two nodes are exchanging information, the nodes around them are unable to send data until the conversation is between two ends. To address this shortcoming of current SWSNs, the multichannel mode is being investigated. Nodes in an MWSN's vicinity have access to a variety of channels for sending and receiving data; each channel is associated with a particular set of paired nodes. Increased data gathering rates are possible with MWSNs as a result of a rise in the quantity of data being transferred concurrently. While SWSNs only take into account time slot conflicts, MWSNs take into account both channel and time slot conflicts at the same time. Time slot selection, channel access, and switching are studied in most MWSN research. When neighbor nodes utilize distinct channels, the sending node sends data by shifting to some other node's receptive channel and then returns to its own to receive data. Calibrating and switching stations take time. Time synchronization difficulties arise with channel switching [4]. Channel changing repeatedly for one node-to-node transmission wastes energy and time. Because sensor nodes are energy-constrained, channel shifting may save energy. Multiple channels on various nodes need frequent channel switching, wasting energy and time. MWSNs employ complementary channels and time slots to provide a collision-free schedule. This can be fixed by assigning separate time slots, not numerous channels. Similarly, channel assignments don't need distinct time windows. Nodes must have a minimal number of channels to decrease switching, energy utilization, and data collection time.

Many proposed MAC protocols [5–10] in the WSN space focus on single channels of interaction. Single-channel scenarios where energy efficiency [11], scalability [12], and flexibility [13] are important design goals benefit greatly from the use of these protocols. Z-MAC and Burst-MAC are two examples of single-channel MAC procedures that work well with planned communication and attempt to deliver maximum throughput [13]. Whereas these methods do well in single-channel circumstances, the throughput may be further increased by sending data in parallel across many channels, which eliminates the possibility of congestion and interference. The effects of interference may be mitigated in several ways, including via multi-channel communication [13], transmission power regulation [14], and the development of minimal interference sink structures [15]. So, it is being inferred that multi-channel allocation is one of the prime concerns for good efficiency but there are some other ways by which it can be enhanced like the convergecast which has been highlighted in the next section of the paper.

The organization of the paper is as under. Section 1st of the paper defines the introduction of the paper which is followed by the role of channel allocation and data aggregation in the Convergecast. The similar work which has been done by the other researchers has been defined in the 3rd section. The comparative analysis of the multichannel and aggregation methods pertaining to various factors has been done in the 4th section of the paper. The 5th section refers to the conclusion and future work.

2 Channel Allocation and Aggregation Role in Covergecast

Minimizing delays and ensuring packet delivery are two of the most immediate concerns for data convergecast, along with conserving resources. Delays from beginning to finish are minimized to keep data as current as possible. Furthermore, improved accuracy in monitoring is achieved by assured packet delivery. Due to the enormous amount of sensors in convergecast, collisions are a significant barrier to achieving limited latencies. Indeed, data loss occurs as a result of collisions. Delivery timings of retransmitted packets are unpredictable and increase overall packet delay. Collision-free methods provide constrained latencies, but contention-based methods are inefficient because of back-off and collisions. In reality, these protocols guarantee that a node's transmissions will not conflict with those of other nodes. This is accomplished by strategically assigning channels and intervals to nodes to prevent these interferences. Furthermore, a node is only considered to be "active" when it is either sending data to or receiving information from its parents. If a node is not communicating, it will switch off its antenna to save power.

The fast collection is possible for periodic traffic owing to contention-free MAC methods like Time Division Multiple Access (TDMA), which remove the need for retransmission after a collision and ensure transmission completion time. However, the challenge of designing a conflict-free TDMA process has been known to be NP-complete. Without each and every one of these broadcasts, it is difficult to gather all of the data produced by sensor nodes. Assigning these slots such that all data produced by sensors is given to the sink in a single collection frame while allocating as few slots as possible is a challenging problem.

These convergecast use cases highlight the importance of convergecast's delay efficiency and data accuracy. Similar to data aggregation situations, convergecast applications are useful in times of crisis and need continuous data collection from individual nodes. As an example, consider the practice of monitoring patients. Information acquired from critically ill patients must be sent to a sink node quickly and reliably. Energy-efficient data collection is made more difficult in convergecast because each packet created in a node must be delivered to a sink node without even any content aggregation in the intermediary node. Since the intermediate node frequently combines and averages numerous data streams, consolidation typically uses less power than convergence. Delay efficiency relates to how quickly a node can transmit its detected information back to the sink node, whereas memory, power efficiency, bandwidth, and data correctness are all issues that arise with convergecast. Typical sensor networks have always struggled with memory, energy, and throughput limitations. Since convergent applications care about the data detected by individual nodes, they have the additional challenge of ensuring data accuracy. However, by averaging sensory input from numerous nodes, certain inaccuracies might be hidden in the aggregated data.

3 Literature Work

In this section, the work that has been done in the field of convergecast by other researchers has been defined. The work is categorized into sub-sections about scheduling related, multichannel related, and energy using aggregation related concerning the convergecast in WSN. Some of the most important works have been discussed here.

3.1 Multi Channel MAC Related Work

Delay-sensitive operations, the context of which and the flow of traffic which are often well-known, benefit from fixed channel allocation. The minimal maintenance benefits of a fixed channel allocation are offset by the potential for worse performance in a dynamic operating environment for a WSN. Fixed channel allocation is used by both TMCP [16] and MCRT [17]. The sink in TMCP [16] partitions the network into numerous sub-trees, assigns unique channels to every sub-tree in a receiver-centric manner, and only sends data flows via the sub-tree associated with the sink. The main goal of these protocols is to reduce the time spent waiting for data by eliminating costly and time-consuming channel switches.

Multichannel adaptive allocation algorithms have been added to multichannel WSNs to increase resilience in the face of channel variance and traffic variations. However, without a coordination strategy in place to make sure that two nodes that want to talk to each other set their radios to the same channel, the connection is poor. Most adaptive allocation protocols, like Y-MAC [18] and MuChMAC [19], rely on precise synchronization sensors to get around the connection issue. Still, delay-critical operations may be unable to take advantage of such solutions due to the significant switching delays that are inherent to them.

There are use cases where the freshness of the supplied data is not optimal while switching interfaces often, making dynamic channel allocation undesirable. Many standards, such as MMSN [20] and TACA [21], are part of this movement. Given that it has been shown that loaded channels experience higher transmission time delays and more radio collisions, EMMAC [22] employs a node's permutation channel schedule to shift among channels with the goal of minimizing the load capacity of each channel. If the channel allocation is receiver-dependent, just like in EM-MAC [22], then independent hopping sequences need extra calculation and storage at every sender node. One of the frequency-key hopping's benefits is that it's less susceptible to interference. When the traffic load is high, RMCA [23] and ARM [24] algorithms use probability-based randomized channel selection.

MuChMAC [19] is a protocol that mixes TDMA with asynchronous MAC methods. It's a kind of receiver-based channel allocation, where each node may choose its own switching sequences for the channel it receives depending on its ID and the slot number currently in use. In order to solve the scheduling, channel allocation, and power management problems simultaneously in multi-power WSNs, a new routing technique has been presented by Li et al. [25]. They come up with a heuristic strategy using the random walk technique to determine the best routes in order to reduce both power usage and the end-to-end delay along pathways. Each connection has a different method of dynamically assigning channels. This method uses a minimal set of channels (colors) to cover all sensors (vertices) in a way that ensures no two nodes in a row have the same set of channels.

3.2 *Energy Efficiency Through Data Aggregation Related Work*

The simple graph concept is frequently employed in conventional communication networks. However, the architecture of a large-scale WSN s might include several thousand nodes. Therefore, a lot of control messages are needed to set up channels. However, numerous real-time control messages are required to keep an established route due to the poor dependability of the sensor network and wireless communication. These activities need a lot of additional bandwidth and power. Thus, studying the role of energy is an important concern.

To maximize availability in PSNs used for IoT applications, the authors of [26] suggest Data Collection and Aggregation with Selective Transmission (DGAST) method. Data from the sensors is collected on a regular basis by DGAST, which helps to preserve the sensors' standby time. The DGAST protocol splits the PSN's lifespan into cycles. Every cycle consists of four steps: data collection, data aggregation, selective communication, and sample frequency adjustment at each node in light of dynamic changes in the climate in the perceived environment.

A Systematic Data Aggregation Model (CSDAM) for processing data in real-time is presented in [27]. At first, the network is organized into a cluster, consisting of both awake and inactive nodes, and a Cluster-Head (CH) is chosen based on the sensors' rankings in terms of their current energy status and their distance from the

Base Station (BS). In this case, the CH serves as the aggregator. Second, Aggregation occurs across a three-tiered structure, with the data processing at the third tier being simplified due to the first two tiers' work. If an aggregator's energy drops below a certain level, it is replaced with another. Third, priority should be given to real-time apps over those that aren't.

As discussed by the authors of [28], data aggregation is a useful method for extending the useful life of WSNs and decreasing their energy consumption. By merging and integrating data that is relevant and similar, unnecessary packets may be avoided and redundancy is reduced. The primary goal of their study is to provide an efficient new means of data aggregation that makes use of the open-pit mining concept. Each sensor node, as described by the authors of [29], looks for other sensor nodes that have made identical readings within a certain time period and aggregate them. The mean data of the observations that fall inside the chosen set are conveyed to the CH. Information not found is sent rather than a blank set if sensors are unable to detect any data during intervals. The authors could then filter out any unnecessary information or regional outliers.

In order to improve energy conservation and extend network life, the researchers of [30] have conducted thorough research on this type of interconnected sensor-based network. In [31], the authors suggest an energy-efficient data-gathering format for use in WSNs. The sensors are gathered for probable rendezvous location selection using clustering techniques. To save power, only the RPs necessary for transmission are activated, and the rest are put to sleep based on their maximum weight. This method is used because it maximizes the productivity of data collection in a sensor network. The Optimization method is used to determine the most effective route for the mobile sink to follow.

In [32], the authors suggest a pattern-based Redundancy Elimination Data Aggregation (REDA) technique they term. The suggested pattern is data-driven and makes use of differential data amassed over several iterations at distinct sensor nodes. As a result, at no point throughout the cycle is information sent from sensor nodes in the same group to the corresponding CH that has already been sent. An analysis of REDA's performance demonstrates a 44% reduction in energy usage compared to a protocol lacking data gathering techniques. Adaptive Data Collection (ADC) is a methodology proposed by the authors of [33] for collecting sensor data at regular intervals to increase the lifespan of a Periodic Sensor Network (PSN). The ADaC protocol's lifespan is quantified in cycles. There are two phases to every cycle. The first step is data collection. Second, it adjusts its sample rate in response to changes in a dynamically controllable way based on the similarities across periods within a single cycle that uses a Euclidean distance metric.

The authors offer multi-layer large data aggregation architecture and a Priority-Based Dynamic Data Aggregation (PDDA) approach [34]. Since most current methods focus only on data aggregation at the central server level, the suggested PDDA methodology operates at the bottom layer of sensors. In [35], an effective clustering technique called the Low-Energy Adaptive Clustering Hierarchy (LEACH) is described; in which nodes inside a cluster submit their information to a local CH. In [36], the authors offer a protocol called HEED (Hybrid Energy-Efficient Distributed

Clustering) that chooses CHs on a periodic basis based on a combination of the node's remaining energy and a supplementary variable, such as the node's closeness to its peers or its grade. It provides a fairly consistent distribution of CHs over the network with little message overhead.

Some other latest work has been done by other researchers in the context of convergecast in WSN in the field of scheduling, channel allocation, and energy management which is of prime concern [37–45].

4 Comparative Analysis of Scheduling Techniques

In this section, a comparison of the different ways that convergecast is done in WSN has been made. Channel allocation is also one of the prime concerns that have been noticed through the literary work which also embarks on the efficiency of any convergecast approach. In Table 2, various methods for channel allocation have been evaluated based on the type of allocation of the channel along with their QoS parameters such as robustness, delay, energy, and scalability.

Table 1 illustrates the three broad categories into which channel assignment approaches fall: the static, the semi-dynamic, and the dynamic. Approaches that don't evolve are static, which limits their usefulness. This means they are useful in situations when both the surroundings and the application are stable and well-understood. The semi-dynamic strategy is appropriate when conditions and traffic levels remain constant over an extended period of time.

Throughout the lifespan of a network, sensors will transmit many distinct types of flows, each with unique QoS needs. In contrast, channels are given to traffic patterns that are not limited in any manner so that conditional QoS assurances are

Table 1 Comparative evaluation of channel allocation techniques on WSN

Reference Work	Static Channel	Dynamic Channel	Semi Dynamic Channel	Robust	Delay Efficient	Energy Efficient	QoS	Scalability
[16]	✓	✗	✗	✗	✓	✗	✗	✗
[17]	✓	✗	✗	✗	✓	✗	✓	✗
[20]	✗	✗	✓	✗	✗	✗	✗	✗
[21]	✗	✗	✓	✓	✗	✗	✓	✗
[22]	✗	✗	✓	✓	✗	✗	✗	✗
[31]	✗	✗	✓	✓	✓	✓	✓	✗
[23]	✗	✗	✓	✓	✗	✗	✓	✗
[24]	✗	✓	✗	✗	✓	✓	✗	✓
[18]	✗	✓	✗	✗	✗	✓	✗	✓
[19]	✗	✓	✗	✗	✗	✓	✗	✓

Table 2 Comparative evaluation of data aggregation techniques on WSN

Reference Work	Cluster Based	Tree Based	Homoge-nous Node	Heteroge-neous Node	Mobility	Centralized	Distributed	In-Network
[26]	✓	✗	✓	✗	✗	✗	✓	✗
[27]	✓	✗	✓	✗	✗	✗	✓	✗
[28]	✓	✗	✓	✓	✗	✗	✓	✗
[29]	✓	✗	✓	✗	✗	✗	✗	✓
[30]	✓	✗	✓	✗	✓	✗	✓	✗
[31]	✗	✗	✓	✗	✓	✗	✓	✗
[32]	✓	✗	✓	✗	✗	✗	✓	✗
[33]	✓	✗	✓	✗	✗	✗	✓	✗
[34]	✓	✓	✓	✗	✗	✗	✓	✗
[35]	✓	✗	✓	✗	✓	✗	✓	✗
[36]	✓	✗	✗	✓	✗	✗	✓	✗

maintained. It is also inferred that the semi-dynamic methods are more favorable in terms of robustness and meeting the QoS parameters but they are not taken care of the delay and energy to a large extent. On the other hand, dynamic methods are more scalable and energy efficient but they lack QoS and robustness. Thus the focus on creating fault-tolerant methods with energy efficiency and meeting the QoS is required.

Nodes in WSNs may be either "homogeneous" or "heterogeneous," depending on the network's needs. The latter is distinguished by upgrading, and the ability of greater starting energy is a feature that has been considered here. The topology of a network plays a crucial role in making use of redundant data. It's possible for the networks to include node groups, to be tree-based, or to be clustered.

Table 2 demonstrates that there are three distinct categories of aggregating algorithms: centralized, decentralized, and in-network. In the first, a central node either acts as the data aggregator and coordinator itself or appoints another node to do so. For the latter, the algorithm's decentralized nature ensures that all nodes have an equal chance of being selected as the aggregator given a certain set of criteria. The third kind is also the one in which the algorithm operates locally and determines both the data aggregator and the aggregation method. Initial research often believed that any of the nodes may serve as the aggregator despite the fact that this was not the case. Methods were then introduced in which aggregators are mobile nodes that are not limited by available resources, allowing them to freely roam the network and gather information from a variety of nodes.

In static networks, where no change happens to their architecture over time, the cluster-based technique is quite effective. However, such a method is usually ineffective when used inside a dynamic network. Hence, it would be beneficial to investigate data aggregation strategies for wireless or adaptable networks. Table 2 further suggests that the primary inclination is toward the dispersed nature and that most work has been done on homogeneous nodes rather than heterogeneous nodes. In addition, data aggregation using a tree structure has received less attention.

5 Conclusion and Future Work

There is a wide range of potential uses for WSNs. In this paper, we focus on some of the challenges presented by multichannel communications in convergecast networks, including the selection strategy, periodicity, and manner of assigning channels. The delay and efficiency of the convergecast algorithm improve dramatically when many channels are used. We have compared and contrasted many approaches to data aggregation, taking into account characteristics such as network and method types, node diversity, and the movement of the aggregator nodes. Methods for allocating channels and setting up schedules have also been studied. It has been inferred from the aforementioned literature review, that one of the primary factors that impact the effectiveness of any convergecast strategy is channel assignment. Semi-dynamic approaches are preferable in terms of robustness and achieving the QoS requirements, but they do not adequately address the issues of latency and energy consumption. While dynamic approaches are more flexible and energy-efficient, they fall short in other key areas, including QoS and robustness. Since energy consumption is one of the main challenges in sensor networks, future research will examine methods to save even more energy by using a variety of channels. Also to save energy via the investigation of transmission power management techniques on the nodes and the security of communicated information will be executed.

References

1. Balazka D, Rodighiero D (2020) Big data and the little big bang: an epistemological (r)evolution. Front Big Data 3:345–348. https://doi.org/10.3389/fdata.2020.00031
2. Ibarra-Esquer J, González-Navarro F, Flores-Rios B, Burtseva L, Astorga-Vargas M (2017) Tracking the evolution of the internet of things concept across different application domains. Sensors (Basel) 17(6):1379
3. Kandris D, Nakas C, Vomvas D, Koulouras G (2020) Applications of wireless sensor networks: an up-to-date survey. Appl. Syst. Innov. 3(1):14
4. Chen H, Cui L, Lu S (2009) An experimental study of the multiple channels and channel switching in wireless sensor networks. In: Proceedings of the 4th International Symposium on Innovations and Real-Time Applications of Distributed Sensor Networks (IRADSN), pp. 54–61.
5. Ye W, Heidemann J, Estrin D (2003) An energy-efficient MAC protocol for wireless sensor networks. In: Proceedings of Twenty-First Annual Joint Conference of the IEEE Computer and Communications Societies
6. van Dam T, Langendoen K (2003) An adaptive energy-efficient MAC protocol for wireless sensor networks. In: Proceedings of the First International Conference on Embedded Networked Sensor Systems - SenSys 2003

7. Rajendran V, Obraczka K, Garcia-Luna-Aceves JJ (2006) Energy-efficient, collision-free medium access control for wireless sensor networks. Wirel Netw 12(1):63–78
8. Lu G, Krishnamachari B, Raghavendra CS (2004) An adaptive energy-efficient and low-latency MAC for data gathering in wireless sensor networks. In: 18th International Parallel and Distributed Processing Symposium, 2004. Proceedings (2004)
9. Polastre J, Hill J, Culler D (2004) Versatile low power media access for wireless sensor networks. In: Proceedings of the 2nd International Conference on Embedded Networked Sensor Systems - SenSys 2004
10. Rhee I, Warrier A, Aia M, Min J, Sichitiu ML (2008) Z-MAC: a hybrid MAC for wireless sensor networks. IEEE ACM Trans Netw 16(3):511–524
11. Langendoen K, Halkes G (2005) Energy-efficient medium access control, in Embedded Systems Handbook. CRC Press, Boca Raton
12. Demirkol I, Ersoy C, Alagoz F (2006) MAC protocols for wireless sensor networks: a survey. IEEE Commun Mag 44(4):115–121
13. Ringwald M, Römer KU (2008) BurstMAC-a mac protocol with low idle overhead and high throughput (work in progress). In: IEEE International Conference on Distributed Computing in Sensor Systems, pp. 24–26.
14. ElBatt T, Ephremides A (2004) Joint scheduling and power control for wireless ad hoc networks. IEEE Trans Wirel Commun 3(1):74–85
15. Fussen M, Wattenhofer R, Zollinger A (2005) Interference arises at the receiver. In: 2005 International Conference on Wireless Networks, Communications and Mobile Computing
16. Wu Y, Stankovic JA, He T, Lin S (2008) Realistic and efficient multi-channel communications in wireless sensor networks. In: IEEE INFOCOM 2008 - The 27th Conference on Computer Communications
17. Wang X, Wang X, Fu X, Xing G, Jha N (2009) Flow-based real-time communication in multi-channel wireless sensor networks. In: Roedig U, Sreenan CJ (eds) Wireless Sensor Networks. Springer, Berlin, Heidelberg, pp 33–52. https://doi.org/10.1007/978-3-642-00224-3_3
18. Kim Y, Shin H, Cha H (2008) Y-MAC: an energy-efficient multi-channel MAC protocol for dense wireless sensor networks. In: 2008 International Conference on Information Processing in Sensor Networks (IPSN 2008)
19. Borms J, Steenhaut K, Lemmens B (2010) Low-overhead dynamic multi-channel MAC for wireless sensor networks. In: Silva JS, Krishnamachari B, Boavida F (eds) Wireless Sensor Networks LNCS, vol 5970. Springer, Heidelberg, pp 81–96. https://doi.org/10.1007/978-3-642-11917-0_6
20. Zhou G, Huang C, Yan T, He T, Stankovic JA, Abdelzaher TF (2006) MMSN: multi-frequency media access control for wireless sensor networks. In: Proceedings IEEE INFOCOM 2006. 25TH IEEE International Conference on Computer Communications
21. Wu Y, Keally M, Zhou G, Mao W (2009) Traffic-aware channel assignment in wireless sensor networks. In: Liu B, Bestavros A, Du D-Z, Wang J (eds) Wireless Algorithms, Systems, and Applications. Springer, Berlin, Heidelberg, pp 479–488. https://doi.org/10.1007/978-3-642-03417-6_47
22. Tang L, Sun Y, Gurewitz O, Johnson DB (2011) EM-MAC: a dynamic multichannel energy-efficient MAC protocol for wireless sensor networks. In: Proceedings of the Twelfth ACM International Symposium on Mobile Ad Hoc Networking and Computing, pp. 1–11
23. Yu Q, Chen J, Sun Y, Fan Y, Shen X (2010) Regret matching based channel assignment for wireless sensor networks. In: 2010 IEEE International Conference on Communications
24. Li J, Zhang D, Guo L, Ji S, Li Y (2010) ARM: An asynchronous receiver-initiated multichannel MAC protocol with duty cycling for WSNs. In: International Performance Computing and Communications Conference
25. Li J, Guo X, Guo L (2011) Joint routing, scheduling and channel assignment in multi-power multi-radio wireless sensor networks. In: 30th IEEE International Performance Computing and Communications Conference
26. Rhee I, Warrier A, Min J, Xu L (2006) DRAND: distributed randomized TDMA scheduling for wireless ad-hoc networks. In: Proceedings of the 7th ACM International Symposium on Mobile Ad hoc Networking and Computing, pp. 190–201

27. Wang Y, Henning I (2007) A deterministic distributed TDMA scheduling algorithm for wireless sensor networks. In: 2007 International Conference on Wireless Communications, Networking and Mobile Computing
28. Lee WL, Datta A, Cardell-Oliver R (2008) FlexiTP: a flexible-schedule-based TDMA protocol for fault-tolerant and energy-efficient wireless sensor networks. IEEE Trans Parallel Distrib Syst 19(6):851–864
29. Wu F-J, Tseng Y-C (2009) Distributed wake-up scheduling for data collection in tree-based wireless sensor networks. IEEE Commun Lett 13(11):850–852
30. Lin C-K, Zadorozhny V, Krishnamurthy P, Park H-H, Lee C-G (2011) A distributed and scalable time slot allocation protocol for wireless sensor networks. IEEE Trans Mob Comput 10(4):505–518
31. Yu C, Fiske R, Park S, Kim WT (2012) Many-to-one communication protocol for wireless sensor networks. Int. J. Sens. Netw. 12(3):160
32. Bagaa M, Younis M, Ouadjaout A, Badache N (2013) Efficient multi-path data aggregation scheduling in wireless sensor networks. In: 2013 IEEE International Conference on Communications (ICC)
33. Bagaa M, Younis M, Ksentini A, Badache N (2013) Multi-path multi-channel data aggregation scheduling in wireless sensor networks. In: 2013 IFIP Wireless Days (WD)
34. Zeng B, Dong Y (2014) A collaboration-based distributed TDMA scheduling algorithm for data collection in wireless sensor networks. J Netw 9(9):2319. https://doi.org/10.4304/jnw.9.9.2319-2327
35. Soua R, Minet P, Livolant E (2014) A distributed joint channel and slot assignment for convergecast in wireless sensor networks. In: 2014 6th International Conference on New Technologies, Mobility and Security (NTMS)
36. Younis O, Fahmy S (2004) HEED: a hybrid, energy-efficient, distributed clustering approach for ad hoc sensor networks. IEEE Trans Mob Comput 3(4):366–379
37. Osman M, Nabki F (2021) OSCAR: an optimized scheduling cell allocation algorithm for convergecast in IEEE 802.15.4e TSCH networks. Sensors 21(7):2493. https://doi.org/10.3390/s21072493
38. Ghosh R, Mohanty S, Patnaik PK, Pramanik S (2022) A novel secured method for rapid data accumulation in energy-aware WSN. In: Pramanik S, Sharma A, Bhatia S, Le D-N (eds) An Interdisciplinary Approach to Modern Network Security. CRC Press, Boca Raton, pp 167–187. https://doi.org/10.1201/9781003147176-10
39. John J, Kasbekar GS, Baghini MS (2021) Maximum lifetime convergecast tree in wireless sensor networks. Ad Hoc Netw 120:102564
40. Thekkil TM, Prabakaran N (2021) A multi-objective optimization for remote monitoring cost minimization in wireless sensor networks. Wirel Pers Commun 121(1):1049–1065
41. Sadeq AS, Hassan R, Sallehudin H, Aman AHM, Ibrahim AH (2022) Conceptual framework for future WSN-MAC protocol to achieve energy consumption enhancement. Sensors 22(6):2129. https://doi.org/10.3390/s22062129
42. Zeng B, Liang Z, Gao X (2022) SINR-based slot reuse algorithm for multi-channel wireless sensor networks. Research Square
43. Lu Y, Wang K, He E (2022) Many-to-many data aggregation scheduling based on multi-agent learning for multi-channel WSN. Electronics (Basel) 11(20):3356
44. Vo V.-V, Le D.-T, Kim M, Choo H (2022) Data aggregation latency minimization in multichannel duty-cycled WSNs. In: 2022 International Conference on Information Networking (ICOIN)
45. Mahesh N, Vijayachitra S (2022) Hierarchical autoregressive bidirectional least-mean-square algorithm for data aggregation in WSN based IoT network. Adv Eng Softw 173:103275

Design of Hybrid Energy Storage System for Renewable Energy Sources

Arockiaraj, Suban Athimoolam, and Praveen Kumar Santhakumar

Abstract By integrating an additional storage mechanism with a regular storage device, the developed system proposes to boost the efficiency of energy storage setup for PV systems and prolong the lifetime of the batteries used in them. To achieve higher power regulation performance, it employs a logical controller to manage the power supplied by two DC storage devices battery and a super capacitor. A traditional system stores electricity during surplus generation and supplies it on demand using a single storage unit. However, in order to alleviate power fluctuations produced by high power disturbances, these devices feature a storage device degradation limitation owing to frequent charge/discharge cycles. Various energy storage technologies have been integrated to construct a system of energy storage to sustain power instability in order to fully leverage the potential of renewable energy sources. To eliminate the constraints of employing single storage unit, the suggested strategy combines the advantages of a super capacitor and a battery, with a renewable power generation unit. In addition, the fuzzy controller is employed as a converter to transfer power between energy sources instantly. Simulation with MATLAB/SIMULINK is used to verify the control design. It is designed and tested a prototype of the planned Hybrid Energy Storage (HES) with fly-back converter and closed loop control.

Keywords Solar energy · HES · Fuzzy logic control · Battery · Super capacitor

Arockiaraj (✉) · P. K. Santhakumar
Mepco Schlenk Engineering College, Sivakasi, India
e-mail: arockiarocks@gmail.com

S. Athimoolam
Velammal College of Engineering and Technology, Madurai, India

Arockiaraj · S. Athimoolam · P. K. Santhakumar
Mepco Schlenk Engineering College, Sivakasi, Tamil Nadu, India

N. Hasteer et al. (eds.), *Decision Intelligence Solutions*, Lecture Notes in Electrical Engineering 1080,
https://doi.org/10.1007/978-981-99-5994-5_30

1 Introduction

1.1 Background

Renewable energy is generally considered to be very promising, futuristic and developing area in the field of energy generation in all over the world because of its clean, economy property and to reduce the emission of polluting air such as CO2 [1]. However, the renewable energy sources like solar and wind power generation output are strongly fluctuating and the current weather forecast is not possible to accurately predict the availability of these sources. Different energy storage devices are used to support real power compensation to maintain power availability at the output of renewable energy power sources [2].

To accommodate rising electrical demand, contemporary non conventional power systems are more profoundly loaded greater than ever before [4–7]. The Lead acid battery is most widely used storage device which is now being replaced by lithium-ion batteries because of its high energy density and efficiency [3]. Also, Lithium ion has higher number of life cycle before it drops its power than a lead acid set [8]. Even though it has higher life cycle it also starts to lose its capacity after around 5000 charge discharge cycles.

The efficiency of storage device also depends on the rate at which charging and discharge. For a better performance slow charging and discharging is essential which cannot always be controlled in practical case [9]. So, to improve this performance an additional storage device with higher efficiency can be used to provide power at high demand and thus by regulating charging discharging current for battery.

In this system, during short time power demand, most of the required power is delivered by super capacitor. As a result, the battery's lifespan is extended. A DC/DC fly-back converter which is bidirectional element connects a parallel kind of mixed energy storage system based on super capacitor and lithium battery. The output of renewable energy sources linked through appropriate power system devices based on nature of DC/AC grid system [10]. The proposed system is modeled, designed and simulated with the help of MATLAB and prototype. The hardware controller implementation is also validated with the help of Arduino board.

1.2 Characteristics of Storage Devices

The energy density of a storage device is the amount of energy that a storage device can store within specific weight of the storage device. Higher the energy density lesser the size of the storage device required, so lesser space is required and cost also decreases. The life of a storage device is defined as the number of maximum charge and discharge cycle a storage device can undergo without losing its energy storage capacity [5]. Generally, it is considered to be the number of cycles a storage device undergoes before it degrades to 80% of its initial capacity. The energy efficiency of a storage device is considered as the proportion of the energy extracted from the storage device to the energy received at the input [1]. The remaining energy lost in the form of heat due to the internal resistance losses. The internal resistance of the storage device is due to the internal chemical properties of the storage device this varies for each type of device [3].

2 System Design and Modelling

The systems taken for this case is a standalone PV power system connected with a constant load. The PV array considered here is 800 W maximum output power panel. The DC bus voltage is 60 V. It consists of 2 PV panels in series and 2 in parallel. The HES is associated to the system at DC terminal of main panel output. A load considered adapts 48 V batteries getting charged as similar to an electric vehicle being charged. The simplified illustration of the projected system is exposed in Fig. 1.

2.1 *Fly Back Converter Design*

The bidirectional fly-back converter is made known in Fig. 2. The parallel capacitors C_1, C_2 are filter capacitors and the resistors Rs and capacitors Cs at both sides form the snubber circuit. The snubber circuit is for protection of MOSFETs from over potential voltage and current caused by leakage inductance during switching. Therefore the equation for calculating inductance and voltage are given below.

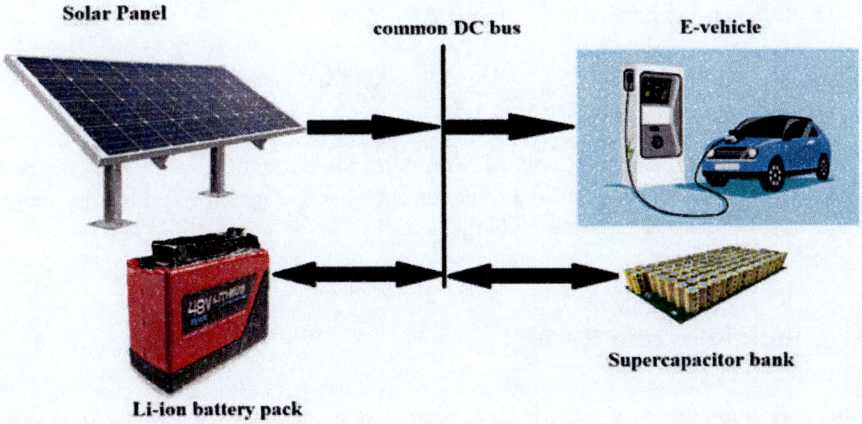

Fig. 1 System considered for study

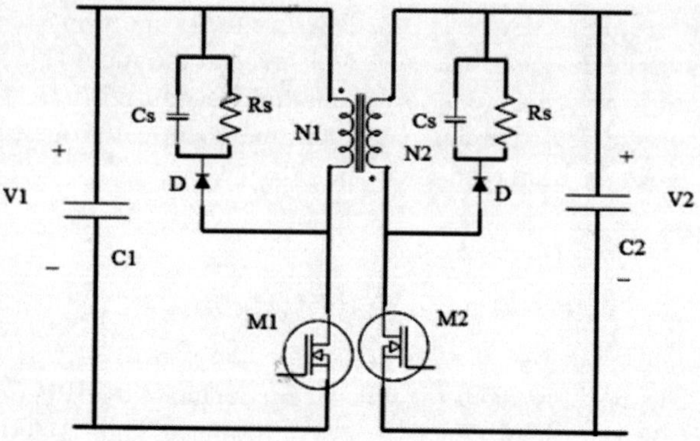

Fig. 2 Fly-back converter under bi-directional mode

$$V2 = \left(\frac{Ton}{Toff}\right)\left(\frac{N2}{N1}\right)V1 \tag{1}$$

$$V1 = \left(\frac{Ton}{Toff}\right)\left(\frac{N1}{N2}\right)V2 \tag{2}$$

$$L1 = \frac{D_{max} \times V1^2_{max}}{2 \times P \times F} \tag{3}$$

$$L2 = \frac{D_{max} \times V2^2_{max}}{2 \times P \times F} \tag{4}$$

where V1, V2 are primary and secondary side voltages, D_{max} is the maximum allowable duty cycle, F is frequency of modulated signal based on pulses, P defines the power transfer between primary and secondary side and Ton and T_{off} are on and off time period respectively.

The Lithium-ion battery is also modeled as follows.

$$f_2(i_t, i*, i) = Ea - K\frac{Q_b}{Q_b + i_t}i * -K\frac{Q_b}{Q_b - i_t}i_t + Aexp(-Bi_t) \tag{5}$$

E_a represents the constant voltage. K denotes the constant based on polarization levels. i* and i are correspondingly the currents due to capacitor and battery under frequency. Q_b denotes maximum charge.

3 Simulations and Results

The corresponding form for the described system is designed using MATLAB/SIMULINK software. For solar panel connection, already available block is used with our desired characteristic. The load is connected through buck converter model and the two bidirectional fly back converter for two power storage devices. These are coupled on common DC bus to load converter and associated solar panel output. The battery and super capacitor model are available in SIMULINK and the properties be changed to our system description.

3.1 Fly Back Converter

The waveforms in the Figs. 3 and 4 show the simulation result of fly back converter. The winding's output waveforms show the voltage is boosted from 12 to 60 V. The converter is controlled using a closed loop PI controller to maintain a constant voltage at output. The Fig. 3 shows the waveforms of the fly back converter.

The equivalent model for the described components is intended using MATLAB software. The load is connected through buck converter model and the two bidirectional fly back converter for two power storage devices are allied to the common DC bus. Here the super capacitor participates more for power share. Because its initial condition is considered to be almost full and also only lesser amount is discharged and charged, also in battery, the depth of discharge is lesser, so the degradation of battery capacity is completely reduced.

The Fig. 4 shows the power flow graph that viewing solar panel output power, load power, battery power and super capacitor power for dynamic load for constant irradiance 200 W/m^2. As it can be seen from results, most power is delivered from

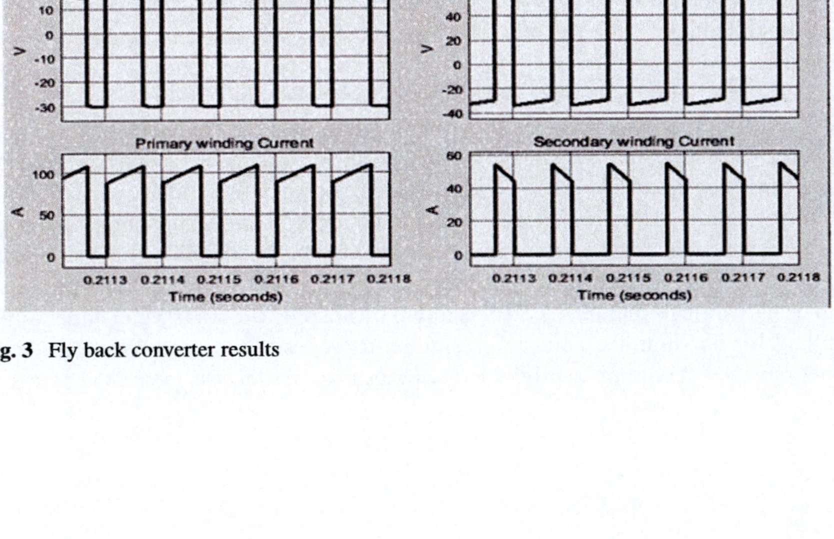

Fig. 3 Fly back converter results

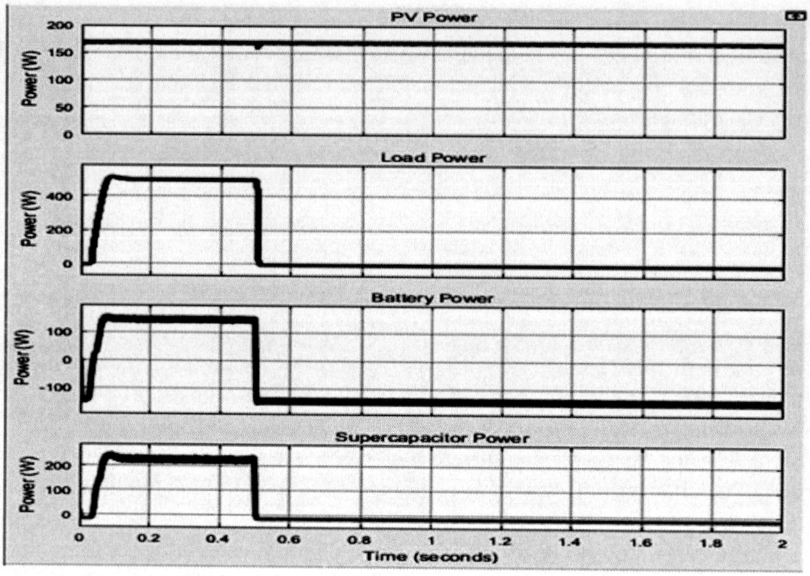

Fig. 4 Power flow graph with irradiance 200 W/m^2 and dynamic load

super capacitor, so the energy losses occur due to internal resistance of the storage device will be lesser. Hence the efficiency of the storage system will be higher.

4 Hardware Implementation

The controller is implemented using Arduino UNO board. The voltage of battery and super capacitor are measured using inbuilt analog to digital converters in Arduino UNO and the necessary PWM signal needed for controlling the bidirectional fly back converter is generated using PWM generator in Arduino UNO. The converter will control the voltage at the DC bus. The Fig. 5 reveals the complete hardware implementation of proposed scheme. The configuration as follows: Battery bi-directional fly back converter turn ratio N1:N2 = 10:20, Super capacitor bi-directional fly back converter turn ratio N3:N4 = 5:20, Switching frequency of fly back converter, f = 32 kHz, Nominal battery voltage = 48 V, Capacity of battery = 200 Ah, Maximum super capacitor voltage = 48 V, Capacity of super capacitor = 500 F.

Identification: (1)-solar panel (2)-Lithium battery (3)-Super capacitor (4)-battery bidirectional fly back converter (5)-super capacitor bidirectional fly back converter (6)-Arduino UNO (7)-LCD (8)-Load fan.

The hardware implemented HES is tested by keeping the solar panel under direct sunlight for maximum irradiance and hiding it for less irradiance. When the solar panel receives maximum sunlight and without load, the excess generated power is

Fig. 5 Hardware implementation

directed towards both the storage devices. For a initial condition of super capacitor at 50% of charge and battery is at approximately 50% of charge, the duty cycle to super capacitor with fly back converter is higher than battery with fly back converter. Once the super capacitor voltage started reaching toward maximum value, the power starts to flow towards battery.

For load connected model, solar panel is hidden from sunrays; the duty cycle of super capacitor fly back converter is higher than battery. After a period of time, the energy stored in capacitor decreases and also voltage across terminal decreases, so the duty cycle to super capacitor fly back converter decreases and battery fly back converter increases. The output voltages are given in the Fig. 6 and 7 which corresponds to super capacitor and battery.

The super capacitor voltage and battery voltage are continuously measured using Arduino UNO and the converter is operated only if these voltages were in the operating region otherwise the device is cutout of the system by making duty cycle to the converter as zero.

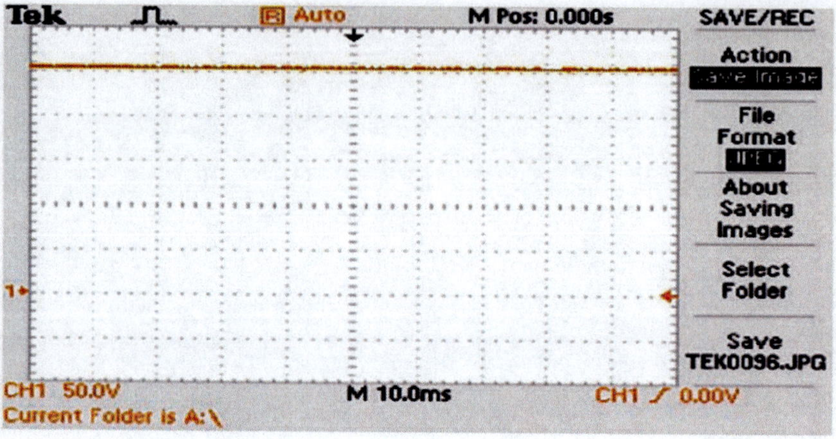

Fig. 6 Hardware result for output voltage of battery

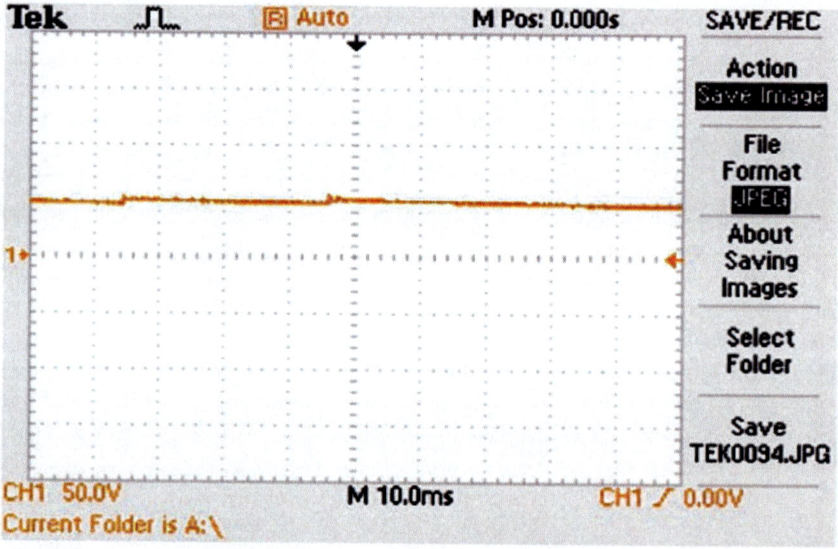

Fig. 7 Hardware result for output voltage of super capacitor

5 Conclusion

The proposed system improves the efficiency of total system of storage during charging period and even discharging. The prototype system with solar panel and a DC load is implemented using fly back converter and Arduino UNO for close loop control. This hybrid system reduces the voltage and current stress on battery and during short duration of load demand. This also lowers number of charge and discharge cycles for the system of storage hence extending its existence time by slowing degradation of capacity of battery.

References

1. Bahloul M, Khadem Sk (2018) Impact of power sharing method on battery life extension in hess for grid ancillary services. IEEE Trans Energy Convers:1–11
2. Xu D, Cen H (2020) A hybrid energy storage strategy based fuzzy control to suppress power fluctuation of grid-connected photovoltaic power system. In: Asia energy and electrical engineering symposium, pp 1–6
3. Joshi KD, Chandrakar V (2017) Power oscillation damping using ultra capcitor and voltage source based FACTS controllers. Electric. Instrum Commun Eng: 1–6 (2017)
4. Shengzhe Z, Kai W, Wen X (2017) Fuzzy logic-based control strategy for a battery/supercapacitor hybrid energy storage system in electric vehicles. Chin Autom Congress (CAC): 1–4
5. Li , Tseng, K (2015) Energy efficiency of lithium-ion battery used as energy storage devices in micro-grid: 005235–005240, 1–6. https://doi.org/10.1109/IECON.2015.7392923
6. Sakthisudhursun B, Muralidharan S, Arockiaraj S (2020) Implementation of nearest level control modulation technique for multilevel inverter using Arduino. Int J Adv Sci Technol 29(7):1096–1102
7. Arockiaraj S, Manikandan BV (2021) Novel transient stability analysis of grid connected hybrid wind/PV system using UPFC. Int J Electric Eng Educ: 1–19. https://doi.org/10.1177/00207209211003268
8. Weimin Z, Weibing Y, Sun K, Li C, Jiang W, Xia Y (2018) An multipurpose optimization equivalent filter in hybrid energy storage systems for output fluctuation suppression of photovoltaic generation systems. In: IEEE 23rd international conference on digital signal processing (DSP), Shanghai, China, pp 1–5
9. Lei L, Shengtie W, Guizhen T (2016) Grid power quality improvement with statcom/hess for wind turbine with squirrel-cage induction generator. In: IEEE 11th conference on industrial electronics and applications (ICIEA), pp 1–6
10. Arockiaraj S, Manikandan BV, Sakthisudhursun B (2018) Intensive analysis of sub synchronous resonance in a DFIG based wind energy conversion system (WECS) connected with smart grid. In: Communications in Computer and Information Science, vol 844, no 19, pp 242–253

Efficient Net V2 Algorithm-Based NSFW Content Detection

Aditya Saxena, Akshat Ajit, Chayanika Arora, and Gaurav Raj

Abstract Youngsters today are highly influenced by the explicit content available online on websites, apps, and various social media platforms. Explicit content such as NSFW (Not safe for work) data, nude images, or images showing private parts and vulgarity are a threat to society and can have a negative impact. Children have free access to easily accessible unfiltered online content, which increases the possibility of viewing such content. Such exposure and online shared material that contains unfiltered content are not appropriate for the community, especially children. In today's world, technology is much more advanced and there is a possibility of children becoming victims of NSFW content because they do not know the difference between good and bad content. The authors of this paper aim to review and analyze the methodologies used for protecting the young generation and people of all age groups from pornographic & NSFW content. This paper also aims to implement an image and content-based approach using research review for filtering pornographically related data (images & text). This research paper aims to provide filtered and clean content for the safety of children and the community in general. In the model analysed by the authors, the algorithms Efficient Net V2L, Efficient NET V2M, and CNN algorithms have been used to detect NSFW content as well as objectionable and adult content that is dangerous for both children and teenagers. These algorithms are researched and analyzed by the authors and further, the best one is utilized for the classification of nude and non-nude on different epoch values.

Keywords Convolution Neural Network (CNN) · Not Suitable for Work (NSFW) · Efficient NetV2L · Efficient NetV2M · Batch Size

A. Saxena · A. Ajit · C. Arora (✉) · G. Raj
Sharda University, Greater Noida, Uttar Pradesh, India
e-mail: arorachayanika1@gmail.com

A. Saxena
e-mail: adi12saxena@gmail.com

N. Hasteer et al. (eds.), *Decision Intelligence Solutions*, Lecture Notes in Electrical Engineering 1080,
https://doi.org/10.1007/978-981-99-5994-5_31

1 Introduction

The young generation is easily influenced and affected by the inappropriate content available on the internet today. Controlling, blocking, and analyzing this sensitive content such as pornographic images, and texts are necessary [2, 3] and can be implemented using ML and Dl models. The dataset used in this paper consists of 2174 images. These images dataset is further divided into training and testing datasets.

The training dataset consists of 1325 images, whereas the validation dataset has 250 images, and the testing dataset has 128 images. The authors use CNN and SVM algorithms for image and text identification of explicit content and detection with better performance. This paper majorly studies and deals with algorithms to create a suitable working environment for children and individuals [5]. Not suitable for work content detection and analysis would also be discussed in depth by the authors of this paper. Table 1 represents the findings and uses cases of the research work implemented.

Table 1 Findings from previous literature

Algorithm	Findings	Use Cases	Accuracy
CNN	CNN, also known as convnet, is a type of artificial neural network architecture that is primarily used for deep learning algorithms. It is a subset of machine learning that is particularly useful for image classification, object detection, and other applications in the field of deep learning	Computer vision Image recognition and classification Learned the object's attribute in consecutive flow Feature extraction	85%–97% [2, 6]
Efficient Net V2L	Belongs to the Efficient Network algorithms family A scaling and faster training algorithm Works on the neural networks in-depth Enhancement of CNN algorithm itself Uses mobilenetv2 CNN architecture Smaller,,faster and better at working in layers	Useful in scaling, increasing efficiency, and accuracy	85%–91% [1, 3]
Efficient Net V2M	Scaling and faster training algorithm Works on the neural networks in-depth Scales the dimensions. Gives better results and accuracy Forms a baseline network	Used for a faster training process, speeding up parameters efficiency and accuracy Useful in image classification, identification enhancement	93–96% [1, 3]

2 Literature Review

Kanwal Yousaf and team conducted research to develop an EfficientNet-BiLSTM that achieved a remarkable accuracy rate of 95.66%. This model outperformed the attention mechanism-based EfficientNet-BiLSTM framework, which had an accuracy rate of 95.30%. However, adding too many layers of BiLSTM can increase network complexity and lead to slower training. It was determined that training the model with a single fully connected layer using a pre-trained CNN architecture was insufficient [1].

Mahmoud Mohammed Taha et al. discussed the concept of filtering animated cartoon characters implemented to check the working algorithm in their paper. Evaluations like Convolutional Neural Network (CNN), Long Short Term Memory (LSTM), Global Fit Ellipse (GFE), and the Local Fit Ellipse (LFE) were discussed. The authors cover up the limitations in the already existing models and study them briefly and improve the accuracy of the system [2].

Nishtha Ahuja et al. in their research study identified nude images and studied various methods and algorithms already in use including the Explicit Content Detection System. A Machine Learning Solution for Detecting Explicit Images and Automatic Detection of images containing Nudity and identify Not Suitable for Work (NSFW) images. In the future, the authors need to avoid the situation and plan to scrape various porn websites and identify the various patterns and confirm the results [3].

In their research, André Tabone and his team developed a model capable of locating and labeling sexual organs in images to detect pornography, and extended the model to perform image classification by providing 19 semantically meaningful descriptors of the content. Additionally, the model was applied to the domain of Child Sexual Abuse (CSA). The team faced the challenge of reusing information with different heads to perform multiple tasks in a single model and increasing the overall system efficiency by using a strong image feature extraction base [4].

Huynh et all researched about an organization skilled in detecting pornographic videos with high speediness and correctness. Mainly 3 primary components are used to create with end-to-end encryption: extraction of mainframes from a video, selecting the best Tensorflow object detection API model to locate and classify objects in videos, with the ultimate goal of classifying the entire video. While the model was able to achieve success with smaller datasets, it has the potential to be extended to categorize other types of videos, such as those related to politics, sports, and violence [5].

Gajula and Rasoul Banaeeyan developed a model using a supervised learning-based Support Vector Machine (SVM) algorithm to determine whether an image is safe or unsafe. The model blurs or colors the exposed skin area completely. The proposed automated censorship system can be implemented on a low-cost embedded system. The LRF-ELM model, which consists of 30 feature maps and 12 layers, achieved an average accuracy of 82.9%. When tested on a newly-collected large-scale dataset, the proposed method achieved an accuracy of approximately 91%, highlighting its effectiveness [6].

Chen and his team developed a method to extract textual and visual features from HTML source code to detect explicit data on websites. The method uses spatial bag-of-visual words to learn visual features. The team also compared and contrasted various methods used for detecting explicit data and found that URL-based approaches are not as effective as they used to be. Text-based detection is vulnerable to the "curse of dimensionality." [7].

Westlake et latest conducted research using a custom-built web-crawler designed to extract freely available exploitation websites. They used a hybrid model that distinguished networks based on image hash values and keywords. However, further investigation is needed before drawing a firm conclusion on the effectiveness of their methodology in translating to peer-2-peer networks and Usenet groups [8].

3 Methodology

3.1 Research Gap Identification

The pornographic sensitive content available online can have hazardous effects on individuals, especially the younger generation of today. This not only distracts their mental peace and well-being but also troubles their parents [9]. The authors of this paper aim to review and analyze the methodologies used for protecting the young generation and individuals from pornographic & NSFW content.

Developing an approach using Research Review on NSFW for content-based filtering of pornographically related data (images & text) is the main goal. Implementation and analysis of the CNN algorithm and Efficient Nets algorithms (V2L and V2M) is done by the authors of this research paper for better performance and accuracy. The Batch size is taken as 32 for attaining better accuracy and results. Table 2 represents the various training accuracy, testing accuracy, average precision, average loss, and average F1 score of the CNN algorithm, Efficient Net V2L, and Efficient Net V2M.

Table 2 Comparative analysis of different algorithms with batch size taken as 16

Algorithm Name	Training Accuracy	Testing Accuracy	Average Precision	Average Loss	Average Recall	Average F1 Score
CNN	69.51	9609	82.8	48.735	82.2	52.99
Efficient Net V2L	97.74	96.88	98.31	6.735	98.31	81.725
Efficient Net V2M	92.91	98.44	95.675	16.835	95.675	83.375

3.2 Objectives of the Research

The paper aims at resolving the following objectives by the means of various algorithms. The authors of this paper analyze and work on the Efficient Net V2 family and CNN. The authors of this paper aim to review and analyze the methodologies used for protecting the young generation and people of all age groups from pornographic & NSFW content. This paper also aims to implement an image and content-based approach using Research Review on NSFW for content-based filtering of pornographically related data (images & text). Implementation and Analysis of methodology (CNN) implementation to protect websites and to develop an environment safe for all.

The CNN Algorithm utilizes the neural layers and functioning inherited from DL algorithms. CNN is useful for image classification, object detection, and interpreting data. CNN helps in the working of Efficient Net V2 Algorithms and other classification and detection algorithms. The primary benefit of utilizing CNN is that it can detect significant features without any human supervision, as evidenced by its ability to learn the distinctive characteristics of dogs and cats from images on its own [1].

Additionally, the EfficientNet V algorithm family is utilized for creating compact models that can be trained quickly. This returns a Keras image classification model and optionally loads pre-trained imageNet weights. This Efficient Net family has main 2 classifications: Efficient Net v2M and Efficient Net v2L. Both have minor differences but majorly are the same and belong to this family. It is basically a type of convolutional neural network. EfficientNet is an approach to developing models that involves a combination of training-aware neural architecture search and scaling, which jointly optimizes training speed. The approach focuses on engineering and scale and has been demonstrated to produce excellent results with reasonable parameters [4]. The key benefit of EfficientNet is that it enables achieving top results by carefully designing the architecture. The approach is illustrated by a graph that plots ImageNet accuracy against model parameters (Fig. 1).

4 Implementation and Analysis

The dataset comprises 2174 Images in total, with 1325 images in training, 250 in validation, and 128 in testing phases respectively. The three algorithmic models utilized by the authors as researched.

4.1 The Efficient Net V2M Algorithm

The Efficient Net V2M belongs to the Efficient Net algorithm family. It is used on the dataset of images classified into nude and non-nude categories. Each image is resized

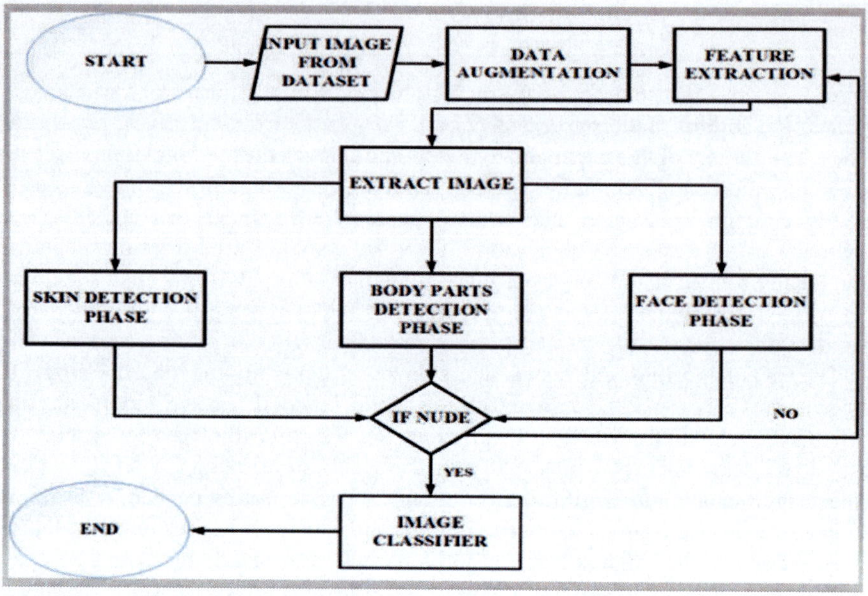

Fig. 1 Flowchart of the CNN implementation done by the authors of this research paper

to 150 × 150 pixels size and batch size values are changed by the authors to enhance the accuracy in each epoch value. This algorithm is an enhancement of CNN which scales and reduces the parameter size. The main advantage and improvement from the previous algorithms are it trains the model faster and provides the best accuracy and performance. MobileNetV2 concept is also used along with compound coefficients. The authors of this paper import the libraries and split the training, testing, and validation dataset. Followed by this the label mapping and plotting of training is implemented. Data augmentation is performed and following this a pretrained Efficient Net V2M model is developed. The activation functions 'relu' and 'softmax' are used in the input and output layers respectively. Adam optimizer is used for optimization purposes. Model metrics is compiled based on the training data, epoch values, and batch size. The last step of implementation is testing the model on testing dataset and prediction model development. The average accuracy is 97.1800% for training and 96.8775% for testing data.

4.2 *The Efficient Net V2L Algorithm*

The Efficient Net V2L algorithm is used on the dataset of images classified into nude and non-nude categories. Each image is resized to 150 × 150 pixels size and batch size values are changed by the authors to enhance the accuracy in each epoch value. This algorithm is an enhancement of CNN which scales and reduces the parameter size. It uses a compound coefficient for scalarity and better performance and accuracy. It works faster than the CNN network, and the MobileNetV2 concept is also used. The authors of this paper import the libraries and split the training, testing, and validation dataset. Followed by this the label mapping and plotting of training is implemented. Data augmentation is performed and following this a pre trained Efficient Net V2L model is developed. The activation functions 'relu' and 'softmax' are used in the input and output layers respectively. Adam optimizer is used for optimization purposes. Model metrics are compiled based on the training data, epoch values, and batch size. The last step of implementation is testing the model on the testing dataset and prediction model development. The average accuracy is 98.280% for training and 98.050% for testing data.

4.3 The CNN Algorithm

CNN algorithm is used on the dataset of images classified into nude and non-nude categories. Each image is resized to 150 X 150 pixels size and batch size values are changed by the authors to enhance the accuracy in each epoch value. The CNN algorithm works in deep neural network layers depending on batch size and epoch values. This process is followed by label mapping and plotting sample training data by the authors to analyze the outputs. A simple convolution 2D model is developed using Keras. Convolutional Neural Network has three main layers of operation namely: CNNs are composed of three fundamental layers: the convolutional operation layer, the fully connected layer, and the pooling layer [10]. The pooling layer performs one of two operations: Average pooling, which calculates the average of each kernel, or MAX pooling, which computes the maximum value of the kernel. In this case, we use the MAX pooling layer. Additionally, the activation functions 'relu' and 'softmax' are employed in the input and output layers, respectively.CNN algorithm gives various accuracy values at different batch sizes and epoch values. The average accuracy is 76.961% for training data and 92.772% for testing.

5 Result and Discussion

The accuracy of the model used for NSFW image detection and censorship is approximate 97.1800% for training and 96.8775% for testing in the Efficient Net V2M Algorithm, 97.8725% for training and 97.8975% for testing Efficient Net V2L, and for 75.1627% for training and 91.88% for testing CNN Algorithm. This has been achieved by using the features in the dataset. The overall dataset comprises 2174 images in total and the dataset has been preprocessed and augmented. The dataset was divided into 1325 images constituting training data, 250 images for validation data, and 128 images constituting the testing data. Efficient Net V2M is a scalable and faster training algorithm working on deep neural networks but at a faster pace. Efficient Net V2M is a better version of CNN. It changes the dimensions and gives better results and accuracy thanks to this feature. This algorithm forms the underlying network and scales each neural layer. Efficient Net V2M is used for a faster training process, acceleration of efficiency, and accuracy of parameters. It is useful in image classification, identification, and enhancement. Giving an accuracy of 97.1800% for the training dataset and 96.8775% for the testing dataset (Tables 3, 4, 5).

Efficient Net V2L belongs to the Efficient Network family of algorithms. It is a scalable and faster training algorithm working on deep neural networks but at a faster pace. Efficient Net V2L is an improvement on the CNN algorithm itself. It uses the

Table 3 Comparative analysis of the results for the Efficient Net V2M Algorithm

Epoch Values	Training Accuracy	Testing Accuracy	Average Loss	Average Precision	Average Recall	Average F1 Score
5,30	93.213	96.09	7.595	97.23	97.23	85.82
10,30	97.000	96.88	11.095	95.025	95.025	84.08
15,30	98.720	97.660	10.041	98.580	98.580	74.605
20,30	100.000	96.880	7.065	98.44	98.44	74.605
Total Average	97.180	96.889	8.692	97.318	97.318	84.096

Table 4 Comparative analysis of the results for the Efficient Net V2L Algorithm

Epoch Values	Training Accuracy	Testing Accuracy	Average Loss	Average Precision	Average Recall	Average F1 Score
5,30	95.020	97.660	5.675	98.715	98.715	50.227
10,30	98.420	98.440	5.732	98.440	98.440	92.710
15,30	99.770	97.440	5.675	98.818	98.818	52.356
20,30	99.920	98.050	4.515	99.180	99.180	87.260
Total Average	98.280	98.050	5.398	98.786	99.786	70.638

Table 5 Comparative analysis of the results for the CNN Algorithm

Epoch Values	Training Accuracy	Testing Accuracy	Average Loss	Average Precision	Average Recall	Average F1 Score
5,30	62.260	97.660	5.461	79.960	43.890	43.890
10,30	76.680	93.750	3.199	87.000	87.000	69.150
15,30	84.750	89.620	3.542	86.905	86.905	72.235
20,30	84.150	90.620	3.362	87.385	87.385	73.260
Total Average	76.961	92.772	3.891	86.062	86.062	64.633

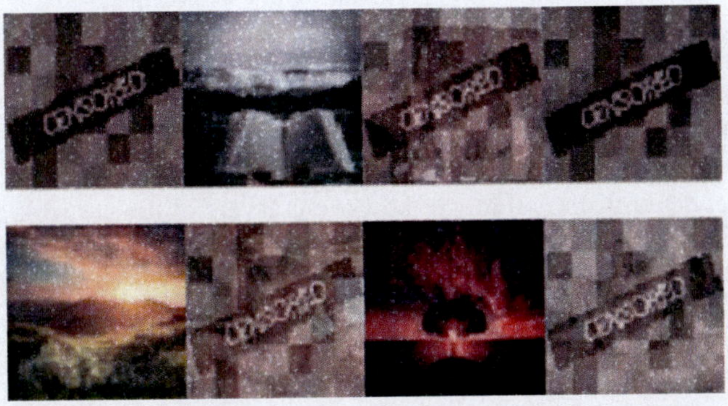

Fig. 2 The output image of Efficient Net V2M Algorithm with a batch size of 32 and epoch values of 5 and 30

MobileNetV2 CNN architecture. It's smaller, faster, and works better in layers. These are useful in scaling up, and increasing efficiency and accuracy. Giving an accuracy of 98.280% for the training dataset and 98.050% for the testing dataset. A CNN algorithm is a subset of ML and DL where various deep neural networks are utilized. It is majorly useful in object detection, facial recognition, image classification, and

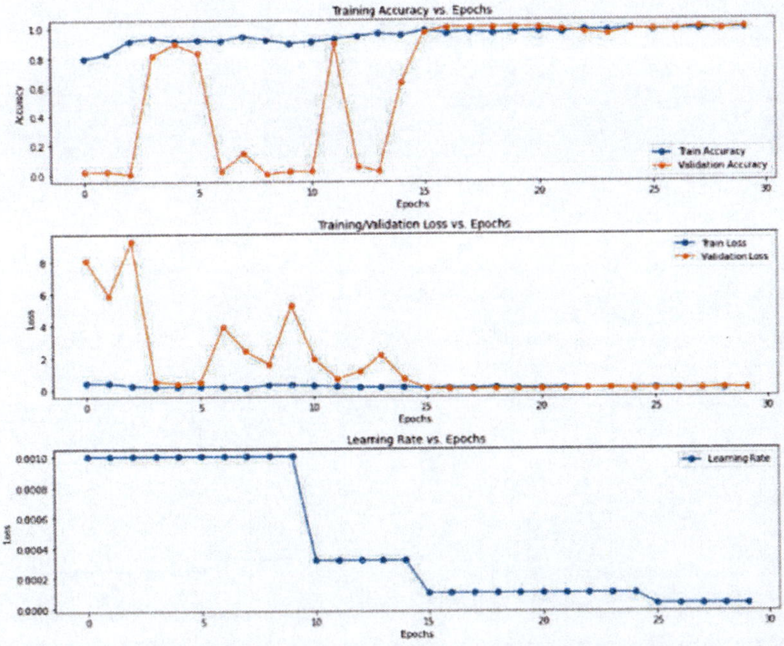

Fig. 3 Graphs plotted between training epochs and accuracy in 2.1, graphs plotted between training loss epochs and learning rate in 2.2 for the Efficient Net V2L Algorithm

real-time analysis of data. The authors of this paper analyze the CNN algorithm and work on the dataset. The accuracy attained is 76.961% for the training dataset and 92.772% for the testing dataset in the work. CNNs are highly valuable for their ability to deliver accurate results, especially when working with extensive datasets. The key advantage of CNNs is their ability to learn object properties in a step-by-step process as the object data traverses through the various layers of the network. The images are classified as sensitive-nude and non-sensitive non-nude images and censored and blurred as shown in Fig. 2.

Figures 2 represent the images of training and testing classified as nude and non-nude for all three algorithms Efficient Net algorithm. Each of the algorithms is used at an epoch value of 5 and 30 respectively. A batch size of 32 is chosen as constant for accurate and better results accuracy and performance (Fig. 3, 4 and 5).

As the authors analyze through the graphs and dataset of training and testing, Efficient Net V2L Algorithm gives an accuracy of 98.280% and 98.050% for training and testing respectively. This algorithm gives the best accuracy and also attains the best results for other parameters. Followed by this, Efficient Net V2M gives 97.180% and 96.889% for training and testing respectively. CNN gives the least accuracy

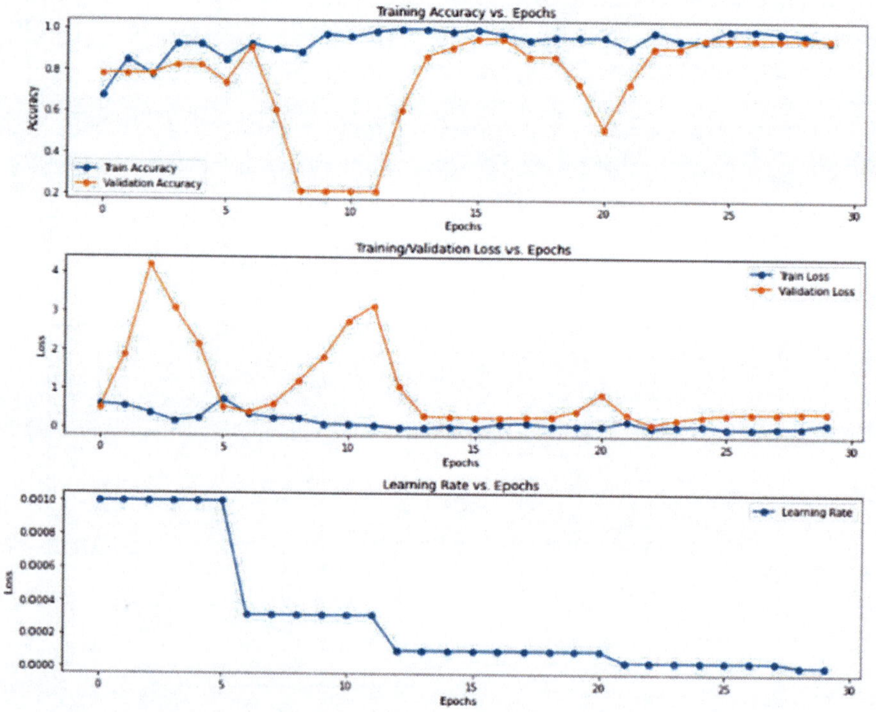

Fig. 4 The training vs epochs and Accuracy, Training loss vs epochs and Learning rate vs epoch's graphical analysis of Efficient NetV2M Algorithm

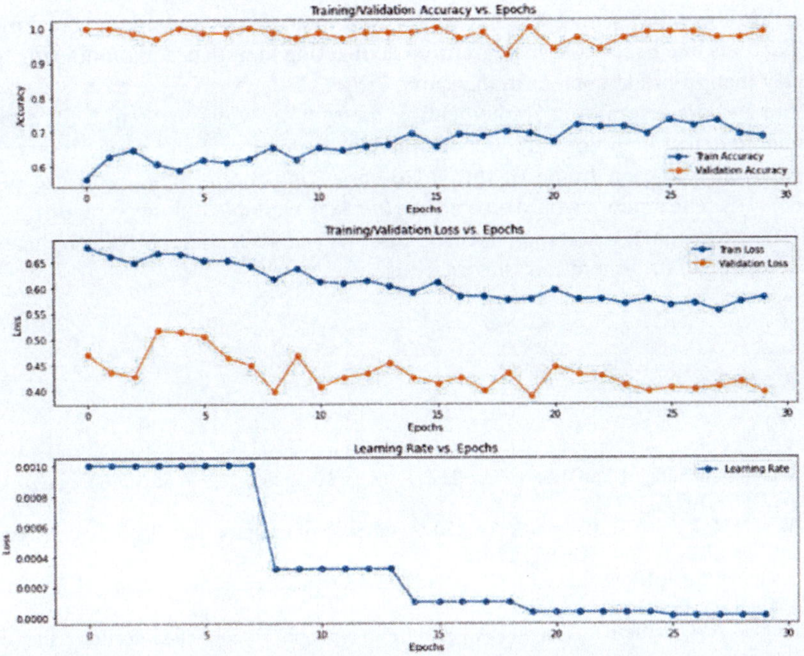

Fig. 5 The training vs epochs and Accuracy, Training loss vs epochs, and Learning rate vs epochs graphical analysis of CNN Algorithm

values of 76.961% and 92.772%. Thus the authors utilized the Efficient Net V2L Algorithm and analyzed NSFW content.

6 Conclusion and Future Scope

The authors of this paper implemented Efficient Net V2M, Efficient Net V2L, and CNN algorithms. For the algorithm Efficient Net V2M, the average training accuracy attained is 97.1800%, and that for testing is 96.8775%. Similarly, for the algorithm Efficient Net V2L, the average training accuracy attained is 98.280%, and that for testing is 98.050%. The CNN algorithm achieved the least accuracy and performance with a training average accuracy of 76.961% and a testing accuracy of 92.772%.

The current research analysis focuses only on images and textual data, a limitation the authors will try to overcome in the near future. Another limitation is that the dataset used classifies the image-nude and not specific parts displaying explicit content.

In the near future, the authors of this research paper would like to enhance the system's accuracy. The authors would be trying to implement this system in schools, enhancing it further in universities and industry traveling buses and vehicles. Along with that, the authors suggest carrying forward their research work on social media applications like Facebook, Instagram, etc. collecting their data and monitoring them for any inappropriate content in the same.

The authors also aim at implementing the same in video detection for explicit content to detect inappropriate video content along with text and image filters. This would work for each frame of the video enhancing the accuracy and use of the model [10]. The authors would also utilize the data available online on platforms like YouTube, Snapchat, Instagram, Twitter, Google, and such social media applications for detection, data generation, and analysis.

References

1. Yousaf K, Nawaz T (2022) A deep learning-based approach for inappropriate content detection and classification of YouTube videos. IEEE Access 10:16283–16298. https://doi.org/10.1109/ACCESS.2022.3147519
2. Taha MM, Zaky AB, Alsammak AW (2022) Filtering of inappropriate video content a survey. Int J Eng Res Technol (IJERT) 11(02)
3. Ahuja N, Shah RR, Shukla J (2021) Nudity detection indraprastha institute of information technology, New Delhi
4. Tabone A, Camilleri K, Bonnici A (2021) Pornographic content classification using deep-learning
5. Huynh V-N, Nguyen HH (2021) Fast pornographic video detection using deep learning, 1–6. https://doi.org/10.1109/RIVF51545.2021.9642154
6. Gajula G, Hundiwale A, Mujumdar S, Saritha LR (2020) A machine learning based adult content detection using support vector machine, 181–185. https://doi.org/10.23919/INDIACom49435.2020.9083700
7. Chen Y, Zheng R, Zhou A, Liao S, Liu L (2020) Automatic detection of pornographic and gambling websites based on visual and textual content using a decision mechanism. Sensors 20:3989. https://doi.org/10.3390/s20143989
8. Westlake B, Bouchard M, Frank R (2012) Comparing methods for detecting child exploitation content online. In: Proceedings - 2012 European intelligence and security informatics conference, EISIC 2012. https://doi.org/10.1109/EISIC.2012.25
9. Karamizadeh S, Chaeikar S, Jolfaei A (2022) Adult content image recognition by Boltzmann machine limited and deep learning. Evol Intel. https://doi.org/10.1007/s12065-022-00729-8
10. Nian F, Li T, Wang Y, Xu Ml, Wu J (2016) Pornographic image detection utilizing deep convolutional neural networks. Neurocomputing 210:283–293. https://doi.org/10.1016/j.neucom.2015.09.135
11. Nguyen QH, Tran HL, Nguyen TT, Phan DD, Vu DL (2020) Multi-level detector for pornographic content using CNN models. In: 2020 RIVF international conference on computing and communication technologies (RIVF). IEEE, pp 1–5
12. Cifuentes J, Orozco ALS, Villalba LJG (2021) A survey of artificial intelligence strategies for automatic detection of sexually explicit videos. Multimed Tools Appl, 1–18
13. Tran HL, Nguyen QH, Phan DD, Nguyen TT, Vu DL et al (2020) Additional learning on object detection: a novel approach in pornography classification. In: International conference on future data and security engineering. Springer, pp 311–324
14. Wehrmann J, Simões GS, Barros RC, Cavalcante VF (2018) Adult content detection in videos with convolutional and recurrent neural networks. Neurocomputing 272
15. Shamila Ebenezer A, Deepa Kanmani S, Sivakumar M, Jeba Priya S (2022) Effect of image transformation on EfficientNet model for COVID-19 CT image classification. Mater Today

Proc 51:2512–2519. https://doi.org/10.1016/j.matpr.2021.12.121. Epub 2021 Dec 13. PMID: 34926175; PMCID: PMC8666302

16. Lin X, Qin F, Peng Y, Shao Y (2021) Fine-grained pornographic image recognition with multiple feature fusion transfer learning. Int J Mach Learn Cybern. 12. https://doi.org/10.1007/s13042-020-01157-9
17. Aruleba I, Viriri S (2021) Deep learning for age estimation using EfficientNet. https://doi.org/10.1007/978-3-030-85030-2_34
18. Yin H, Yang C, Lu J (2022) Research on remote sensing image classification algorithm based on EfficientNet. In: 2022 7th international conference on intelligent computing and signal processing (ICSP), pp 1757–1761. https://doi.org/10.1109/ICSP54964.2022.9778437
19. Touvron H et al (2022) ResMLP: feedforward networks for image classification with data-efficient training. IEEE Trans Pattern Anal Mach Intell. https://doi.org/10.1109/TPAMI.2022.3206148
20. Chubachi K (2020) An ensemble model using CNNs on different domains for ALASKA2 image steganalysis. In: 2020 IEEE international workshop on information forensics and security (WIFS), pp 1–6. https://doi.org/10.1109/WIFS49906.2020.9360892

A Survey of Green Smart City Network Infrastructure

Shraddha Gupta and Ugrasen Suman

Abstract Smart cities are offering world-class services with significant use of resources. Thus, we need a huge power supply. Green smart city is minimizing energy utilization and maximizes eco-friendly techniques. Therefore, we need green energy and limit power wastage. Recent studies recommended use of sensors, machine learning (ML), internet of things (IoT), and cloud computing for optimizing power utilization. In this paper, we provide a review of green smart city and key areas of services explained. Additionally, recent development and enhancement at the different levels of smart cities to save energy are explained. The finding shows an improvement in conventional services and the use of ML in smart city applications that it can help in reducing the power requirements. Finally, the key objectives are proposed for future studies.

Keywords Smart city · Network · Proactive scheduling · Machine learning · Green infrastructure

1 Introduction

A smart city is a modern urban area that uses different methods, and sensors to collect specific data. The data is collected from citizens, devices, buildings, and resources. The collected data will be used to manage resources and services for citizen welfare. The aim is to monitor and manage different city activities such as, traffic systems, power plants, water supply, crime, hospitals, and others that becomes feasible by integration of information and communication technology (ICT) [1]. ICT is an umbrella which includes all communication devices like radio, television, cell phones, computer and network, satellite and more. Additionally, includes various

S. Gupta (✉) · U. Suman
School of Computer Science and IT, Devi Ahilya University Indore, Indore, India
e-mail: shraddhagupta22@yahoo.com

U. Suman
e-mail: ugrasen123@yahoo.com

N. Hasteer et al. (eds.), *Decision Intelligence Solutions*, Lecture Notes in Electrical Engineering 1080,
https://doi.org/10.1007/978-981-99-5994-5_32

services such as video conferencing and distance learning. The ICT also optimize smart city operations and services [2]. The ICT allows city administrators to interact with community and city infrastructure. Additionally, offers techniques to manage the urban flow and real-time responses during different critical issues like disasters, terror, and others [3]. There is a need of significant and continuous energy supply to execute these services. The conventional sources of energy for developing smart cities are also responsible for the emission of CO2 and other toxic gases. Therefore, there is a need to create sustainable smart cities. Sustainable smart cities aim to minimize the consumption of non-renewable energy. The sustainable smart city raises the need for green information technology (IT), low-powered hardware, and software applications. However, recycling will reduce need for new devices and will reduce energy consumption. Additionally, emission of CO_2 can be controlled by decentralized systems [4]. On the other hand, rapid urbanization in the entire world and employment of IT services to achieve comfort, increasing risk of global warming [5]. Thus, there is a need to evolve new techniques and technologies to minimize the environmental losses and offers the balance between society modernization and environment [6]. 5G technology is useful to deliver services efficiently. It provides higher speed, lower latency, and greater capacity. That enables quicker download, lower lag, and an impact on work, and play. The applications of 5G in smart city can be recognized in different areas such as, energy, industrial, health care, media and entertainment, and public transportation [7].

In this paper, we provide two key insights; first, an understanding of the sustainable smart city, its components, and areas of considerations. Second, an overview of different technologies, methods, algorithms, and optimizations contributed recently to achieve green smart city development. Finally, we discuss conclusion of the work done in this paper and provide the future study plan to preserve the energy of smart city infrastructure network.

2 Sustainable Smart City

The sustainable and green smart city infrastructure has multiple components, which are aimed at better utilization of available natural and artificial resources. Additionally, helps in reducing resource wastage. Figure 1 demonstrates the different components of green sustainable smart city and is described as follows [33]:

1. *Sustainable energy:* The conventional source of energy has a significant impact on the environment. Therefore, there is a need to adapt and develop new techniques for the generation of green energy.
2. *Waste management:* Sustainability requires life-saving resources in waste management and recycling. Proper waste management helps us to generate power, recycle goods, and create new products that are low-cost and have low carbon footprints.

Fig. 1 Sustainable smart city infrastructure [33]

3. *Sustainable transportation:* Sustainable transportation empowers infrastructure and the development of eco-friendly transportation systems that emit less carbon and improve fuel efficiency. In addition, finding an alternate method of transportation and promoting e-vehicles.
4. *Green building:* One of the innovative concepts is to generate energy through solar power. It also reduces energy consumption with low-power and smart devices.
5. *Natural resource management:* The optimal use and preservation of natural resources are one of the keys to green smart cities.
6. *E-government:* E-government monitors the activity of the city for different management perspectives.
7. *Smart health:* The primary goal of the smart city is the development of smart healthcare systems and remote monitoring of patients to support and save lives. This includes the development of smart health infrastructure.
8. *Water management:* Water is one of the essential resources for human sustainability, but due to changing environmental conditions, we start facing water issues. The reduction and recycling of water is an essential objective of a green smart city.
9. *Social responsibility:* A sustainable and green smart city requires the attention of the government. It is also a social responsibility to maintain and adapt the policies created for sustainable smart cities.
10. *Smart education:* Reduction in the use of paper, promotion, and adoption of digital devices in education will help to reduce cutting down trees.

The utilization of sustainable sources of energy helps to transform the smart city into green smart city, therefore, discovery and development of sustainable sources of energy is the main concern [9]. The existing techniques for energy-saving include traditional methods and utilize machine learning techniques. Among both them, ML- based techniques become much more popular due to the ability to deal with a large volume of data and proactive resource management. The ML-based techniques formulate the issue of resource management as an optimization problem, prediction, classification, and clustering [10]. Recently, several ML techniques are being utilized in applications such as, medical science, agriculture, engineering, business, banking, and many more [11]. These techniques can also be used to analyze and obtained trends in data for improving the Quality of Service (QoS) of smart city applications.

3 Energy Saving Strategies

The aim of green smart cities is to promote and use green sources of energy, reduce energy consumption and prevent energy wastage. Therefore, significant efforts have been made in this direction by different researchers. Some essential contributions are highlighted in following subsections:

3.1 Energy Saving in Smart Cities

Energy consumption can be managed and reduced in smart cities with several techniques [13]. Some of the popularly used techniques are as follows-

- An intelligent lighting system for saving energy.
- Renewable energy is used for electricity.
- Preserving and utilizing energy efficiently.
- Replacement of old lighting with LEDs to reduce energy consumption and cost
- The forecast to provide number of lamps required which also provide the cost-efficiency.
- Understand the demand and supply of energy.

A Hybrid Appliance Load Monitoring System (HALMS) is essential for developing smarter grid and energy infrastructure. There is a need to introduce the smart plug which utilizes IoT for monitoring. The use of smart plugs enhances energy management [14]. On the other hand, a study highlights the smart plug, relevant activities, and limitations. It describes various services, applications, and the role of ICT. They discuss the features of 5G and how 5G could be best for implementation. This lists down a few Smart City use cases [15].

5G offers new ways of business for energy providers and also enhances communication with smart grids. Therefore, a review of the 5G for smart grids has been discussed. They discuss wireless technologies and architectural changes to understand the user- based analysis of 5G from the grid perspective [16]. Similarly, there is a focus on designing an IoT-based energy management system. This technique utilizes deep reinforcement learning. They first discussed IoT-based energy management and then a model based on software. Finally, provide a scheduling technique for energy efficiency [17]. A 5G PPP-compliant software developed [18], which includes:

- Providing scalable, lock-in free, and trusted plug 'n' play technique.
- Inclusion of 5G devices' shows massive Machine Type Communications, critical MTC and Extended Massive Broad Band communications with distributed, secure, scalable and energy-efficient solution.
- Extended Mobile Edge Computing to reduce load, enhance network capacity and delay reduction.

The energy-saving affects the energy efficiency of a 5G base station. The difficulty also found in operational necessities, and has practicality [19]. Similarly, in a study authors implement and test the optimal minimum- energy settable-complexity bandwidth manager (SCBM) [20]. The features are:

- Implementation complexity is acceptable depending on energy consumption and implementation complexity trade-off.
- It reduces the energy requirement of network using a constraint of downtimes and migration times.
- It responds faster for mobility and fading.
- It maintains energy versus delay comparisons.

In a real-world example is China's new 5G infrastructure, which is taking attention. But, deployment of 5G base stations is consuming a large amount of energy which cannot be overlooked. Additionally, the overall power consumption is also increase. Some base stations in idle state cause power waste. Therefore, base station data can be used to train the ML algorithm like LSTM, for future trend prediction. During this prediction if the predicted workload is below than a threshold, it will be closed to reduce power wastage. By implementing the power saving strategy, the energy consumption is reduced by 18.97% [21].

3.2 Machine Learning Based Power Saving

In smart city transformation the "Green AI" is a core contributor. The green AI is focused on preserving energy by using the advance computational techniques. It provides the opportunity for developing the energy efficient solution by utilizing the technology [22]. The aim is two-fold;

- Discussing the limitations of theoretical and practical use of AI.
- Discuss the requirement of AI techniques for implementing the green AI.

In a review, VOS-viewer was used and thirty-nine articles were identified based on Data Mining (DM) and Machine Learning, and subjects are identified where smart cities usage the DM and ML. According to the findings Predictive techniques and the methods based on mobility and environment are the major area of investigation [23]. Further, a method named Home Energy Management as a Service (HEMaaS) was introduced. That is including a neural network-based Q-learning algorithm. The Q- learning is an adaptive learning process which is used for flexible and energy-efficient solution design. This technique provides a rapid and adoptable solution for minimizing the demand and save energy [24].

In the online environment of the smart city, we can transfer optimization problem into task partition and functional task. To reduce the cost of energy and fast execution, a workload allocation technique is used based on multi-action deep reinforcement learning [25]. In similar study, the ML methods can work as an optimization tool for WSN-IoT placement in smart cities. The results show that 61% of methods used supervised learning as compared to 27% reinforcement learning and 12% unsupervised learning [26]. Similarly, to offer and enable better traffic monitoring and real time updates, a less expensive vehicle detecting sensors are employed between the roads. IoT is used to collect traffic data and communicated to the server for analysis [27]. A deep learning model like LSTM can be used to forecast air quality. The results are found promising and can be utilized for different prediction related problems in smart cities [28]. Similarly, a framework is studied that integrates the IoT with ML, which extracts and utilizes the data for different applications [29].

This includes two cases:

- Vehicle's pollution detection and pattern of traffic.
- Demonstrates scalability and capability to employ for large urban regions.

3.3 Case Studies

Andalusian city: The aim is to study and identify the relationship among ML technology, smart cities and employment of ML techniques. The relationship is identified for achieving the sustainability. The sustainability areas and domains are studied, and Sustainable Development Goals (SDGs) are explained. They develop a model cities machine learning sustainability (EARLY) which is used to handle the issues of health and environment. The ML methods, SGD, applicability, and scope are discussed [30].

China: Authors review on ML to predict air pollution based on sensor data [31]. They collected relevant papers and review them, which provides the following highlights:

- In place of utilizing the normal ML need to apply sophisticated and advanced ML models.
- China was mostly used for case study.
- The matter with 2.5 μm was mostly predictable.
- In 41% articles the prediction is done for next day.
- In 66% of the work hourly rate data was used.
- 49% articles are utilizing open dataset.
- For accurate air quality prediction need to consider the factors like spatial properties, weather, and temporal features.

Central London U.K.: An ML-based multi- level automatic fault detection system (MLe- AFD) was introduced which utilizing the remote HVAC fan coil unit (FCU) behavior analysis. The two-stage sequential clustering was used for recognizing the abnormal behavior. The performance is tested using statistical techniques and cross-verified by experts. The experiments were conducted in central London's commercial building, that method can recognize three types of faults remotely and notify the event to staff [32].

4 Observations

In near future, the global warming and energy crisis will become one of the major concerns. There is a need to promote sustainable and eco-friendly life style and to replace conventional and inefficient methodologies of providing the services. Therefore, there is a need to identify most expensive energy consumers in different areas of services in smart cities and optimize them to reduce the energy consumption and energy wastage. In this context, there is also need to explore the possibilities and applicability of ML techniques to analyses, improve and optimize existing infrastructure.

5 Discussions

Green computing in green AI is the main area of study in green smart city infrastructure. The term green indicates the methods and techniques involved to preserve the non-renewable energy sources and create the eco-friendly surroundings. And the next term i.e., computing indicates the use of computing equipment and devices to preserve energy. AI and smart city indicate the domain where we utilize the green labeled techniques. The collected literature can be categorized in three main parts:

1. We have performed the survey, reviews, and tutorials for preserving and optimizing the energy consumption.
2. The employment of ML techniques in serving different real-world applications in smart cities. This section shows how we can make use of ML for improving urban life while, accomplishing a sustainable, energy-efficient, and green lifestyle.
3. The real-world applications where we analyze the data and activities of cities for improving the Quality of service of smart city applications and preserve energy resources.

The studies are also helpful to understand required resources and algorithms. We overserved that the supervised learning algorithms are most frequently used algorithm category for energy saving problems in smart city network. The usage of machine learning techniques in problem solving is presented in Fig. 2. According to the literature, several ML techniques are used in applications of energy conservation.

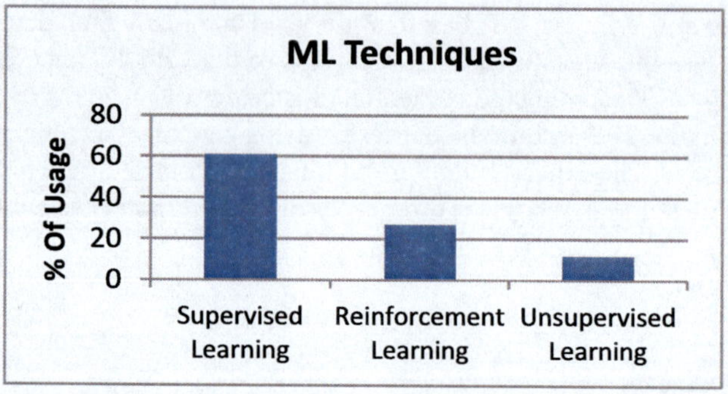

Fig. 2 Types of ML Algorithm

Fig. 3 Percentage of Methods Allows Prediction

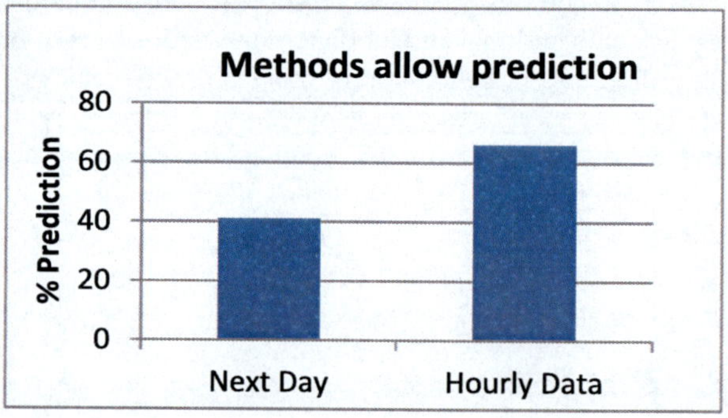

These machine learning approaches performs predictions for the next day data values and hourly data values proliferating in regular interval. The contributions of these prediction-based techniques are demonstrated in Fig. 3.

The current smart city infrastructure is a combination of smart sensing devices and network. A comparative highlight of literature is present in Table 1.

The table consists of sensor and ML technology utilized to serve in smart city. In this context the cellular network will work as backbone of the communication for exchanging the information between city and city administration. There is a need to establish more and more cellular base stations to increase the reach and high-speed communication. These cellular base stations helpful in scaling the network and smart city services to larger area. But the ideal base stations are consuming a constant amount of energy. That indicates the power wastage in smart city network infrastructure. In this context, we utilize the ML techniques for limiting the consumption and wastage of energy. Therefore, for future studies we proposed to utilize ML techniques to understand and optimize the utilization of network devices and resources to minimize the wastage of energy [12]. The following objectives are proposed in future work:

1. Propose proactive energy-saving techniques.

Table 1 Comparing the literature based on employed technology

Ref	Application	Use of ML	Use of sensors
[13]	Intelligent lighting system	Yes	Yes
[14]	smarter grid and energy infrastructure	Yes	Yes
[15]	Smart City use cases	No	Yes
[16]	5G for smart grids	Yes	No
[17]	IoT-based energy management	Yes	Yes
[18]	5G PPP-compliant software	No	No
[19]	energy-saving of a 5G base station	Yes	No
[20]	minimum-energy settable-complexity bandwidth manager	No	No
[21]	5G base stations energy management	Yes	No
[22]	fundamental shortfalls in AI system	Yes	Yes
[23]	where smart cities usage the DM and ML	Yes	Yes
[24]	Home Energy Management as a Service	Yes	Yes
[25]	green workload placement	Yes	Yes
[26]	optimization tool for WSN-IoT nodes in smart cities	Yes	Yes
[27]	traffic monitoring and real time updates	Yes	Yes
[28]	predict future air quality	Yes	Yes
[29]	pollution detection and pattern of traffic	Yes	Yes
[30]	Relationship among ML technology, smart cities and employment of ML techniques	Yes	Yes
[31]	predict air pollution	Yes	Yes
[32]	ML-based multi-level automatic fault detection system	Yes	Yes

2. Design local and central power-saving techniques.
3. Design real-time power optimization and scheduling technique.
4. Forecast the resource schedule in advance.

6 Conclusion and Future Work

Smart cities are offering different levels of services and comfort for their citizens. The city administrators are making plans and developing applications to support the smart city activities and processes. But, to execute these services a significant amount of power is required. Additionally, idle devices and energy wastage in smart city network is a major concern. In this context, we need to develop and design strategies to minimize the power requirements and wastage. In this paper, we are focused on understanding the existing technologies which are transforming smart cities into green smart cities. According to our findings scheduling and optimization techniques are helpful for conserving energy. Additionally, utilization of machine learning for preparing proactive scheduling techniques is also an essential technique that can be utilizing for monitoring the power requirements in different city areas and also helpful to understand the upcoming demand for energy.

References

1. Daniel S, Doran MA (2013) geoSmartCity: geomatics contribution to the smart city. In: The proceedings of the 14th annual international conference on digital government research, pp. 65–71
2. Bawany NZ, Shamsi JA (2015) Smart city architecture: vision and challenges. IJACSA 6(11)
3. Lea R, Blackstock M (2014) Smart cities: an IoT-centric approach. In: Proceedings of the 2014 international workshop on web intelligence and smart sensing, pp 1–2
4. Naim A (2021) New trends in business process management: applications of green information technologies. Br J Environ Stud, 12–23
5. Avasalcai C et al (2021) Resource management for latency-sensitive IoT applications with satisfiability. IEEE Trans Serv Comput
6. Hancke GP et al (2013) The role of advanced sensing in smart cities. Sensors 13(1):393–425
7. Gohar A, Nencioni G (2021) The role of 5G technologies in a smart city: the case for intelligent transportation system. Sustainability 13(9):5188
8. Gielen D et al (2019) The role of renewable energy in the global energy transformation. Energy Strategy Rev 24:38–50
9. Jayachandran M et al (2021) Operational planning steps in smart electric power delivery system. Sci Rep 11(1):1–21
10. Shurdi O et al (2021) 5G energy efficiency overview. Eur Sci J 17(3):315–327
11. Groshev M et al (2021) Towards intelligent cyber-physical systems: digital twin meets artificial intelligence. IEEE Commun Mag 59(8):14–20
12. Patel P et al (2019) A survey on intelligent transportation system using internet of things. In: Emerging research in computing, information, communication and applications, pp 231–240
13. Bachanek KH et al (2021) Intelligent street lighting in a smart city concepts-a direction to energy saving in cities: an overview and case study. Energies 14:3018
14. Suryadevara NK, Biswal GR (2019) Smart plugs: paradigms and applications in the smart city-and- smart grid. Energies 12(10):1957
15. Rao SK, Prasad R (2018) Impact of 5G technologies on smart city implementation. Wireless Pers Commun 100(1):161–176
16. Sofana RS et al (2019) Future generation 5G wireless networks for smart grid: a comprehensive review. Energies 12(11):2140
17. Liu Y et al (2019) Intelligent edge computing for IoT-based energy management in smart cities. IEEE Netw 33(2):111–117
18. Leligou HC et al (2018) Smart grid: a demanding use case for 5G technologies. In: IEEE international conference on pervasive computing and communications workshops, pp 215–220
19. Chang KC et al (2020) Energy saving technology of 5G base station based on internet of things collaborative control. IEEE Access 8:32935–32946
20. Baccarelli E et al (2018) Fog-supported delay-constrained energy-saving live migration of VMs over MultiPath TCP/IP 5G connections. IEEE Access 6:42327–42354
21. Wang Y (2021) Energy-saving scheme of 5G base station based on LSTM neural network. In: 2nd international conference on applied physics and computing. J Phys Conf Ser 2083(3):032026
22. Yigitcanlar T et al (2021) Green artificial intelligence: towards an efficient, sustainable and equitable technology for smart cities and futures. Sustainability 13:8952
23. deSouza JT et al (2019) Data mining and machine learning to promote smart cities: a systematic review from 2000 to 2018. Sustainability 11:1077

24. Mahapatra C et al (2017) Energy management in smart cities based on internet of things: peak demand reduction and energy savings. Sensors 17(12):2812
25. Zhou Q et al (2020) Learning-based green workload placement for energy internet in smart cities. J Mod Power Syst Clean Energy 10(1)
26. Sharma H et al (2021) Machine learning in wireless sensor networks for smart cities: a survey. Electronics 10:1012
27. Sharif A et al (2017) Internet of things- smart traffic management system for smart cities using big data analytics. IEEE, pp 281–284
28. Kok İ et al (2017) A deep learning model for air quality prediction in smart cities. In: International Conference on Big Data (BIGDATA). IEEE, pp 1983–1990
29. Zhang N et al (2016) Semantic framework of internet of things for smart cities: case studies. Sensors 16(9):1501
30. Heras ADL et al (2020) Machine learning technologies for sustainability in smart cities in the post-COVID era. Sustainability 12(22):9320
31. Iskandaryan D et al (2020) Air quality prediction in smart cities using machine learning technologies based on sensor data: a review. Appl Sci 10(7):2401
32. Dey M et al (2020) A case study based approach for remote fault detection using multi-level machine learning in a smart building. Smart Cities 3(2):401–419
33. https://social-innovation.hitachi/en-in/knowledge-hub/collaborate/smart-sustainable-cities/.

Leveraging Machine Learning Algorithms for Fraud Detection and Prevention in Digital Payments: A Cross Country Comparison

Ruchika Gupta, Priyank Srivastava, Harish Kumar Taluja, Sanjeev Sharma, Shyamal Samant, Sanatan Ratna, and Aparna Sharma

Abstract The COVID-19 outbreak has brought unprecedented shifts for the global financial system and has hastened the adoption of digital payments. More insidious fraud schemes have risen as a result of these shifts, creating fertile ground for all sorts of financial fraud. Therefore, this paper presents a complete review of the fraud environment in digital payments, as well as the various approaches used by regulatory agencies in various nations throughout the world. This paper further describe how machine learning algorithms may be used to detect and prevent digital payment fraud in the post-pandemic period. Finally, some of the major obstacles and promising possibilities are discussed in order to give future inspiration for intelligent payment fraud detection.

Keywords Machine learning algorithms · ML · Financial frauds · Digital payments · Fraud detection

R. Gupta (✉)
ABES Business School, Ghaziabad, UP, India
e-mail: guptaruchika81@gmail.com

P. Srivastava · S. Sharma · S. Samant · S. Ratna
Amity University, Noida, UP, India

H. K. Taluja
HR Institute of Technology, Ghaziabad, UP, India

A. Sharma
Noida International University, Noida, UP, India

N. Hasteer et al. (eds.), *Decision Intelligence Solutions*, Lecture Notes in Electrical Engineering 1080,
https://doi.org/10.1007/978-981-99-5994-5_33

1 Introduction

Financial fraud has wreaked havoc on the global economy over the last few decades, jeopardizing capital markets' efficiency and stability. To make matters worse, the coronavirus pandemic (COVID-19) breakout in early 2020 caused severe disruptions in global financial markets. Because of the health and financial issues due to pandemic, people are more vulnerable, and fraudsters profit from it.

By sending bogus emails demanding charity donations, hawking counterfeit personal protective equipment (PPE), or advertising a COVID-19 vaccination, bad actors continue to prey on people's fears and anxieties. Phishing attempts and emails directing naïve consumers to harmful websites with suspicious files have been on the rise. Personal information might be stolen, enterprise networks could be accessed without permission, and financial fraud could occur as a result of these attempts [1].

Furthermore, quarantine rules allow for internet banking and distant transactions, but remote banking finds it difficult to gather full information for client identity verification, resulting in a high rate of credit fraud [2]. Furthermore, moving a firm from offline to online exacerbates information asymmetry and makes fraud detection more difficult [3].

To summarize, the COVID 19 epidemic has exacerbated the problem of digital payment fraud. On the one hand, the danger and number of occurrences of payment fraud have skyrocketed; on the other hand, detecting these frauds has become increasingly difficult.

Payment fraud losses have more than quadrupled in the previous decade, according to Industry Analyst. Illicit payments cost the economy $10 billion in 2011. This sum is expected to reach more than $32 billion by 2020, with fraud accounting for 6.83 cents of every $100 spent. In addition, the expenses of payment fraud are likely to rise in the future. By 2027, fraud losses are expected to have increased by another 25%, surpassing $40 billion. Not to mention the chargeback costs, investigation expenses, and fines levied by government and credit card companies [4].

As a result, payment fraud has become a serious problem in the post-pandemic age, with stronger objectives, more subtle forms, and more sophisticated techniques. As a result, this paper gives a thorough review of the payment fraud environment as well as the various techniques used by regulatory organizations in various nations throughout the world. The study goes on to address the use of machine learning algorithms in the post-pandemic age for identification and prevention of digital payments fraud. Finally, some of the most important issues and promising possibilities are discussed in order to give future inspiration for intelligent payment fraud detection.

2 Digital Payments Frauds Landscape: Pre and Post Covid 19

Selecting the metrics for transaction classification performance. No matter the method used to synthesis the models, the best classifier is ultimately chosen because it is impossible to predict the efficacy of a certain model or their ensemble in advance. The philosophy for categorizing the solutions into automatic and manual ways is obviously very different. In manual sorting, the developer's experience and intuition are the primary drivers of finding the optimal answer; in automatic sorting, it's the algorithms and computer power.

Customers resorted to digital channels to substitute or complement face-to-face contacts as the quickly spreading coronavirus prompted lockdowns and ongoing social distancing measures throughout the world. As a result, mobile applications and web outlets have proliferated. Total US online spending in May 2020, during the height of the first wave of lockdowns, hit $82.5 billion, up 77% from the same month a year before, according to Adobe's Analytics Digital Economy Index unit. However, although this historic and quick move to digital provided a virtual lifeline to customers in a time of crisis, it also spurred a multibillion-dollar increase in payment fraud throughout the world.

Consumers have grown increasingly vulnerable to cyber criminals, according to a poll performed by Fintech Company FIS in April 2021 in India, with 34% of respondents reporting financial theft in the previous year. For individuals between the ages of 25 and 29, this percentage climbs to 41%. Consumers were also victims of card scams and skimming, according to the poll. Financial crimes were usually committed through phishing, followed by QR code/UPI scams.

According to another survey by [4], fraud attempts have increased by nearly 35%, indicating that thieves are increasingly active in breaking through digital channels. Independent interviews were conducted with the payment and security executives from 20 countries were performed between January and September 2020 for this study. These countries belong to the given regions: North and South America, Africa, Europe and Asia Pacific.

On comparing the period between March 11, 2020 and March 10, 2021 with the previous year period, [6, 8, 9] revealed that the fraction of suspected fraudulent digital transaction attempts against firms originating in India climbed 28.32%. This analysis is based on data from billions of transactions and more than 40,000 worldwide websites and applications that are part of the company's fraud analytics product package. These websites and applications receive visitors from all over the world. The prepared sample was split into two separate data sets for machine learning training and testing. 1500 users were taken from the blacklist and 1100 individuals were taken from the whitelist as innocent users in order to guarantee that the training data was balanced and the research gave the most value from the blacklist users. The test set also includes the remaining 5000 users. In light of this, 2.4% of the test data were selected to be included in the blacklist. 5000 non-fraudulent users and 120

Table 1 The Size of Datasets

	Training	Test	Total
Real or Non-fraudulent	1100	5000	6100
Fake	1100	120	1220
Total	2200	5120	7320

fraudulent users, respectively, made up the test set. The number of fake and legitimate users in the test and training data sets is displayed in Table 1.

3 Digital Payments Fraud Statistics and Practices Adopted: Cross Country Comparison

The surge in online traffic is indeed one of the key reasons leading to payment fraud. According to a recent research of Australian buyers [7], internet purchases increased by 65% between March 2020 and January 2021, while card-not-present fraud increased by 3.8%. This sort of fraud currently accounts for 90% of all payment fraud in the country, accounting for 0.058% of all credit card spending.

Global payment fraud losses have increased from $9.84 billion in 2011 to $32.39 billion in 2020, according to [9]. The research also projects continuous increase in payment frauds. The projected figures for 2027 are approximately 25% higher from 2020 costing around $40.62 billion. (Fig. 1).

3.1 *European Payments Fraud Statistics*

In 2019, the overall value of fraudulent transactions utilizing SEPA (Single Euro Payments Area) cards collected throughout the world was €1.87 billion. 1.03 billion pounds have been the value for such duplicitous transactions for cards issued in the euro region exclusively. [10] In comparison to 2018, the total value of aggregate card

Fig. 1 Global Payments Fraud Statistics (Source: [9])

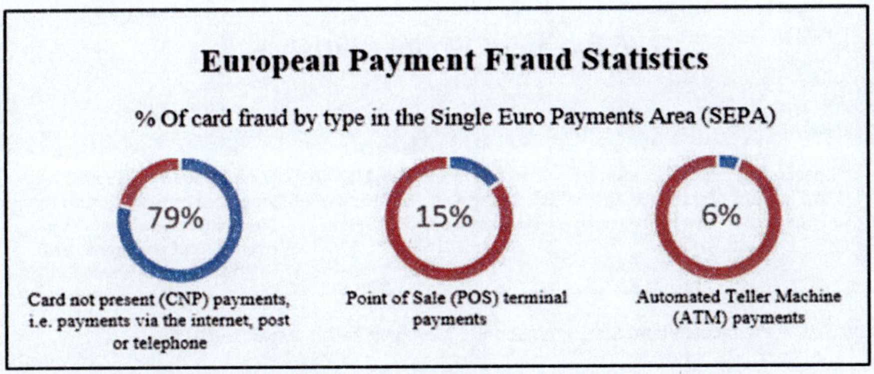

Fig. 2 European Payments Fraud Statistics (Source: [9])

transactions utilizing SEPA-issued and acquired cards climbed by 6.5%, while fraud increased by 3.4% (Fig. 2).

Practices Adopted by Regulatory Bodies in Europe:

- The updated Payment Services Directive (PSD2) requires the use of the following two standards namely Regulatory Technical Standard and Common & Secure Open Standard for strong customer authentication.
- Various preventive measure has been adopted by the concerned service providers. Geo blocking, tokenization of cards, increasing security standards are amongst the few.
- The European Central Bank jointly with the European Banking Authority published certain guidelines in the year 2018. These guidelines pertain to fraud reporting and provide an overview of the kind of statistical data to be collected from the concerned service providers.

3.2 UK Payments Fraud Statistics

UK Finance has observed a 5% increase in APP (Authorized Push Payments) fraud losses over the preceding year, with the number of incidents growing by 22% to over 150,000 in 2020. (Finestra, 2020). The amount of money lost to fraudsters in the UK reached £754 million, up 30% over the same period previous year, and was defined by UK Finance as "at a level that constitutes a national security danger" (Fig. 3). Identity theft complaints increased by 113% from 2019 to 2020, reaching roughly 1.4 million cases, according to the US Federal Trade Commission.

Practices Adopted by Regulatory Bodies in UK

- The UK continues to work on its New Payment Architecture (NPA), that will provide financial institutions (FIs), payment providers, and other stakeholders access to the new payment platform that will allow them to make rapid payments.

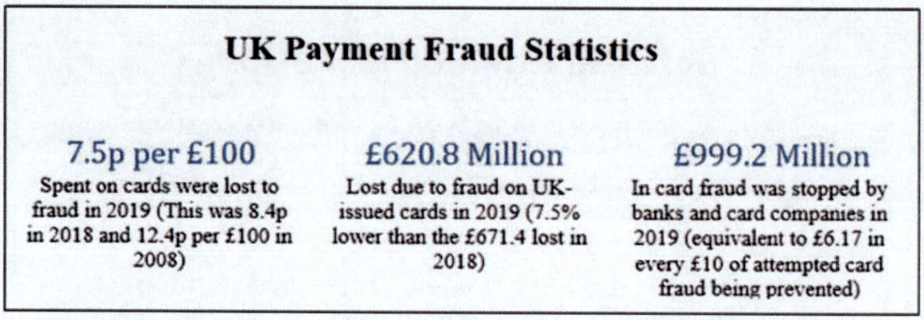

Fig. 3 UK Payments Fraud Statistics (Source: Merchant savvy research (2020))

- The Bank of England is preparing to switch to the ISO 20022 communications standard. The NPA will be easier to use now that it has been updated to the current ISO standard.
- Pay.UK had created a new solution to combat Authorized Push Payments (APP) fraud. It acts as a framework for classifying pertinent customer data, making it easier for both banks engaged in a payment transaction (the sending and receiving banks) to detect a fraudulent transaction.

3.3 US Payments Fraud Statistics

Synthetic identity fraud is the quickest developing sort of financial crime in the United States. The number of incidents of identity theft is considerable. In the United States, about 1.4 million incidents were recorded, with the Federal Trade Commission reporting that the number of cases in 2019 was three times higher than in 2018. According to Juniper Research, between 2018 and 2023, merchants are expected to lose $130 billion in digital CNP (Card-not-Present) fraud (Fig. 4).

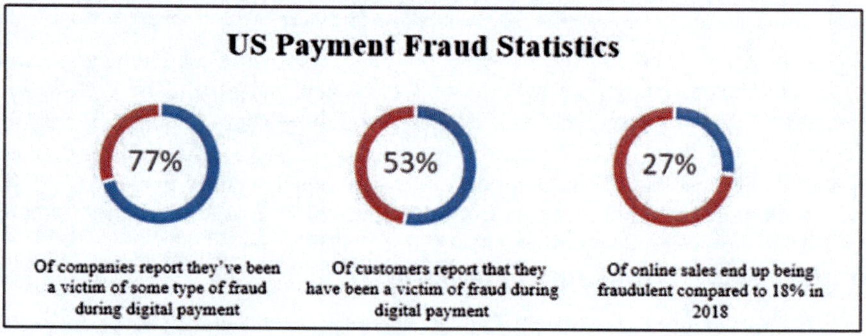

Fig. 4 US Payments Fraud Statistics (Source: Merchant savvy research (2020))

Practices Adopted by Regulatory Bodies in US

- While PCI DSS is not a nationally mandated security standard in the United States, it is imposed by the Payment Card Industry Security Standard Council, which is made up of major credit card companies and keeps it as an industry standard. The norm has been enshrined in the legislation of certain states.
- The US Department of Justice is requesting that the US Congress alter the present legislation to make it unlawful for international criminals to hold, acquire, or sell a stolen credit card provided by a US bank regardless of geographic location.
- In the United States, federal law restricts cardholder responsibility to $50 in the case of credit card theft. This needs to be reported within a time frame of 60 days.

3.4 *Asia Pacific (APAC) Payments Fraud Statistics*

The cost per transaction in Australia is USD3.51; USD3.61 in Hong Kong, USD3.87 in Japan, and USD3.84 in India, which varies with the dollar valuation. When the average of the regions including other APAC markets as well is being calculated, it estimates around 3.40 USD [11] (Fig. 5). We divided financial fraud into four categories based on the results of our investigation, including financial statements, bank fraud, insurance fraud, and others. Data provides information on the quantity of articles evaluated and the sort of fraud. It can be seen that, of the review papers, 51% of the articles deal with bank fraud, 31% with financial statements, 20% with insurance fraud, and the remaining articles deal with other fraud at 14% and 6%, respectively.

Practices Adopted by Regulatory Bodies in Asia Pacific

- In areas such as the Asia–Pacific region, several governments have established policies that favor indigenous payment networks such as UnionPay in China and

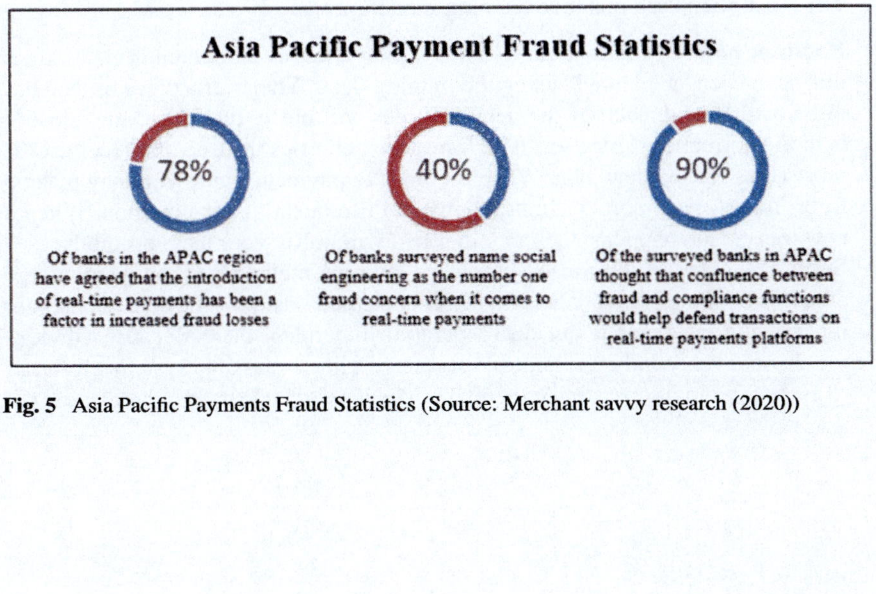

Fig. 5 Asia Pacific Payments Fraud Statistics (Source: Merchant savvy research (2020))

Rupay in India. The goal is to reduce fees and charges while also bringing the unbanked people into the financial mainstream.

- Open banking, cooperation, information sharing, and privacy are all becoming more regulated.
- Safe authentication processes for existing and new cards, SMS or e-mail alerts when transactions occur or an agreed-upon limit is surpassed, and customer authorization in case of using OTP. Consumer recommendations were also provided by the MAS (Monetary Authority of Singapore) in June 2014 to defend against phishing attempts.

4 Application of Machine Learning Algorithms for Fraud Detection and Prevention

ML (Machine learning) is a branch of Artificial Intelligence technology which is based on the algorithm development. These self-generated algorithms analyze the data and discovers the correlation amongst customer behavior and likelihood of their involvement in malicious activities. ML approaches also improve with the size of the database to which they have been fitted, indicating that the more fraudulent operations example they are trained on, the better they spot fraud [5, 12, 14].

A machine learning model must acquire data before it can detect fraud. The model evaluates all of the information obtained, segments it, and extracts the necessary characteristics. Following that, the machine learning model is taught how to estimate the likelihood of fraud using training sets. Finally, it builds machine learning models for fraud detection.

Prominent financial institutions, on the other hand, are already employing machine learning to tackle fraudsters. MasterCard, for example, used AI and machine learning for tracking suspicious transactions wherein the self-generated algorithms review each transaction and report if it finds any suspicious activity.

The most often used machine learning models are:

- **Random Forest:** The random forest is simply a set of individual decision trees that have been "generated" using the training data. Whenever a piece of data has to be categorized, each of the decision trees will notify the forest how close it is to the correct classification. The forest then chooses the tree that received the most votes for the new data. When it comes to payment fraud, you may make a forest out of many sorts of transactions, both frequent and uncommon. When a new transaction happens, the random forest will notify your team promptly.
- **Support Vector Machine:** A supervised learning method that can handle large amounts of complicated data is SVM. This method positions the data on a bigger input region and generates an ideal separation hyperplane there. Because it divides the classes evenly and at the largest distance, this linear classifier system is referred to as the optimum separating hyperplane. The maximum margin hyperplane is the

plane that results in the largest disparity between classes. The data points nearest to the greatest margin hyperplane are represented by support vectors.

- **The K-nearest neighbors (KNN)** method is a practical, uncomplicated supervised machine learning technique that is effective in handling both classification and regression operations. The KNN model typically uses a limited collection of the closest samples to decide the class label. The KNN model is a kind of non-parametric model that can find comparable neighbor hoods in a dataset that are closest to a given sample point and construct a new sample point depending on the distance between two samples of data. It is used for both classification and regression applications. The performance of this approach is probably hampered by imbalanced datasets, despite the fact that it performed well on many datasets. One of the most well-known methods for estimating distance is the Euclidean distance. For the purpose of detecting financial fraud, certain new methods have been established recently. The experimental findings demonstrated that the KNN model is superior for credit card fraud detection.
- **Neural Networks:** Neural networks are mathematical concepts that retain information using learning methods based on the brain. In several applications, neural networks have been used to simulate unknown relationships between various parameters using a huge number of instances (Anuradha et.al 2020). The model is trained on a labelled dataset, and input data is routed via many layers (i.e., sets of mathematical functions). 1–2 hidden layers are used in this sort of model. What makes this challenge (fraud detection) so difficult is because in the actual world, the vast majority of transactions are legitimate, with only a small percentage accounting for fraudulent conduct. This suggests we're dealing with the issue of skewed data.
- **The Genetic Algorithm (GA):** It is a sort of Evolutionary-inspired Algorithm (EA) that is frequently used to solve a variety of optimization problems with a decreased computing cost. Typically, EAs exhibit the following characteristics:

  – Population: The population is a sample of potential solutions maintained by EA techniques.
  – Fitness: A person inside the population is a solution. Each person is distinguished by a gene representation and a fitness metric.
  – Variation: The person evolves through mutations inspired by the evolution of biological genes (Fig. 6).

5 Algorithm for Fraud Detection

1. Load the Initial data set
2. List all the factors that one responsible for fraud detection.
3. Set the Target variable
4. Select the training data set.
5. Compute the Initial population from list step No 2
6. Compute the fitness function.

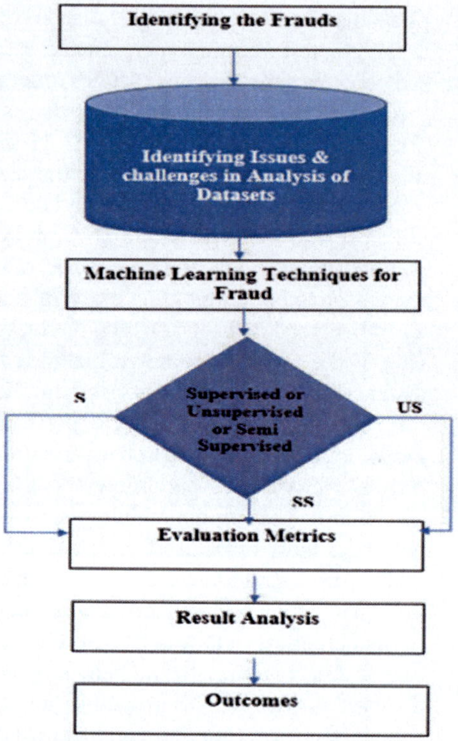

Fig. 6 GA algorithm for fraud detection

7. Conduct the crossover
8. Compute the mutations
9. Calculate the fitness
10. Generate the optimal fitness score, q
11. Update the list B

In this work, the RF technique is utilized as the GA fitness method. In addition, the RF approach is utilized because it eliminates the issue of over-fitting that is sometimes observed when utilizing conventional Decision Trees (DTs). In addition, RF works well with both continuous and categorical variables, it is known to perform optimum on datasets with a class imbalance issue. Tree-based ML techniques are alternatives to the RF. The fitness method is defined as a function that decides if a candidate solution (a feature vector) is ft or not. The fitness metric is determined by the accuracy produced by a certain attribute vector during the testing phase of the RF technique within the GA. The first algorithm gives more information regarding the implementation of RF in the GA (Fig. 7, 8 and Tables 2, 3).

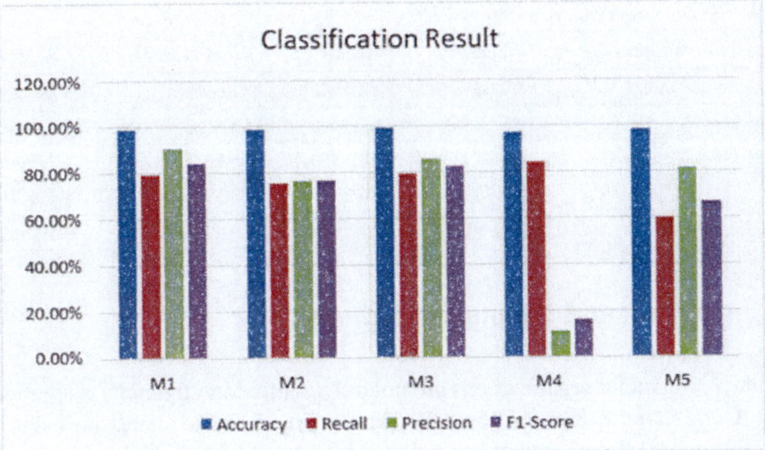

Fig. 7 Classification results for V_1 Set

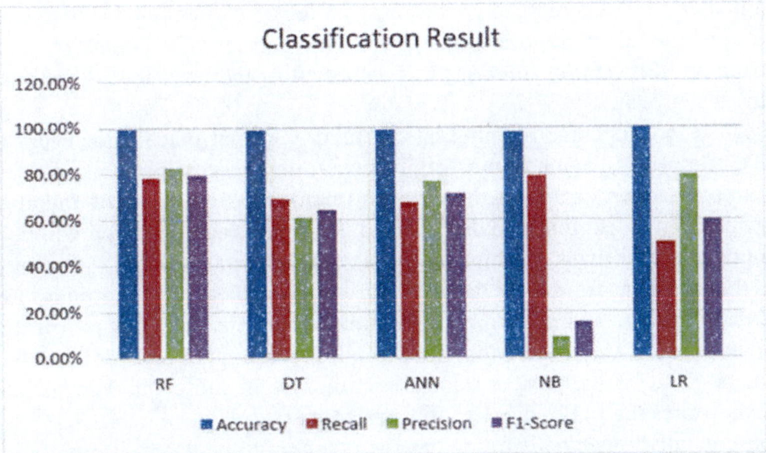

Fig. 8 Classification results for V_2 Set

Table 2 Classification results for V_1

Machine Learning Model	Accuracy	Recall	Final Accuracy	Score-1
M1	98.84%	78.99%	90.69%	83.85%
M2	98.82%	75.22%	76.22%	76.22%
M3	98.94%	78.87%	85.61%	82.10%
M4	97.23%	83.95%	10.83%	15.65%
M5	97.81%	59.52%	81.27%	66.70%

Table 3 Classification results for V_2

Machine Learning Model	Accuracy	Recall	Final Accuracy	Score-1
RF	99.03%	78.10%	82.69%	89.26%
DT	98.57%	69.14%	62.62%	63.06%
ANN	98.81%	67.37%	76.53%	71.09%
NB	97.75%	78.87%	8.59%	15.47%
LR	99.79%	49.78%	79.41%	59.66%

6 Conclusion and Recommendations

Nowadays, Financial organizations are not only confined to financial risk reduction; rather they are more focused towards fraud prevention and detection using latest technologies [13]. This has been necessitated by the pandemic which has been a major culprit in exponential increase in the number of fraudulent transactions. Amongst a wide array of emerging technologies for the purpose of safeguarding systems, machine learning has emerged as the most preferred choice due to its uniqueness in spotting fraudulent activities amongst a wide dataset. The popularity of Fraud Detection and Prevention solutions has increased as the need for AI and machine learning has grown.

In banks too, machine learning has secured a special place since every day, a massive number of payment transactions occur, thus increasing the size of banks' payment transaction databases. Many useful hidden insights may be found in this data, which can be exploited to detect fraud. Payment fraud detection will be faster and more effective thanks to human decision-making and machine learning algorithms that can learn from these datasets. Furthermore, there are a few things to think about when using machine learning algorithms to detect and prevent payment fraud: How much does it cost to overlook true fraud? What impact will the scam have on customers' faith? Is the present fraud-detection system sufficient? How much time and money does the fraud investigation procedure take?

Consequently, machine learning has the potential to be a very useful tool in the identification and prevention of fraud as well. It might help companies create a more secure environment for their clients, and even provide clients the strength to keep embracing fintech. If well- designed data models and efficient operational rules are employed, fraud detection with machine learning may be straightforward and time-saving to implement.

References

1. Gupta R, Agarwal SP (2017) A comparative study of cyber threats in emerging economies. Globus Int J Manag IT 8(2):24–28

2. Rosen LW (2020) COVID-19 and emerging global patterns of financial crime. Congressional Research Service
3. Karpoff JM (2021) The future of financial fraud. J Corp Finan 66:101694
4. Javelin Strategy & Research (2020) The Escalation of digital fraud: global impact of the coronavirus. https://www.javelinstrategy.com/research/escalation-digital-fraud-(2020)
5. Anuradha et al (2020) Calculation and evaluation of network reliability using ANN approach. Procedia Comput Sci 167:2153–2163
6. Credit Bureau TransUnion (2021). https://www.transunion.com/annual-credit-report
7. Crawford A (2021) Card fraud rises with increased online shopping. https://mozo.com.au/credit-cards/articles/card-fraud-rises-with-increased-online-shopping-says-report(2021)
8. Finextra. App Fraud continues to rise as criminals target bank customers. https://www.finextra.com/newsarticle/37751/app-fraud-continues-to-rise-as-criminals-target-bank-customers-online(2021)
9. Merchant savvy (2020) Global Payment Fraud Statistics, Trends & Forecasts. https://www.merchantsavvy.co.uk/payment-fraud-statistics/
10. Card Fraud Report (2021) European Central Bank Report. https://www.ecb.europa.eu/pub/cardfraud/html/ecb.cardfraudreport202110~cac4c418e8.en.html
11. LexisNexis (2021) True Cost of Fraud APAC Study. https://en.prnasia.com/releases/apac/cost-of- fraud-in-Asia-pacific-markets-is-high-according-to-LexisNexis-risk-solutions-study-324904.shtml
12. Gupta A, Gupta R, Kukreja G (2021) Cyber security using machine learning: techniques and business applications. Appl Artif Intell Bus Educ Healthcare 954:385
13. Srivastava P, Khanduja D, Agrawal VP (2018) Mitigation of risk using rule based fuzzy FMEA approach. In: 2018 8th international conference on cloud computing, data science & engineering (confluence). IEEE, pp 26–30
14. Anuradha et al (2020) A heuristic approach for all-terminal network reliability using OANN. Int J Emerg Technol 11(3):721–730

Automatic Data Generation for Aging Related Bugs in Cloud Oriented Softwares

Harguneet Kaur and Arvinder Kaur

Abstract Context: Cloud oriented applications are the online delivery of IT resources which increases the adaptability of the users by sharing computer infrastructure. The cloud oriented software constitute the features such as distributed file system, database storage, MapReduce methodology, BigData Analytics and many more which may introduce technical issues with the long running time. These technical debts decreases system's performance and increases the risk of failure and known as Aging related bugs (ARB). The process cause due to ARBs is addressed as Software Aging. The ARBs are caused due to null pointer exception, memory leakage, overflow data, deadlock state or overuse of the resources in the running system whose effect can be deadly also, therefore it will be great if they get predicted before testing of the software stage. Finding ARBs manually in the dataset using aging related keywords is common but when the dataset is large like cloud oriented software then it becomes a challenging task.

Objective: This is the first paper that presents the automatic generation of dataset for aging in cloud oriented softwares.

Method: ARB reports are automatically generated using the summary/description of bug reports through SEARCH ARB algorithm with the help of aging related keywords.

Results: The Manual Extraction of ARB reports from the dataset is compared with the application of automata algorithm.

Conclusion: The results suggested that manual extraction is comparable to automata in cloud oriented datasets therefore automata is preferable as it saves time and cost.

Keywords Software aging · Automata · Aging related bug · Cloud · Bug Report

---

H. Kaur (✉) · A. Kaur
University School of Information, Communication and Technology, GGSIPU, Dwarka, New Delhi, India
e-mail: harguneet.13316490018@ipu.ac.in

N. Hasteer et al. (eds.), *Decision Intelligence Solutions*, Lecture Notes in Electrical Engineering 1080,
https://doi.org/10.1007/978-981-99-5994-5_34

1 Introduction

Cloud computing applications is a trending technology that provide services in various fields such as art, business, data storage, education, backup services, entertainment, social network etc. These applications allows different group of people to share computer resources easily besides sitting at the remote places. Once the data is stored in the cloud storage area, it is easy to get the access and back up from it. It reduces the maintenance cost for hardware and software both. Nowadays, Cloud computing softwares are available in open source repositories from where it can be easily accessible.

Software aging [7] is the process observed in long running softwares where performance degradation and resource depletion increases with the period of time. The reasons for the software aging are memory leakage, unterminated threads, disk fragmentation, overflow error, unreleased files and locks, uninitialized memory read and dereference of NULL pointer in software. It is the challenging task to uncover ARBs in the software during testing as the replication of these bugs is not easy. Therefore identification of ARB reports in the cloud oriented softwares improve the testing efforts of software quality assurance team in terms of cost and time. As per the knowledge, software aging bug reports are usually extracted by the authors using the aging related keywords while reading the given description of bug report in the repository. But when the data is large like cloud oriented software where it contains thousands of bug reports then there is a need for the automata as manual extraction is time consuming and labour intensive. Manually extraction of aging related bug reports introduce the risk of errors and efficiency i.e. if different researcher perform the same task then it may yield different results. Therefore, automatic extraction of bug reports is preferred with more complex and large datasets. Automatic extraction has greater number of benefits includes less hands on time, decreased risk of errors and increased efficiency.

Machida et al. [11] analysed cloud oriented softwares [2] which stated the existence of aging bugs in it. Therefore in this study the automatic extraction of aging reports is proposed from the cloud oriented dataset [16]. As per our knowledge this is the first study which explains the procedure to extract aging related bug reports from the cloud oriented dataset through automata. It uses the aging related keywords explored in the recent research [4, 6] for prediction of ARBs. In this study aging related bug reports are extracted manually also and then it is compared through our automata. The results are comparable i.e. manual task and automata produces almost equal number of ARBs. In some dataset automata produces more number of ARB therefore it is recommended to use automata for the extraction process. The results shows that there are fewer aging related bug reports in comparison with non aging bug reports which give rise to class imbalance problem discuss in our further studies. The dataset for the prediction [19] of aging related bugs is prepared in this study: automatic identification of ARBs in cloud oriented dataset and the extraction of static code metrics using Understand Tool [15].

The paper is divided in section wise: Section 2 tabulates the Related studies in a detailed manner and Sect. 3 contains the research methodology of our study. Research methodology brief about Aging related keywords, software metrics, Datasets and experimental setup. The results are explained in Sect. 4 and Sect. 5 explain the threats to the validity. Sect. 6 concludes the research performed in this study along with references.

2 Related Work

Software aging is the process in which software starts showing failure with long running span of time. This study focused on automatic extraction of aging related bugs in cloud oriented software. The work is distributed into two parts: 1. Automatic generation of aging related bug reports using aging related keywords.

2. Extraction of software code metrics from the cloud oriented softwares. The objectives of the related studies, Performance Measures used, Datasets explored and the research gap for each study are given in detail in Table 1. ST1 and ST3 manually sorted ARBs in bug reports for the prediction purpose where as the dataset with thousand bug reports, it becomes the task to sort it manually. Lov Kumar et al. [10] studied the relation between source code metrics and ARB which concludes that source code metrics are used for ARB prediction. The concept of software aging was given by Huang in 1995.

Cotroneo [4, 5] presented his empirical work on relating software aging and static code metrics of the software. It predicted ARBs using software complexity metrics and built fault prediction on open source static softwares i.e. MySQL, and Linux. The research gap found in Cotroneo's work is the bug reports which are manually studied to identify ARB's using aging related keywords. Cotroneo et al. [6] discussed rejuvenation (SAR) and aging in softwares and mark the issues towards the same which should be taken into consideration in future. Massimo Ficco [8] worked on apache storm to predict software aging bugs in runtime. ST2, ST5 and ST6 studied aging related bug prediction using transfer learning approach on Static open source softwares but not on large datasets such as cloud oriented softwares [1]. ST4 declared the presence of ARBs in cloud computing datasets by reading the bug report description manually. According to the recent studies [18], the dataset collection is performed manually [12] while finding the aging related keywords in the bug reports whereas this study sorted aging related bug reports through automata. In this study, author explored cloud oriented softwares for the identification of ARBs through automata.

Table 1 Related Studies

SNo.	Author	Objective	Dataset	Performance measure	Research Gap
ST1	Xiaoxue Wu [18]	It identified aging related bug reports in invalid bug reports and then the prediction is performed	HDFS and HBase	Recall, Precision, F1-score and AUC	Manually explored ARBs in all bug reports
ST2	Qin, Fangyun et al. [15]	Proposed transfer learning approach for ARB prediction in cross project	Linux, MySQL, Apache HTTPD	PD (probability of detection), PF (probability of false in alarms),Bal (Balance)	Performed cross project ARB prediction not in cloud oriented softwares
ST3	Lov Kumar et al. [10]	ARB prediction is performed on the basis of static code metrics using machine learning classifiers	Linux, MySQL	Area under ROC, F-measure and accuracy	Manually explored ARBs in all the bug reports
ST4	Fumio Machida et al. [11]	It stated that ARBs exist in five cloud computing open source softwares	Hadoop Mapreduce, Cassandra, Eucalyptus, Memcached and Xen	Classification	It manually performed the identification of ARBs in bug reports
ST5	Qin, Fangyun, et al. [14]	Proposed the transfer learning aging related bug prediction approach (TLAP)	Linux and MySQL	Probability of Detection, PF and Bal	Transfer Learning is applied on MySQL and Linux not on cloud oriented softwares
ST6	Fangyun Qin et al. [13]	The empirical study is performed using cross project approach to predict ARBs	Linux, MySQL, HTTPD	PD, PF and Bal	Linux and MySQL are used in software aging and manual sorting of ARBs in bug reports

3 Research Methodology

3.1 Aging Related Keywords

Aging Related Bugs are observed in the running softwares caused due to the effects of aging. Softwares have limited memory and time which when exhausted lead to deadlock state or hanging or crashing of the software. It generally happens due to

Table 2 Aging Related Keywords

Keywords			
Leakage	Race	Memory	Aging
deplet	Overflow	flush	NPE
null pointer	Buffer Exhausted	deadlock	flush
MEMORY	LEAK		

leakage of memory in threads and data objects or deadlock stage or infinite loop. In our work, ARB reports are sorted through the algorithm using the aging related keywords such as race, memory, deadlock, null pointer exception, leakage, overflow etc. In previous research, the bug reports are manually sorted using aging related keywords. But with cloud oriented datasets when there are thousands of bug reports then it is better to use these keywords in automata. The aging related keywords are given in Table 2 [17].

3.2 *Software Metrics*

Software metrics help in the prediction of bugs therefore in this study, static code metrics are extracted from the cloud oriented dataset. The metrics help in measuring the performance, productivity and planning for the software applications. The software static code metrics are automatically extracted using the Understand tool for analysis purpose [4]. The metrics are given in Table 3. There are two sets of static code metrics collected in this research: McCabe's complexity and Program size. McCabe's Complexity is the measure of cyclomatic complexity that counts the number of paths, number of operands and operators whereas Program size measures the lines of code and files by evaluating the productivity of developer.

4 Datasets

It is difficult to find software aging in the datasets due to its low proportion and limiting condition of reproducibility. Majorly, the benchmark datasets i.e. LINUX, MYSQL are being considered for the study of software Aging but nowadays as the increasing demand for cloud oriented datasets, we are considering the same for the dataset collection. Cloud oriented datasets are also available at open source repositories such as Cassandra, Storm, Hive, Hadoop Mapreduce and Hadoop Hdfs. The detailed description for every dataset used is given Table 4. Datasets are driven from GitHub.

Apache Cassandra is a popular distributed NoSQL database which is designed in the objective to handle large amount of data. It is available in the open source with the characteristics of wide column store and fast support for the clusters which may

Table 3 Software Metrics

S. No.	Metrics	Type
1	AvgCyclomatic, AvgCyclomaticModified, AvgCyclomaticStrict, AvgEssential, Cyclomatic, CyclomaticModified, CyclomaticStrict, Essential, MaxCyclomatic, MaxCyclomaticModified, MaxCyclomaticStrict, MaxEssential, SumCyclomatic, SumCyclomaticModified, SumCyclomaticStrict, SumEssential	McCabe cyclomatic complexity
2	AvgLine, AvgLineBlank, AvgLineCode, AvgLineComment, Count-ClassBase, CountClassDerived, CountDeclClass, CountDeclClass-Method, CountDeclClassVariable, CountDeclExecutableUnit, CountDeclFile, CountDeclFunction, CountDeclInstanceMethod, CountDeclInstanceVariable, CountDeclInstanceVariablePrivate, CountDeclInstanceVariableProtected, CountDeclInstanceVariablePublic, CountDeclMethod, CountDeclMethodAll, CountDeclMethodDefault, CountDeclMethodPrivate, CountDeclMethodProtected, CountDeclMethodPublic, CountInput, CountLine, CountLine Html, CountLine Javascript, CountLine Php, CountLineBlank, CountLineBlank Html, CountLineBlank Javascript, CountLineBlank Php, CountLineCode, CountLineCode Javascript, CountLineCode Php, CountLineCodeDecl, CountLineCodeExe, CountLineComment, CountLineComment Html, CountLineComment Javascript, CountLineComment Php, CountOutput, CountPath, CountPathLog, CountSemicolon, CountStmt, CountStmtDecl, CountStmtDecl Javascript, CountStmtDecl Php, CountStmtExe, CountStmtExe JavascriptCountStmtExe Php, MaxNesting, RatioCommentToCode	Program Size

belong to multiple data-centers. It is purely java based with Management Extensions (JMX).

Hadoop MapReduce generates datasets with the help of distributed parallel clustering algorithm. Mapreduce is basically the formation of two processes: Map and Reduce. Map filters and sort the data whereas Reduce helps in executing the function. It has the characteristics of high scalability and high fault tolerance. It works

Table 4 Dataset

Dataset	Version	Total Files
Cassandra	3	2764
Hadoop Mapreduce	0.23.0	1412
HDFS	0.20.0	521
Hive	3.1.0	7112
Storm	2.3	2371

on processing of large dataset based on clustering. This framework works on large applications.

HDFS() It is a Hadoop Distributed File System which works on Master-slave architecture and stores the large amount of data. It is composed of 2 parts: Metanode and n Datanodes. Metanode contains the meta data which behaves as a Master node and datanode contains the data that behaves as slaves. It basically used for streaming and parallel processing.

Hive is a SQL data warehouse used for writing, reading and managing huge datasets. It is available at Apache foundation and implements queries using SQL under Java platform. It works without using low level Java-API. It is developed by Facebook and other multinational companies like Amazon, Netfflix and other web services.

Storm is real-time software more dominant than Hadoop to process huge data using clustering. It process data in seconds and worn by Twitter, NaviSite and Wego for the processing of data in fault tolerance manner. It is user-interactive and reliable which can be easily used by small working areas to large industries.

4.1 Experimental Setup

Software aging is the process of performance degradation and consumption of software resources over the long running time. This process is caused by Aging Related Bugs (ARB) which gets accumulated with running time and these are characterised as Unreleased File locks, deadlock state, over utilisation of memory, unterminated threads and disk fragmentation. In the recent research, Aging related bugs are identified in the bugs data manually by finding the keywords of aging in the description of bug report. If any bug report is related to aging then it is marked as Aging related bug report. On the basis of recent studies, aging related keywords are explored and are listed in a list such as deadlock, memory leak, deallocating, overflow buffer, dereference, improper use of lock, null pointer exception, socket leak, etc. In this research, manual task is converted into automata through text mining implemented in python for the ease of researcher to work upon large datasets for building fault prediction models. Manual search can be applied when there is small dataset but when the count of datasets is more then to save cost and time its better to implement automata than manual task. Manual task may produce inconsistent in the number of bugs if different technical researcher reproduce the bugs in the same datasets. Therefore it is highly recommended to use automata for consistent results. The SEARCH ARB algorithm, given in Algorithm 1, is the automata process for the extraction of aging bug reports in cloud oriented dataset. This automata process can be replicated using any other dataset. It is being performed for the first time therefore we have done manually also to cross check the results with automata. The results are comparable as shown in Table 5, although the automata results are better than manual. Thus, the automata implemented in this research is efficient and reliable. For example, in Storm, manual inspection produces 82 and automata produces 84 aging related bugs. Similarly in Hive through manual the count is 517 and in automata it is found to be 601. It has been concluded that bug reports extracted through automata and manual are almost same. Therefore it is always recommended to use automata process than manual as manual can produce different results every time.

The implementation for the extraction of Aging related bug reports is explained in Figure 1 for the cloud oriented softwares. Aging related files are the files which are contaminated with the software aging problem. JIRA repository is used to collect the bug reports of cloud oriented datasets. There is a REST API which collects the 1000 bug reports at each circle. However cloud oriented datasets have thousands of bug reports therefore REST API is used where with some changes all the bug reports gets collected. This is how the bug reports are collected. Now the automata step i.e. SEARCH ARB is implemented to extract only aging related files from the bug data. The algorithm is implemented in python where the ARBs are resulted in tabular form. Static code metrics are driven from the Understand tool The source code of all five open source cloud oriented datasets are collected from Github repository. The formation of dataset is performed with the help of static code metrics and the ARB metric which says '1' for the java file with the presence of aging and '0' for the absence of aging. Implementation of this study is explained thoroughly in Algorithm 1 - SEARCH ARB where the algorithm is executed in python-Eclipse to extract only the aging related bug reports from thousand of bug reports of the dataset. The implementation justification can be access from: https://github.com/Harguneet/Search keyword.

Table 5 Manual comparison with automata for aging related bugs

S. No	Dataset	Total Bugs	Manual ARB collection	Automata ARB extraction
1	Cassandra	7869	601	677
2	Hadoop Mapreduce	2541	192	190
3	Hadoop HDFS	5125	377	344
4	Storm	1316	82	84
5	Hive	9598	517	601

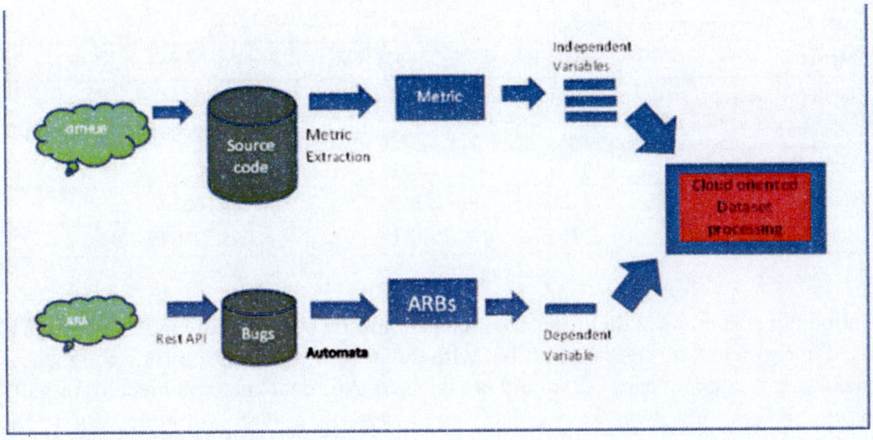

Fig. 1 Framework for ARB extraction

5 Results

Aging related bugs are caused with the long runtime due to the memory or internal changes inside the software. Generally rejuvenation i.e. rebooting of the software is performed if the system hang's out or crash but is not always the solution as the software may loose important information. Therefore if the aging is predicted in the software before hand then the software can be protected from the failure. Aging related bugs are not easily reproducible which is the main hindrance of this phenomenon. To avoid major crashes or damage, it should be the important task to locate ARBs in datasets. It is observed that percentage of ARB files in the dataset is low in comparison to non ARB files shown in Table 6. For instance, total java files are 2371 in Storm dataset out of which 131 are only the aging related files. The percentage found out to be 5.52% only which is less than 50% the requirement for the balance dataset. The percentage of aging related files of Cassandra is noticed to be 13% which is again the mark for the imbalance dataset. Therefore we can say that software aging is the imbalance problem [3] where minority files are less than the majority files. If the prediction is performed then it is necessary to balance the dataset as minority files may be left in the presence of majority files. In our case, minority files are aging related files and majority files are other java files of the dataset.

6 Threats to Validity

Open source data are easily accessible and reusable but it may be effected by unwanted attacks therefore Industrial datasets are being preferred to validate the hypothesis. In this study, cloud oriented datasets are considered which are available

Table 6 ARB proportion in dataset

Dataset	Total Files	ARB Files	Proportion of ARB
Hadoop HDFS	521	170	32.62956
Cassandra	2764	385	13.92909
Hadoop Mapreduce	1412	161	11.40227
Hive	7112	367	5.160292
Storm	2371	131	5.525095

online but it is advised that industrial data should be considered. Understand tool is used to extract static code metrics but with the industrial limitation Halstead metrics and aging related metrics [15] could not be used. Automata is implemented for gathering ARB reports using keywords. Some keywords may be not added due to the time limitation.

Algorithm 1: SEARCH_ARB

Input: The Excel File is imported Containing Bug Key, Bug Id and Bug description/summary of bug reports extracted for each cloud dataset.

1. Excel File is converted into list 'AL BUG' holding the complete information of Bugs
2. Create New List of Aging Related Keywords 'ARK LST'-'null pointer' 'race' 'memory' 'deadlock' ' leak' 'Overflow' 'Buffer exhausted' 'NPE' with all the posibilities of capitalisation of the word.
3. **while** *From the keywords list 'ARK LST', check if these keywords are there in the description of bug reports from AL BUG list* **do**
   - A new list 'FIN AR' is formed contains ARB reports;
4. If any Duplicate then it is removed in FIN AR if any aging related bug is repeated using Set conversion.
5. Convert FIN AR into dataframe and then export it to excel file AR FILE.
6. Repeat this process for each dataset.

Output: AR EILE is exported retaining aging related bug reports

7 Summary

Five different cloud oriented datasets are collected using GitHub repository and then static code metrics are derived from Understand Tool. Bug reports of the cloud oriented datasets are found in Jira repository and with the help of REST API thousand of bug reports are extracted. This study replaces manual classification of aging related bug reports with the automata by implementing the python algorithm. The algorithm used aging related keywords for the classification. According to the author's knowledge, automata is being provided for the first time in extracting aging related bug reports in cloud oriented datasets from the bug repository. Therefore in large datasets, the SEARCH ARB algorithm in python is applied to find aging related bug reports from the bug data [9]. The other observation by the author is that there are very few number of aging related bug reports than other bug reports in datasets therefore aging is the class imbalance problem.

8 Future Work

In this work, static code metrics are used for the collection of cloud oriented datasets but the collection of aging related metrics and Halstead metrics are planned in future. More cloud oriented datasets and industrial datasets are in the process which may strengthen the results for the prediction of ARBs.

References

1. Andrade E, Pietrantuono R, Machida F, Cotroneo D (2021) A comparative analysis of software aging in image classifiers on cloud and edge. IEEE Trans Dependable Secur Comput (2021)
2. Araujo J, Matos R, Maciel P, Matias R (2011) Software aging issues on the eucalyptus cloud computing infrastructure. In: 2011 IEEE international conference on systems, man, and cybernetics. IEEE, pp 1411–1416
3. Chouhan SS, Rathore SS (2021) Generative adversarial networks-based imbalance learning in software aging-related bug prediction. IEEE Trans Reliab 70(2):626–642
4. Cotroneo D, Natella R, Pietrantuono R (2010) Is software aging related to software metrics? In: 2010 IEEE second international workshop on software aging and rejuvenation. IEEE, pp 1–6
5. Cotroneo D, Natella R, Pietrantuono R (2013) Predicting aging-related bugs using software complexity metrics. Perform Eval 70(3):163–178
6. Cotroneo D, Natella R, Pietrantuono R, Russo S (2014) A survey of software aging and rejuvenation studies. ACM J Emerg Technol Comput Syst (JETC) 10(1):1–34
7. Dohi T, Trivedi KS, Avritzer A (2020) Handbook of software aging and rejuvenation: fundamentals, methods, applications, and future directions. World Sci (2020)
8. Ficco M, Pietrantuono R, Russo S (2018) Aging-related performance anomalies in the apache storm stream processing system. Futur Gener Comput Syst 86:975–994
9. Kaur H, Kaur A (2022) An empirical study of aging related bug prediction using cross project in cloud oriented software. Informatica 46(8) (2022)
10. Kumar L, Sureka A (2018) Feature selection techniques to counter class imbalance problem for aging related bug prediction: aging related bug prediction. In: Proceedings of the 11th innovations in software engineering conference, pp 1–11
11. Machida F, Xiang J, Tadano K, Maeno Y (2012) Aging-related bugs in cloud computing software. In: 2012 IEEE 23rd international symposium on software reliability engineering workshops. IEEE, pp 287–292
12. Parri J, Sampietro S, Scommegna L, Vicario E (2021) Evaluation of software aging in component-based web applications subject to soft errors over time. In: 2021 IEEE international symposium on software reliability engineering workshops (ISSREW). IEEE, pp 25–32
13. Qin F, Wan X, Yin B (2019) An empirical study of factors affecting cross-project aging-related bug prediction with tlap. Softw Qual J: 1–28

14. Qin F, Zheng Z, Bai C, Qiao Y, Zhang Z, Chen C (2015) Cross-project aging related bug prediction. In: 2015 IEEE international conference on software quality, reliability and security. IEEE, pp 43–48
15. Qin F, Zheng Z, Qiao Y, Trivedi KS (2018) Studying aging-related bug prediction using cross-project models. IEEE Trans Reliab 68(3):1134–1153
16. Shruthi P, Cholli NG (2020) An analysis of software aging in cloud environment. Int J Electric Comput Eng (IJECE) 10(6):5985–5991
17. Tan L, Liu C, Li Z, Wang X, Zhou Y, Zhai C (2014) Bug characteristics in open source software. Empir Softw Eng 19(6):1665–1705
18. Wu X, Zheng W, Pu M, Chen J, Mu D (2020) Invalid bug reports complicate the software aging situation. Softw Q J: 1–26 (2020)
19. Yan Y, Li Y, Cheng B (2021) Predicting software aging with a hybrid weight-based method. J Inf Technol Res (JITR) 14(4):58–69